Ed Bee

SPATIAL PROCESSES
MODELS & APPLICATIONS

SPATIAL PROCESSES
MODELS & APPLICATIONS

A D Cliff & J K Ord

ρ Pion Limited, 207 Brondesbury Park, London NW2 5JN

© 1981 Pion Limited

All rights reserved. No part of this book may be reproduced in any form by photostat microfilm or any other means without written permission from the publishers.

ISBN 0 85086 081 4

Printed in Great Britain by Page Bros (Norwich) Limited

Preface

When *Spatial Autocorrelation* was published in 1973, the literature on spatial and space–time processes was scant indeed. Since then, interest in the field has grown rapidly, and large numbers of papers on theoretical developments and empirical applications have appeared. In *Spatial Autocorrelation* we stressed hypothesis testing about spatial processes. Although this approach clearly remains important as a basis for inductive theory construction in the field sciences, part of the literature growth reflects a shift in emphasis from this focus to models of spatial processes. As a result *Spatial Autocorrelation* had, in our view, become dated, both in terms of its restriction to hypothesis testing and in the literature reviewed. The question therefore arose as to whether to allow the first book to fade quietly away, or whether to update it in some manner. Rightly or wrongly, we opted for the second course. This posed the further issue of whether simply to produce a second edition of *Spatial Autocorrelation* with an emphasis on hypothesis testing, or to extend it in some way. We have taken the latter course. Material taken from *Spatial Autocorrelation* comprises about one-third of the current book and this material has been heavily revised; the rest of the book, which is now some 60% longer, is new. Thus we have included chapters on point pattern analysis (chapter 4), spatial correlograms and model identification (chapter 5), simultaneous and conditional autoregressive and moving-average models in the plane (chapter 6), the effect of autocorrelation upon tests of hypotheses (chapter 7), and estimation of spatial models (chapters 6 and 9). The new title of the book reflects this increased coverage.

One of the difficulties we faced in producing the first book was finding enough empirical applications to illustrate the use of the methods suggested. The present volume has been greatly strengthened by the work of others in terms of applications, and we have been able to draw upon a variety of sources for examples. Specific debts are acknowledged elsewhere. We would, however, like to thank Mark Billinge and Derek Gregory for providing the data on cholera epidemics in Victorian London used in chapter 1.

In organising the material in the book, we have been conscious that the book's potential audience will contain readers with varied mathematical backgrounds and empirical interests. Therefore we have tried to interleave fairly substantial applications of the methods described at regular intervals through the book to enable those whose chief concern is practical rather than theoretical to follow the main lines of argument presented. Chapters 2 and 6 employ more advanced statistical methods than is assumed in the rest of the book. A glossary of principal notation used is provided as pages 243–250. Tables, figures, and equations are numbered sequentially within each chapter, and for cross-referencing purposes the first digit is always the chapter number.

Some of the work described in this book was prepared as part of a project on population and epidemiological forecasting involving ourselves and Peter Haggett of the University of Bristol. This project was funded by

SSRC, and our thanks are due to them for their support. In addition, J K Ord was funded by an SRC Senior Visiting Fellowship to enable him to visit the University of Iowa during autumn 1978. Without this assistance, preparation of the manuscript would have been much delayed.

Personal thanks tend to be regarded as the most important part of a preface by their authors, and the least necessary by those who receive them. The third man of the old firm would probably agree with the second half of this sentence as far as he is concerned, but he has been, as always, a good and kind friend, and a constant source of encouragement and advice. We thank him with affection.

Other bouquets we should like to present go to Terri Moss for her careful conversion of a tangled manuscript into a clear typescript, and to Roy Versey for the production of truly excellent diagrams.

As many colleagues will know, the writing of any book is something of a family affair, with our children in ages from hit parade to playgroup. In a multitude of ways, Margaret and Janice have provided us with support and quiet havens for our work, and we dedicate this book to them with our love.

A D Cliff, J K Ord
Gibbet Hill
University of Warwick
Hallowe'en 1980

Acknowledgements

We should like to thank the following publishers and organisations for permission to reproduce copyright material.

The Association of American Geographers, Washington, DC: figure 4.8
Butterworth, Sevenoaks, Kent: figure 3.6
Cambridge University Press, Cambridge: figure 5.1; table 5.1
Charles Griffin, High Wycombe, Bucks: table 4.1
Economic Geography Clark University, Worcester, Mass: figure 1.8
Edward Arnold, London: figures 1.5, 4.3, 4.6, 7.1, 7.3
Geographical Analysis Ohio State University Press, Columbus, Ohio: table 5.2
Heinemann Medical Books, London: figures 1.3, 1.4
The Institute of British Geographers, London: figure 6.8; tables 6.1, 6.2
International Cooperative Publishing House, Fairland, Md: figures 4.4, 4.11, 4.12; table 4.7
Methuen, London: figure 1.1
Sociological Methodology The American Sociological Association, Pittsburgh, Pa: table 9.1
The Trustees of *Biometrika*, London: figure 4.10
University of California Press, Los Angeles: figure 3.7

Contents

1	**Measures of autocorrelation in the plane**	
1.1	Introduction	1
1.2	Concepts of autocorrelation	7
	1.2.1 Spatial autocorrelation	7
1.3	Basic spatial autocorrelation measures	10
	1.3.1 Measures for nominal data	11
	1.3.2 Measures for ordinal and interval data	13
	1.3.3 Choice of test statistic	14
	1.3.4 Limitations of measures	15
1.4	The weighted coefficients	16
	1.4.1 The form of the coefficients	16
	1.4.2 Structure of the weights	17
1.5	Tests of significance	19
	1.5.1 General procedures	19
	1.5.2 Interpretation of the results	21
1.6	Space–time interactions	22
	1.6.1 Clustering in time and space	22
	1.6.2 Tests of significance	24
1.7	Examples	24
	1.7.1 Map pattern analysis	24
	1.7.2 Diffusion on graphs; measles in Cornwall	24
	1.7.3 Space–time interactions; cholera in the metropolis	27
1.8	Conclusions	33
2	**Distribution theory for the join-count, I, and c statistics**	
2.1	Introduction	34
	2.1.1 General expressions for the moments	34
2.2	The join-count statistics	36
	2.2.1 Mean and variance of BB and BW	38
	2.2.2 Higher moments	40
2.3	The Moran and Geary statistics	42
	2.3.1 The moments of I, given assumption N	42
	2.3.2 The moments of I under assumption R	45
2.4	The distributions of the test statistics	46
	2.4.1 The I and c statistics	47
	2.4.2 The join counts under free sampling	51
	2.4.3 Asymptotic normality under assumption R	52
	2.4.4 The Poisson limit for BB joins	52
2.5	Evaluation of the distribution functions	53
	2.5.1 Monte Carlo study for I	54
	2.5.2 Monte Carlo study for c	56
2.6	Monte Carlo studies for the join-count statistics, $k = 2$	56
	2.6.1 The experimental procedure	56
	2.6.2 Analysis of the results	59
	2.6.3 The Poisson approximation	61
	2.6.4 The other join-count statistics	62
2.7	Monte Carlo tests	63
2.8	Conclusions	65

3	**Map comparison with application to diffusion processes**	
3.1	Introduction	66
3.2	Map comparison	66
	3.2.1 The test	66
	3.2.2 Application to diffusion processes	68
3.3	Empirical example	71
	3.3.1 The Hägerstrand model	71
	3.3.2 Analysis of simulation results	74
	3.3.3 Interpretation of the results	81
3.4	Other methods	84
3.5	Conclusions	85
4	**The analysis of spatial point patterns**	
4.1	Introduction	86
	4.1.1 Random or mapped data?	87
4.2	The method of quadrat counts	87
	4.2.1 The Poisson process	87
	4.2.2 True or apparent contagion?	90
	4.2.3 A study of settlement patterns	92
	4.2.4 Implications for settlement geography	99
4.3	The study of pattern by using distances	99
	4.3.1 The Poisson and other processes	100
	4.3.2 Tests using distances	104
	4.3.3 Reflexive pairs	106
	4.3.4 Spacing of towns along the Mississippi	107
	4.3.5 Estimation of intensity	109
4.4	Analysis of mapped data	111
	4.4.1 First-order (or moment) methods for intensity surfaces	111
	4.4.2 Distance-based methods	112
	4.4.3 Second-order analysis	115
	4.4.4 Spectral analysis	116
	4.4.5 Analysis in space and time	117
4.5	Conclusions	117
5	**Spatial correlograms and related methods**	
5.1	Introduction	118
5.2	Spatial correlograms	118
	5.2.1 Theory	118
	5.2.2 A regional application: measles in Cornwall	120
5.3	The spatial scale of a process	123
	5.3.1 Greig-Smith's method	123
	5.3.2 Other approaches	125
	5.3.3 The variogram	126
5.4	Scale and correlation between processes	127
	5.4.1 Model formulation	127
	5.4.2 Negative estimates	130
	5.4.3 An alternative approach	131
	5.4.4 Empirical application: land use in London	131

5.5	Model identification	134
	5.5.1 The partial correlogram	134
	5.5.2 Computational aspects	135
	5.5.3 Spatiotemporal correlograms	136
	5.5.4 Correlograms and partial correlograms for the London land-use data	137
5.6	Estimation of the weights	139
5.7	Conclusions	140
6	**Models for spatial processes**	
6.1	Introduction	141
	6.1.1 Reaction or interaction?	141
6.2	Model specification	142
	6.2.1 Trend-surface analysis	145
	6.2.2 Autoregressive models	145
	6.2.3 Whittle's model	146
	6.2.4 The conditional approach	148
	6.2.5 Moving-average models	149
	6.2.6 Regionalised variables	151
	6.2.7 Generation of random variables	152
6.3	Estimation procedures	153
	6.3.1 First-order conditional autoregression	154
	6.3.2 First-order simultaneous autoregression	159
	6.3.3 Higher order autoregressive schemes	160
	6.3.4 Moving-average models	160
	6.3.5 Regionalised variables	162
6.4	Choice of tests for spatial autocorrelation	163
	6.4.1 Asymptotic relative efficiency	164
	6.4.2 The likelihood ratio test	166
	6.4.3 Comparison of the I and c statistics	167
	6.4.4 The join-count statistics	170
6.5	Empirical comparisons	173
	6.5.1 Comparisons using regression residuals	173
	6.5.2 Monte Carlo studies	174
6.6	Conclusions	178
	Appendix on the major theorems used in chapter 6	179
7	**Autocorrelation and inferential statistics**	
7.1	Introduction	184
7.2	Comparison of means	184
	7.2.1 The two-sample case	184
	7.2.2 A modified t test	187
	7.2.3 Extension to k samples	189
7.3	Correlation and regression	189
	7.3.1 Correlation	189
	7.3.2 Regression analysis	190
7.4	Spatial variate differencing	192

7.5	Tests using join counts	194
	7.5.1 Expectations under H_0	194
7.6	Conclusions	196
8	**The analysis of regression residuals**	
8.1	Introduction	197
8.2	The linear model and alternative hypotheses	198
	8.2.1 The classical regression model	198
	8.2.2 Consequences of autocorrelated errors	199
	8.2.3 Choice of test statistic	199
	8.2.4 Correlation among the sample residuals under H_0	200
8.3	The moments of the I statistic	200
	8.3.1 Mean of I	201
	8.3.2 Variance of I	202
	8.3.3 Evaluation of the variance	203
8.4	The BLUS procedure	203
	8.4.1 Derivation of the estimates	203
	8.4.2 Moments of I'	204
	8.4.3 Relative efficiency of BLUS procedures	205
8.5	Distributions of the test statistics	205
	8.5.1 Asymptotic normality of the test statistics	205
	8.5.2 A random permutations procedure	206
8.6	Eire: the economic effects of road accessibility	206
8.7	Road accessibility in developing countries	215
8.8	Trend-surface analysis: agricultural land values in Iowa, 1977–8	222
8.9	Conclusions	228
	Appendix	229
9	**Models containing components both regressive and autoregressive**	
9.1	Introduction	231
9.2	Regression models with autoregressive components	231
	9.2.1 A regression model with autoregressive terms	231
	9.2.2 Spatiotemporal processes	233
	9.2.3 Regression with autocorrelated errors	234
9.3	The Huk rebellion in the Philippines	236
9.4	Consumption of agricultural output in Eire	237
9.5	Residual autocorrelation in autoregressive schemes	240
9.6	Conclusions	240
	Appendix	241
Glossary of notation		243
References		251
Index		263

Measures of autocorrelation in the plane

1.1 Introduction
It is often necessary to consider the spatial distribution of some phenomenon, whether in the 'counties' or 'states' of a 'country', in the 'quadrats' of a study area, or as a map of (point) locations of occurrences. Two basic questions may then be asked: (1) is the spatial pattern displayed by the phenomenon significant in some sense and therefore worth interpreting? If it is, (2) can we obtain any information on the processes which have produced the observed pattern from an analysis of the mapped distribution of the phenomenon?

To illustrate some of the issues involved, let us consider the series of diagrams given as figures 1.1-1.4. These diagrams have the common theme of showing the spatial incidence of various diseases. Figure 1.1 describes the changing distribution of local authority areas in southwest England reporting cases of measles for a four-week sequence, 1969-70, during the course of a major measles epidemic in the region. Local authority areas reporting measles cases for the first time in a given week, t, are colour coded black. Local authorities reporting cases in week t, and which had also reported cases in immediately preceding weeks, are cross-hatched ('existing outbreak'). Those authorities which had ceased to report cases in week t, but which had reported cases in week $t-1$, comprise the category 'fade out' of the epidemic on the maps. The consistent abutting of new to existing outbreak areas is apparent. Indeed, for measles cases reported in the Southwest, 1966-70, of the local authorities which were at risk and which reported measles cases in a given week, 1 in 8 was contiguous to a local authority area that had reported cases in the preceding week. The corresponding figures for local authorities two and three steps away from existing outbreaks were 1 in 21 and 1 in 31. The maps therefore raise the question of whether measles epidemics spread by a spatially contagious diffusion process, with highly localised mixing of those at risk and those carrying the disease.

Figure 1.2(a) shows the death rate per 10000 persons from cholera in London, 1848-9. It was but one of several visitations of the disease to the metropolis in the middle of the nineteenth century, and provoked great public concern about the causes and manner of spread of the disease. It is now known that contaminated water supplies are a primary cause of transmission of cholera, and that survival times of the vibrio in water are increased as the chemical composition of the water increases in basicity (Jusatz, 1977). Figure 1.2(a) therefore raises the general issue of whether we can relate areas of high/low incidence of cholera to varying quality of the public water supply. Figure 1.2(b) shows the areas of the metropolis served by the several water companies in 1849 and also gives the locations of their waterworks and reservoirs. The water-company boundaries are

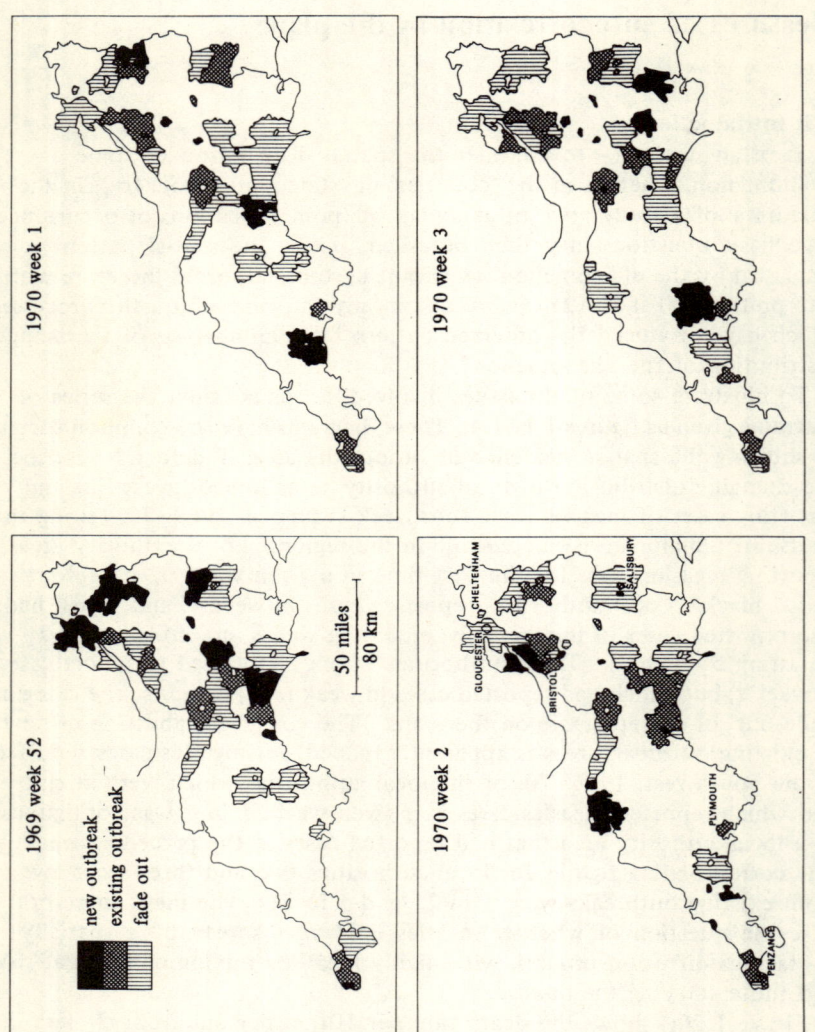

Figure 1.1. Changing distribution of measles notifications in General Register Office (GRO) areas in southwest England, 1969 (week 52)–1970 (week 3). Source: Haggett (1972, page 310).

reproduced as the heavy lines on figure 1.2(a). The high incidence of deaths in the area served by the Southwark and Vauxhall Water Company is apparent. Table 1.1 gives the chemical composition of the waters of the companies as determined by the public inquiry at the time. The Chelsea, and Southwark and Vauxhall Companies have the highest levels of organic matter in their waters, together with the highest levels of soluble lime (base) $Ca(OH)_2$. However, the Chelsea Company has much the highest level

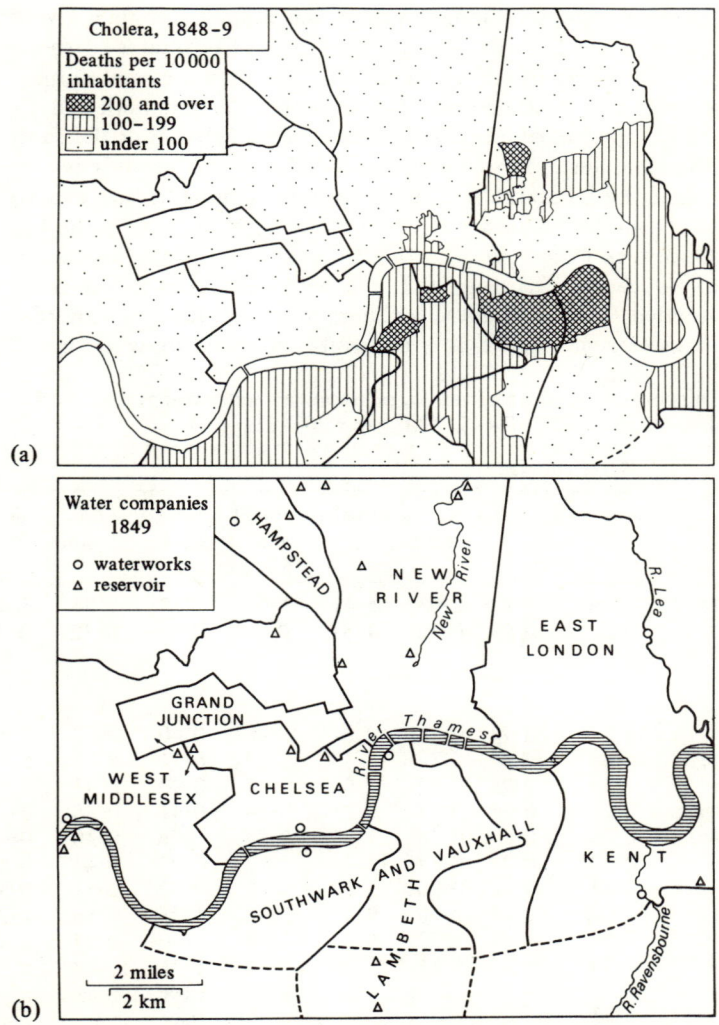

Figure 1.2. (a) Death rates per 10000 inhabitants from cholera in the metropolis, 1848-9. Based on data from PP 1850, XXI page 201. (b) Areas served by the metropolitan water companies, 1849. Source: PP 1850 XXII, facing page 2.

of sulphuric acid, which would result in the Chelsea Company's water being more acidic than that from the Southwark and Vauxhall. This would reduce the effects of the higher level of organic matter in the Chelsea water. We conclude that, on balance, not only is the Southwark and Vauxhall water high in organic matter, but that it also has the highest degree of basicity, thus favouring the sustained survival of the vibrio[1]. The high organic content of the Chelsea, and Southwark and Vauxhall waters was caused by the practice of discharging raw sewerage into the Thames from whence these companies drew the supplies for their riverside waterworks [see figure 1.2(b)]. The particularly evil character of the Vauxhall and Southwark water is revealed in the following quotation:
"When we take into consideration the facts that the Lambeth and Southwark Companies supply contiguous houses throughout the same district, that the elevation of their sites in relation to the river is identical, that the inhabitants supplied by both companies are exactly similar with regard to means and yet the mortality in houses supplied by the Southwark Company exceeded by nearly 2,000 the deaths that would have occurred if cholera had only been as fatal as it was in the houses supplied from the Lambeth Company ..., without viewing the subject in an extreme aspect, it is impossible to avoid drawing the

Table 1.1. Composition of water of various metropolitan water companies, 1854, grains per gallon. Source: PP 1854-5, XXI, pages 208-211.

Chemical	Water company							
	Lambeth	Grand Jn	West Middx	Chelsea	V'hall and S'wark	New River	East London	Kent
Organic matter	1·39	1·92	2·08	5·41	3·26	2·33	1·94	1·48
Silica	0·35	0·09	0·52	1·51	0·24	0·18	0·32	0·42
Sesquoxide of iron, alumina, and phosphates	0·22	0·73	0·46	0·64	0·46	0·40	0·52	0·13
Insoluble lime	5·68	4·97	5·56	5·35	5·99	6·71	6·72	5·34
Soluble lime	0·94	0·98	0·87	2·65	2·37	0·92	0·45	1·25
Insoluble magnesia	0·28	0·34	0·34	0·21	0·24	0·41	0·35	0·10
Soluble magnesia	0·26	0·23	0·16	1·28	0·89	trace	0·10	0·40
Sodium	0·38	0·37	0·64	11·71	5·97	0·93	0·44	0·34
Potassium	0·33	0·25	0·26	1·30	1·09	0·32	0·31	0·52
Chlorine	1·02	0·98	1·16	19·56	12·16	1·43	0·86	1·24
Sulphuric acid	1·60	1·65	1·50	6·04	2·98	1·40	0·84	2·34
Carbonic acid	9·55	8·56	9·11	8·86	9·94	11·44	11·34	8·62
Nitric acid	–	–	–	–	6·50	trace	trace	trace
Ammonia	0·23	–	–	–	0·30	trace	trace	trace

[1] We wish to thank Dr J Paterson, Research Fellow in Chemistry, Christ's College, Cambridge, for providing us with this qualitative interpretation of table 1.1.

conclusion, that the Southwark water must at least have had a more predisposing influence in the production of the disease than that of the Lambeth Company." (Source: PP 1854-5, XXI, page 185).

Figure 1.3(a) categorises the standardised mortality ratios for bronchitis in males in the twenty-nine metropolitan boroughs, 1959-63. The high mortality in the northern and East End boroughs, where socioeconomic status is lowest and, bearing in mind the period covered by the data, air pollution is highest, is striking. Related to this diagram is figure 1.4, which shows the spatial pattern of average annual deaths, 1959-63, in the Welsh

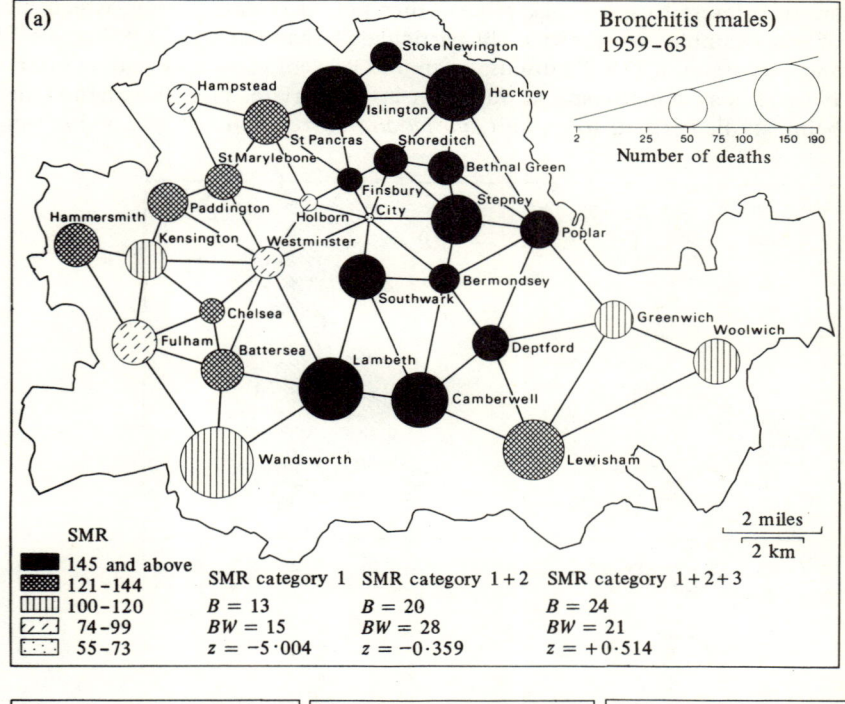

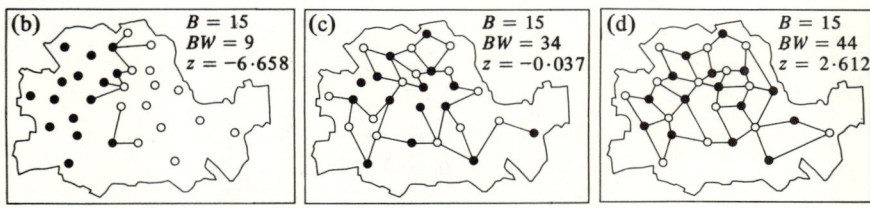

Figure 1.3. (a) Standardised mortality ratios (SMRs) for bronchitis in males in the metropolitan boroughs of London, 1959-63. (b), (c), and (d) show representative clustered, random, and uniform patterns, and corresponding standard normal deviate for the black-white join-count statistic, BW. Source: Cliff and Haggett (1980, forthcoming).

counties for males from (a) tuberculosis of the respiratory system and
(b) bronchitis. Counties with death levels above the crude average for
Wales (left-hand maps) and for the United Kingdom (right-hand maps) are
shaded. On figure 1.4(b), the concentration of excess mortality in the
South Wales industrial and coalfield area is as striking as it is for the same
disease in the northern and East End boroughs shown in figure 1.3(a). In
figure 1.4(a), there appears to be a coastal/noncoastal dichotomy in
mortality rates.

In all the examples given, we are scanning the maps for spatial pattern
in the variation of values of some variate across the map. Such questions
are of interest to many disciplines other than geography (from which most
of our examples are drawn). In particular, Sokal and Oden (1978a; 1978b)
and Jumars et al (1977) discuss examples of ecological and genetic interest.
If there is systematic spatial variation in the variate, then the phenomenon
being studied is said to exhibit *spatial autocorrelation*. It is a major aim

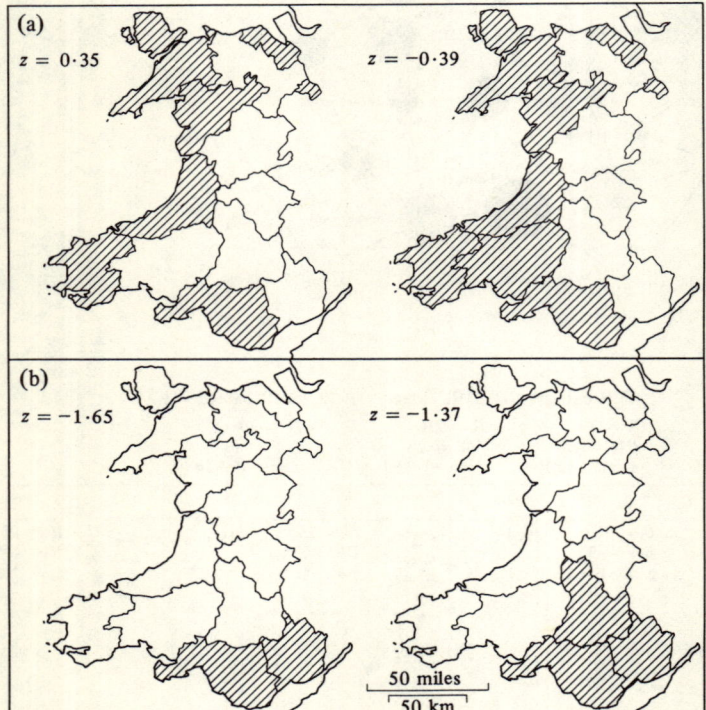

Figure 1.4. Spatial pattern of average annual deaths for males from (a) tuberculosis of
the respiratory system and (b) bronchitis, 1959-63, for Wales by county. Counties
above the crude average for Wales (left-hand maps), and for the United Kingdom
(right-hand maps) are shaded. Symbol z denotes the value of the standardised normal
deviate for the black-white join-count statistic, BW, for each map.

of this book to define various measures of spatial autocorrelation in geographically located variate values. For simplicity of discussion, we shall consider principally the formulation of 'counties' within a 'country', but it is important to remember that any lattice of regions or set of points at which the value of a variate has been obtained could be considered using the methods to be discussed. What the measures tell us is whether the spatial pattern displayed is significant or not—that is, they provide some answer to question (1) on page 1 of whether the pattern is worth interpreting. Having determined this, it is natural to try to establish the nature of the processes that have produced the spatial pattern in the variate values. Our ability to do this will depend upon the problem in hand. The statistical methods provide no hard and fast answers, but we shall see that inferences based upon the circumstantial evidence of mapped patterns form a basis of inductive theory building in subjects such as geography and ecology.

In this chapter, the spatial autocorrelation problem is defined formally in section 2, and some measures of autocorrelation proposed in the literature are reviewed in sections 3 and 4. Clearly, autocorrelation patterns can occur both in space and time, and we consider space–time interactions in section 6. The chapter is concluded in section 7 with a series of examples.

1.2 Concepts of autocorrelation
1.2.1 Spatial autocorrelation
Consider a study area which has been exhaustively partitioned into n nonoverlapping subareas, as in any of the figures 1.1–1.4. Suppose that a random variable, X, has been measured in each of the subareas, and that the value of X in the typical subarea, i, is x_i. X could describe either (1) a single population from which repeated drawings are made to give the x_i; or (2) n separate populations, one for each county; or (3) a partition of a finite population among the n counties. The underlying population model used will depend upon the problem in hand. For example, in figure 1.1 we might regard the number of measles cases reported by each local authority in a given week as part of a finite population comprising the total number of cases recorded in the whole Southwest [population model (3)]. However, in figure 1.2, since each water company draws its water from separate sources, it might be more plausible to treat the number of cholera deaths in each water-company area under population model (2). Finally, in figure 1.3 we could regard the number of deaths from bronchitis in the metropolitan boroughs as generated primarily by a single-factor process—perhaps the level of air pollution—and use population model (1). It is important to note that the choice of population model does not affect the derivation of the measures of spatial autocorrelation, nor the method of analysis. However, as we shall see in chapter 6, it does affect the inferences that can be made.

A basic property of spatially located data is that the set of values, $\{x_i\}$, are likely to be related over space. This idea underlies the concept of the region in geography, and has been graphically summarised by various authors as follows:

Galton (1889), commenting on a paper by Tylor, argued that the historical links between societies, through migration and cultural diffusion, induced dependence among observations.

"It was extremely desirable for the sake of those who may wish to study the evidence for Mr. Tylor's conclusions, that full information should be given as to the degree in which the customs of the tribes and races which are compared together are independent. It might be, that some of the tribes had derived them from a common source, so that they were duplicate copies of the same original. Certainly, in such an investigation as this, each of the observations ought, in the language of statisticians, to be carefully 'weighted'."

Tobler (1970) has referred to "the first law of geography: everything is related to everything else, but near things are more related than distant things."

Gould (1970, pages 443-444) stated

"Why we should expect independence in spatial observations which are of the slightest intellectual interest or importance in geographic research I cannot imagine. All our efforts to understand spatial pattern, structure, and process have indicated that it is precisely the lack of independence—the interdependence—of spatial phenomena that allows us to substitute pattern, and therefore predictability and order, for chaos and apparent lack of interdependence—of things in time and space."

Finally Hepple (1978) gives the following remark by the statistician, F F Stephan (1934), "data of geographic units are tied together like bunches of grapes, not separate like balls in an urn."

If the $\{x_i\}$ display interdependence over space, we say that the data are spatially autocorrelated. The following formal definition may be made.

If, for every pair of counties i and j in the study area the drawings which yield x_i and x_j are uncorrelated, then we say that there is no spatial autocorrelation in the county system on X.

A detailed consideration of models for spatial autocorrelation among variate values is undertaken in chapter 6, but one model of the spatial interdependence among the $\{x_i\}$ is the scheme

$$X_i = \rho \sum_j w_{ij} X_j + \epsilon_i , \qquad i = 1, 2, ..., n . \tag{1.1}$$

Here, the $\{\epsilon_i\}$ are independent and identically distributed variates with common variance, σ^2. The set of weights, $\{w_{ij}\}$, are any set of constants that specify which j subareas in the study area have variate values directly spatially related with X_i. The constant, ρ, is a measure of the overall level of spatial autocorrelation among the $\{X_i X_j\}$ pairs for which $w_{ij} > 0$. For example, we might put $w_{ij} = 1$ (unscaled) if j is physically continuous to i, and $w_{ij} = 0$ otherwise. More general sets of weights may, however, be constructed and are considered in detail in section 1.4.2. Thus Tobler's first law of geography quoted above might be captured by

$$w_{ij} = (c + d_{ij})^{-\alpha}, \qquad (1.2)$$

where d_{ij} is the distance between points or areas, i and j, and α is a 'friction of distance' parameter as used in many gravity and interaction models, and c is a constant ($c > 0$). Finally, when $\rho > 0$ in model (1.1), we say that there is *positive* spatial autocorrelation among the $\{X_i\}$ whereas $\rho < 0$ implies *negative* spatial autocorrelation. The former case is characterised by similar $\{x_i\}$ values in areas with nonzero $\{w_{ij}\}$ values, and the latter by very different (+/−) relationships. Thus in figure 1.3(b), we have arranged 15 of the metropolitan boroughs colour coded black, B, and 14 white, W, in such a way as to maximise positive spatial autocorrelation. Figure 1.3(d) maximises negative spatial autocorrelation. If $\rho = 0$ in model (1.1), there is said to be no spatial autocorrelation in the study area on X, and the variate values are randomly mixed as are the B and W units in figure 1.3(c).

1.2.2 Space-time autocorrelation

Model (1.1) is the spatial equivalent of the time-series model for X_t given $X_{t-k} = x_{t-k}$, namely

$$X_t = \rho x_{t-k} + \epsilon_t, \qquad t = 1, 2, ..., T, \qquad (1.3)$$

where terms are as defined in section 1.2.1 and t indexes the typical time period. Econometricians have long recognised that economic time series display temporal autocorrelation. But the fundamental difference between models (1.3) and (1.1) is that dependence in time can only extend backwards, whereas in the spatial case the dependence is multidirectional. Thus Whittle (1954) has remarked

> "At any instant in a time series, we have the natural distinction of past and future, and the value of the observation at that instant depends only upon past values. That is, the dependence extends only in one direction: backwards ... [In] the more general two dimensional case of [say] a field, a dab of fertilizer applied at any point in the field will ultimately affect soil fertility in *all* directions."

That is, interregional interactions are multilateral.

Very frequently, however, we have to interpret the processes which have produced regional patterns in economic data. As Bennett (1974) has noted, we are not concerned solely with the analysis of cross-sectional (spatial) data. Our task is to unravel the complex patterns of autocorrelation in both time *and* space to gain some insight into the functional dependencies between areas implied by the presence of autocorrelation. Thus we might postulate the dependence of X_{it} upon past values as

$$X_{it} = f(x_{i,t-k}, x_{j,t-k}), \qquad k = 1, 2, \ldots; \quad j \neq i. \tag{1.4}$$

If our time-space matrix is as shown in figure 1.5, model (1.4) would imply a complex 'cone' of dependencies between regions, going back through time. In addition, the time interval between our data recording points is often sufficiently long, compared with the rate of operation of the geographical process, that what appear as simultaneous effects between regions may occur. The dependency of X_{it} on $x_{i,t-1}, \ldots, x_{i,t-k}$ in model (1.4) is, as shown in figure 1.5, a purely *temporal* autoregressive element; the simultaneous effects represent a purely *spatial* autoregressive component; the remaining terms in the historical part of the 'cone' of figure 1.5 represent general space-time covariances. The cone of dependencies which stretches from the present into the past will also project into the future. That part of the cone shows which regions, j, will be affected by area i in the future as multiplier effects work themselves through the time-space system. Although the measures to be developed in the next section are defined in the context of *spatial* autocorrelation, we shall see that they are analogous to time-series measures of autocorrelation.

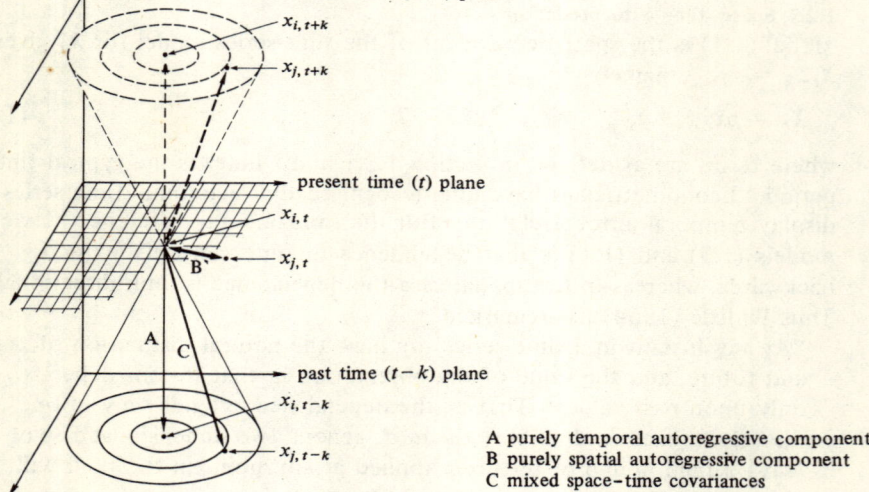

A purely temporal autoregressive component
B purely spatial autoregressive component
C mixed space-time covariances

Figure 1.5. Patterns of dependencies between regions in time and space. Source: Haggett et al (1977, page 355).

1.3 Basic spatial autocorrelation measures
In the remainder of this chapter we assume that the $\{x_i\}$ are raw data rather than residuals from a calculated regression or trend surface. Testing for spatial autocorrelation among such residuals presents special problems and these are discussed in chapter 8. The measures of spatial autocorrelation

which have been proposed in the literature are discussed conveniently according to the kind of data (nominal, ordinal, or interval scaled) to which they may be applied. Fortunately this also coincides with the historical order of development of the measures.

1.3.1 Measures for nominal data

The simplest nominal scale is a binary classification. In each of the n counties we note whether a given event has or has not occurred. If it has, the county is colour coded black B, and if it has not, the county is colour coded white W. If two counties have a boundary of positive nonzero length in common, they are said to be linked by a *join*. A join may link two B counties, two W counties, or a B and a W county. These joins are called BB, WW, and BW joins respectively. To determine whether events in neighbouring counties are spatially autocorrelated or not, we count the numbers of BB, BW, and WW joins which occur in the county system, and compare these numbers with the expected numbers of BB, BW, and WW joins under the null hypothesis, H_0, of no spatial autocorrelation among the counties. Intuitively it can be appreciated that a 'lot' of BB joins, compared with the expected number under H_0, implies clustering of the B counties in the plane, whereas a 'lot' of BW joins implies some sort of alternating pattern of B and W counties as, for example, along the rows and columns of a chessboard.

Put formally, the method of analysis is as follows (Moran, 1948). Let $\{\delta_{ij}\}$ be a connection matrix in which $\delta_{ij} = 1$ if the ith and jth counties are joined, and $\delta_{ij} = 0$ otherwise. Let $x_i = 1$ if the ith county is B, and $x_i = 0$ if the ith county is W. The observed number of BB joins in the county system is then given by

$$BB = \tfrac{1}{2} \sum_{(2)} \delta_{ij} x_i x_j \,, \tag{1.5}$$

and the observed number of BW joins is given by

$$BW = \tfrac{1}{2} \sum_{(2)} \delta_{ij} (x_i - x_j)^2 \,, \tag{1.6}$$

where

$$\sum_{(2)} = \sum_{\substack{i=1 \\ i \neq j}}^{n} \sum_{j=1}^{n} .$$

The observed number of WW joins is given by

$$WW = A - (BB + BW) \,, \tag{1.7}$$

where A is the total number of joins in the county system. The definition of a join given at the beginning of this section implies that $\delta_{ij} = \delta_{ji}$ for all i and j; that is $\{\delta_{ij}\}$ is symmetric. The factor of $\tfrac{1}{2}$ in equations (1.5) and (1.6) enables us to retain the interpretation of BB and BW as counts of the number of joins of each type. BW also reduces to the 'number of runs' statistic in one dimension (Siegel, 1956, pages 52-58) and provides a test of randomness in a time series (see Kendall, 1976).

The usual method employed to determine whether BB, BW, and WW depart significantly from random expectation is to use the fact that these join-count statistics are asymptotically normally distributed (see chapter 2), and to assume that these results hold approximately for moderate sized lattices. The first two moments of the coefficients are then used to specify the location (μ) and scale (σ^2) parameters of the normal distribution. The early work on these measures was carried out for rectangular lattices, although an interesting extension of these results by Freeman (1953) allows for vacant cells on the lattice. Using this approach, we note that the moments of the coefficients may be evaluated under either of two assumptions:
(1) free sampling (or sampling with replacement), where we suppose that the individual counties are independently coded B or W with probabilities p and $q = 1-p$ respectively;
(2) nonfree sampling (or sampling without replacement), where we assume that each county has the same probability, *a priori*, of being B or W, but coding is subject to the overall constraint that there are n_1 counties coloured B and n_2 coloured W, and $n_1 + n_2 = n$.

To illustrate the differences between the two assumptions, suppose that we have a map of a shopping area with shops designated either as food (coded B) or as nonfood (W) stores. Thus the BB count would denote the total number of adjacent pairs of food stores in the area. The null hypothesis describes a situation where shops locate independently of one another, whereas the alternative hypothesis might suggest either clustering (of convenience-good shops?) or regular spacing (to avoid excessive competition?). Detailed studies of such processes are given in Rogers (1974). When we test the null hypothesis under free sampling, we do not fix n_1 and n_2, so that we are comparing our realisation with all possible configurations of n shops. This might correspond to 'no-planning', where no constraints are placed upon the mix of shops in the area. Conversely, when sampling without replacement, we take n_1 and n_2 to be fixed and then look at their spatial pattern conditional upon the numbers observed. This reduces the range of possibilities considerably and we find that the variances for nonfree statistics are usually much lower than those for the corresponding free sampling statistics.

The moments of the join counts were first obtained by Moran (1948). They are given in section 1.5.1 for the weighted coefficients, of which the present statistics are a special case.

Quite commonly the nominal scale will have classes ($k > 2$) rather than the simple binary classification discussed above. Each class may then be assigned one of k distinct colours, and each county is called after the colour of the class into which it falls. Conventionally, the analysis then proceeds by counting the number of joins between counties of (1) the same colour, (2) two different colours, and (3) all counties of different colours.

The statistic for case (3) will be designated J_{tot} (total number of joins) and will be written as

$$J_{\text{tot}} = \tfrac{1}{2} \sum_{(2)} \delta_{ij} y_{ij} , \qquad (1.8)$$

where $y_{ij} = 1$ when counties i and j are different colours, and $y_{ij} = 0$ when the two counties have the same colour. This statistic was first studied by Krishna Iyer (1949) and is very similar in form to a statistic suggested by David (1971a) to measure the spatial diversity of plant species.

Suppose that the commonest species at location i is species r (colour r). This could be a single plant or the most common plant in the area. Then if contiguous locations also have species r as the commonest species, the spatial diversity is said to be low. Conversely, when neighbouring locations all have different most common species, it is apparent that there is a considerable variety of species in the study area. It is evident that low spatial diversity corresponds to low J_{tot} and that high spatial diversity corresponds to high J_{tot}. David's (1971a) measure differs from statistic (1.8) only in its treatment of locations on the boundary of the study area. More complex measures, which do not restrict attention to the commonest species, are described in David (1971b). An alternative measure of spatial diversity, using entropy, is given by Pielou (1975, chapter 5). The first two moments of the distribution of J_{tot} under free and nonfree sampling are given in section 1.5.1.

1.3.2 Measures for ordinal and interval data

If X is ordinal scaled (ranked) or interval scaled, we could group the range of X into k classes, such as quartiles or deciles, and use the colour lattice tests described above; in this case, a loss of information occurs. We now define two further coefficients which assess the degree of spatial autocorrelation between the $\{x_i\}$ in joined counties, where x_i is either the rank of the ith county (ordinal data) or the value of X in the ith county (interval data). Individual county values are therefore retained and the loss of information which occurs if the join-count statistics are employed is avoided.

The first coefficient was proposed by Moran (1950) and is given by

$$I = \frac{n}{2A} \frac{\sum_{(2)} \delta_{ij} z_i z_j}{\sum_{i=1}^{n} z_i^2} , \qquad (1.9)$$

where $z_i = x_i - \bar{x}$ in addition to previously used notation. The second coefficient has been suggested by Geary (1954). Geary's statistic c is defined as

$$c = \frac{(n-1) \sum_{(2)} \delta_{ij} (x_i - x_j)^2}{4A \sum_{i=1}^{n} z_i^2} . \qquad (1.10)$$

Note that both I and c take on the classic form of any autocorrelation coefficient: the numerator term in each is a measure of covariance among the $\{x_i\}$ and the denominator term is a measure of variance.

In terms of temporal autocorrelation, we note that I reduces in one dimension to the familiar serial correlation coefficient; c corresponds in form to (a) the Durbin and Watson d statistic (Durbin and Watson, 1950; 1951; 1971) used to search for temporal autocorrelation in regression residuals, and (b) the von Neumann ratio (von Neumann, 1941).

It is shown in the next chapter that both I and c are asymptotically normally distributed as n increases. As with the join-count statistics, this result is assumed to hold approximately for lattices of moderate size, and I and c are tested for significance as standard normal deviates. The moments of I and c may be evaluated under either of two assumptions: (1) assumption N, normality. Here we assume that the $\{x_i\}$ are the results of n independent drawings from a normal population (or populations); (2) assumption R, randomisation. Whatever the underlying distribution of the population(s), we consider the observed value of I or c relative to the set of all possible values which I or c could take on if the $\{x_i\}$ were repeatedly randomly permuted around the county system. There are $n!$ such values.

As for the join-count statistics, these two assumptions correspond either to allowing drawings from a wider population (1) or to conditioning upon the observed values (2). The restriction to normality simply recognises that the analysis is intractable for other distributions; spatial autocorrelation measures are not alone in this respect! Fortunately, the results in chapter 2 seem to hold for reasonable departures from the normal assumption. Further, the randomised moments may be interpreted as unbiased estimates of the moments for *any* population.

The first two moments of the I and c statistics are given in section 1.5.1.

1.3.3 Choice of test statistic [2]

When the researcher wishes to examine a data set for spatial autocorrelation, he will have to decide which of the coefficients defined in sections 1.3.1 and 1.3.2 to use as his test statistic. The following guidelines are intended to help him make that choice.

(1) With binary (0, 1) data, the join-count statistics may be used. Alternatively, I or c could be employed by putting, say, $x_i = 1$ if an event has occurred in the ith county and $x_i = 0$ otherwise. However, with binary data, I and c reduce, apart from constants, almost exactly to the BW statistic. Thus there is little point with binary data in evaluating I or c rather than the join-count statistics. If the join counts are used, the researcher has the choice between the free and nonfree sampling models. Strictly, free sampling may only be used if p is known *a priori* (exogenously). If p is estimated from the data by n_1/n, then only estimates of the moments are available. It is not known whether this would induce a

---

[2] A full discussion of this topic is deferred until chapter 6.

serious inferential error, but we would suggest that in these circumstances the nonfree model may be more appropriate anyway.

(2) With ranked or interval scaled data, I and c are preferred to the colour lattice approach. Recall that in order to use the colour lattice approach with these data, the $\{x_i\}$ must be grouped into classes, which results in loss of information. I and c preserve the individual x values and so avoid this problem. The choice between I and c will be considered in detail in chapter 6. However, results given in Cliff and Ord (1969, page 45) suggest that the variance of I is less affected by the distribution of the sample data than is the differences-squared form used in Geary's c. This is because the b_2 term in the variance of the Geary statistic has a coefficient of order n^{-1}, whereas for the Moran statistic the coefficient of the b_2 term is of order n^{-2} (see section 1.5.1).

1.3.4 Limitations of measures

The join-count statistics, I and c have two important limitations. First, they suffer from what Dacey (1965, page 28) has called the problem of 'topological invariance'. That is, once the connection matrix $\{\delta_{ij}\}$ has been specified, the size and shape of counties in the system, and the relative strength of links between counties (road and rail links, for example) are completely ignored. The measures are, therefore, invariant under certain transformations of the underlying county structure. Dacey illustrates this point as follows: Consider a county map P_0, with a connection matrix $\{\delta_{ij}\}$, an assigned set of values $\{x_0\}$, and an index value V_0 (any one of the join-count statistics, I or c). Without changing $\{\delta_{ij}\}$ or $\{x_0\}$, it is possible to transform P_0 structurally and to produce a new county map P_1, with $V_1 = V_0$, which ensures the measures of spatial autocorrelation are unchanged. For example, P_1 may be constructed from P_0 by the rule that county boundaries are shortened if the counties separated by the boundary have values of opposite sign, and are increased in length if the counties have values of the same sign. Alternatively, the area of counties may be changed so that the area of, say, counties with positive values is increased, and the area of counties with negative values is decreased. All the coefficients described are invariant over such transformations because the only element of the underlying county structure which they incorporate is the connection matrix. To reinforce this point, figure 1.6 illustrates three different county systems with the same join structure.

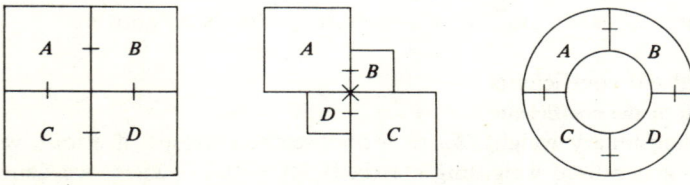

Figure 1.6. Three county systems with the same join structure.

To overcome this difficulty, Dacey (1965) suggested a measure of spatial autocorrelation I', where

$$I' = \frac{n}{2A} \frac{\sum_{(2)} \delta_{ij} \alpha_i \beta_{i(j)} z_i z_j}{\sum_{i=1}^{n} \alpha_i z_i^2} , \qquad (1.11)$$

and in which

$$\alpha_i = \frac{a_i}{\sum_{i=1}^{n} a_i} , \qquad \beta_{i(j)} = \frac{b_{ij}}{\sum_{j \in J} b_{ij}} .$$

Here, a_i equals the area of the ith county, b_{ij} denotes the length of the common boundary between the ith and jth counties, and $j \in J$ directs the summation over all j counties joined to i. Unfortunately, as is discussed in chapter 2, it is not possible to express the moments of I' in a usable form, and so no test of significance based on I' is readily available.

The second limitation is one of usage. As defined, joins exist solely between physically contiguous counties. With $\{\delta_{ij}\}$ thus specified, the measures search for spatial autocorrelation only between counties which are first nearest neighbours. Thus correlogram analysis, to determine how the autocorrelation function decays over space, was not attempted with these measures. There is nothing in the structure of the tests which prevents this kind of analysis; see chapter 5. For example, we could define 'joins' to exist between counties which are second, rather than first, nearest neighbours. Two counties, i and k, might be called second nearest neighbours if they have no common boundary of positive nonzero length, but there exists a county j such that i and j are contiguous, and j and k are contiguous. Then $\delta_{ik} = 1$ for such pairs of counties, and $\delta_{ij} = 0$ otherwise. Generalisation of the concept of a join to second and higher order neighbours in this fashion is easily performed using graph theoretic methods (see Haggett et al, 1977, pages 319–320). Even if this were done, however, all joins would still be given equal weight; and in some studies we might wish to give strong links between counties which are not contiguous, and weak links between contiguous counties. Similarly an individual's perception of geographical space is discontinuous and, in perception studies, the researcher might wish to take this into account when defining a connection matrix.

We now give versions of the join count, I, and c statistics which employ general weights to overcome the two limitations discussed above.

1.4 The weighted coefficients
1.4.1 The form of the coefficients
Instead of using binary weights δ_{ij} to formalise the concept of a join, we now define a generalised weighting matrix W, $W \equiv \{w_{ij}\}$, where we denote the effect of county j on county i by the weight w_{ij}. The weighted join-

count statistics are given by

$$BB = \tfrac{1}{2}\sum_{(2)} w_{ij} x_i x_j ,\qquad (1.12)$$

$$BW = \tfrac{1}{2}\sum_{(2)} w_{ij}(x_i - x_j)^2 ,\qquad (1.13)$$

and

$$WW = \tfrac{1}{2} S_0 - (BB + BW) ,\qquad (1.14)$$

where $S_0 = \sum_{(2)} w_{ij}$. The generalisation of Moran's statistic is

$$I = \frac{n \sum_{(2)} w_{ij} z_i z_j}{S_0 \sum_{i=1}^{n} z_i^2} ,\qquad (1.15)$$

and the generalised Geary coefficient is defined as

$$c = \left(\frac{n-1}{2 S_0}\right) \frac{\sum_{(2)} w_{ij}(x_i - x_j)^2}{\sum z_i^2} .\qquad (1.16)$$

1.4.2 Structure of the weights

The use of a generalised weighting matrix W, as opposed to a binary connection matrix, allows the investigator to choose a set of weights which he deems appropriate from *prior* considerations. This allows great flexibility in defining the structure of the county system, and permits items such as natural barriers and county size to be taken into account. Further, if different hypotheses are proposed about the degree of contact between neighbouring areas, alternative sets of weights might be used to investigate these hypotheses. It is important to stress that care must be used in the choice of weights if spurious correlations are to be avoided. The factors which are most important will depend upon the study in hand. For example, the amount of interaction between any two counties may depend upon the distance between their geographical or demographic centres, the length of common boundary between the counties, and so on. In urban areas the contact between two zones may depend on the frequency of public transport services.

As a particular example, suppose that it is decided that the relevant variables which measure the amount of interaction between any pair of counties are distance between county centres and length of common boundary between counties. Let the distance between the centres of counties i and j be d_{ij}, where distance may be defined by the Euclidean or other appropriate metric (for example, the 'city block' metric in perception studies). Further, suppose that the proportion of the perimeter of county i which is in contact with j is $\beta_{i(j)}$. We exclude from the perimeter of i those parts which coincide with the boundary of the study area, and note

that
$$\sum_{j \in J} \beta_{i(j)} = 1 , \qquad (1.17)$$

where J is the set of counties contiguous to county i. The weighting system is then defined as some function g of d_{ij} and $\beta_{i(j)}$, that is,
$$w_{ij} = g[d_{ij}, \beta_{i(j)}] . \qquad (1.18)$$

Thus, provided $d_{ij} > 0$, we might put
$$w_{ij} = d_{ij}^{-a}[\beta_{i(j)}]^b , \qquad (1.19)$$

where a and b are parameters. Positive values of a and b give greater weights to pairs of counties which have shorter distances between their centres, and which have long common boundaries. It should be noted that equation (1.19) gives positive weights only to counties which are contiguous.

Other forms could be used for equation (1.18). Thus, an exponential function might replace the Pareto form used for distance in that equation. The choice of functional form for the w_{ij} must lie with the investigator. When testing hypotheses, the form of the weights will be determined by the nature of the alternative hypothesis specified, as we shall show in chapter 6. Alternatively, we may adopt the approach of Kooijman (1976) and find the set of weights which, subject to certain basic restrictions, maximises the value of the test statistic; see section 4.6. From the estimation viewpoint, we may construct correlograms to consider correlation at different spatial scales; see chapter 5.

It may happen that the investigator chooses weights such that $w_{ij} \neq w_{ji}$. When testing hypotheses, nothing is changed if we redefine the weights such that
$$w'_{ij} = w'_{ji} = \tfrac{1}{2}(w_{ij} + w_{ji}) ,$$

so that only symmetric weights need be considered. In some theoretical models (see section 6.2) only symmetric weights are allowable, but we shall continue to admit asymmetric weights since the investigator may wish to scale his weights in some way.

For example, we might define the weights such that
$$\sum_{j \in J} w_{ij} = 1 , \qquad i = 1, 2, ..., n , \qquad (1.20)$$

where J is the set of counties contiguous to county i. When the scaling (1.20) is used, $S_0 = n$, and the quantity z_i^*,
$$z_i^* = \sum_{j \in J} w_{ij} z_j , \qquad (1.21)$$

represents a value for z_i 'suggested' by the counties contiguous to i.

Although the example we have given assigns positive weights only to contiguous counties, we could have defined the link between *any* two counties i and k as, say,

$$w_{ik} = d_{ik}^{-a}.$$

This makes it clear that we do not restrict the $\{w_{ij}\}$ to be greater than zero only when the counties are contiguous, as has happened with $\{\delta_{ij}\}$.

1.5 Tests of significance
1.5.1 General procedures

It will be shown in chapter 2 that, when generalised weights are employed, the join-count, I, and c statistics are still asymptotically normally distributed as n increases. An approximate test of significance is therefore provided, as with binary weights, by evaluating the coefficients as standard normal deviates. We denote the mean by μ and the variance by σ^2.

The values of the moments for each of the statistics described so far are given below; the detailed derivations are deferred until chapter 2. The following notation is used:

p_r denotes the probability that a county is colour r,
n_r the number of counties of colour r,

$$S_0 = \sum_{(2)} w_{ij},$$

$$S_1 = \tfrac{1}{2}\sum_{(2)}(w_{ij}+w_{ji})^2, \quad \text{and} \tag{1.22}$$

$$S_2 = \sum_{i=1}^{n}(w_{i.}+w_{.i})^2, \tag{1.23}$$

where

$$w_{i.} = \sum_{j=1}^{n} w_{ij} \quad \text{and} \quad w_{.i} = \sum_{j=1}^{n} w_{ji}. \tag{1.24}$$

Join counts, $k \geqslant 2$.

Free sampling

Joins between counties of the same colour (equivalent to BB joins for $k = 2$)

$$\mu = \tfrac{1}{2}S_0 p_r^2, \tag{1.25}$$

$$\sigma^2 = \tfrac{1}{4}[S_1 p_r^2 + (S_2 - 2S_1)p_r^3 + (S_1 - S_2)p_r^4]. \tag{1.26}$$

Joins between counties of two different colours (equivalent to BW joins for $k = 2$)

$$\mu = S_0 p_r p_s, \tag{1.27}$$

$$\sigma^2 = \tfrac{1}{4}[2S_1 p_r p_s + (S_2 - 2S_1)p_r p_s(p_r + p_s) + 4(S_1 - S_2)p_r^2 p_s^2]. \tag{1.28}$$

Total number of joins between counties of different colours ($k \geqslant 3$; when $k = 2$, this case is equivalent to BW joins)

$$\mu = S_0 \sum_{r=1}^{k-1} \sum_{s=r+1}^{k} p_r p_s, \tag{1.29}$$

$$\sigma^2 = \frac{1}{4}\left[S_2 \sum_{r=1}^{k-1} \sum_{s=r+1}^{k} p_r p_s + (2S_1 - 5S_2) \sum_{r=1}^{k-2} \sum_{s=r+1}^{k-1} \sum_{t=s+1}^{k} p_r p_s p_t \right.$$
$$\left. + 4(S_1 - S_2)\left(\sum_{r=1}^{k-1} \sum_{s=r+1}^{k} p_r^2 p_s^2 - 2 \sum_{r=1}^{k-3} \sum_{s=r+1}^{k-2} \sum_{t=s+1}^{k-1} \sum_{u=t+1}^{k} p_r p_s p_t p_u\right)\right]. \tag{1.30}$$

Nonfree sampling
Joins between counties of the same colour

$$\mu = \frac{S_0 n_r^{(2)}}{2n^{(2)}}, \tag{1.31}$$

$$\sigma^2 = \frac{1}{4}\left[\frac{S_1 n_r^{(2)}}{n^{(2)}} + \frac{(S_2 - 2S_1)n_r^{(3)}}{n^{(3)}} + \frac{(S_0^2 + S_1 - S_2)n_r^{(4)}}{n^{(4)}}\right] - \mu^2. \tag{1.32}$$

Joins between counties of two different colours

$$\mu = \frac{S_0 n_r n_s}{n^{(2)}}, \tag{1.33}$$

$$\sigma^2 = \frac{1}{4}\left[\frac{2S_1 n_r n_s}{n^{(2)}} + \frac{(S_2 - 2S_1)n_r n_s(n_r + n_s - 2)}{n^{(3)}}\right.$$
$$\left. + \frac{4(S_0^2 + S_1 - S_2)n_r^{(2)} n_s^{(2)}}{n^{(4)}}\right] - \mu^2. \tag{1.34}$$

Total number of joins between counties of different colours

$$\mu = S_0 \sum_{r=1}^{k-1} \sum_{s=r+1}^{k} \frac{n_r n_s}{n^{(2)}}, \tag{1.35}$$

$$\sigma^2 = \frac{1}{4}\left\{\left[\frac{S_2}{n^{(2)}} - \frac{4(S_0^2 + S_1 - S_2)(n-1)}{n^{(4)}}\right]\sum_{r=1}^{k-1}\sum_{s=r+1}^{k} n_r n_s \right.$$
$$+ 4\left[\frac{(S_1 - S_2)}{n^{(4)}} + \frac{2S_0^2(2n-3)}{n^{(2)}n^{(4)}}\right]\sum_{r=1}^{k-1}\sum_{s=r+1}^{k} n_r^2 n_s^2$$
$$+ \left[\frac{2S_1 - 5S_2}{n^{(3)}} + \frac{12(S_0^2 + S_1 - S_2)}{n^{(4)}} + \frac{8S_0^2}{n^{(3)}(n-1)}\right]\sum_{r=1}^{k-2}\sum_{s=r+1}^{k-1}\sum_{t=s+1}^{k} n_r n_s n_t$$
$$\left. - 8\left[\frac{(S_1 - S_2)}{n^{(4)}} + \frac{2S_0^2(2n-3)}{n^{(2)}n^{(4)}}\right]\sum_{r=1}^{k-3}\sum_{s=r+1}^{k-2}\sum_{t=s+1}^{k-1}\sum_{u=t+1}^{k} n_r n_s n_t n_u\right\}. \tag{1.36}$$

Interval scaled measures

The subscripts N and R are used to denote the assumptions of normality and randomisation respectively.

The coefficient I

$$E_N(I) = E_R(I) = -(n-1)^{-1}, \qquad (1.37)$$

$$E_N(I^2) = \frac{n^2 S_1 - n S_2 + 3 S_0^2}{S_0^2(n^2-1)}, \qquad (1.38)$$

$$E_R(I^2) = \frac{n[(n^2 - 3n + 3)S_1 - nS_2 + 3S_0^2] - b_2[(n^2 - n)S_1 - 2nS_2 + 6S_0^2]}{(n-1)^{(3)}S_0^2}. \qquad (1.39)$$

The coefficient c

$$E_N(c) = E_R(c) = 1, \qquad (1.40)$$

$$\text{var}_N(c) = \frac{(2S_1 + S_2)(n-1) - 4S_0^2}{2(n+1)S_0^2}, \qquad (1.41)$$

$$\begin{aligned}\text{var}_R(c) = \{&(n-1)S_1[n^2 - 3n + 3 - (n-1)b_2] \\ &- \tfrac{1}{4}(n-1)S_2[n^2 + 3n - 6 - (n^2 - n + 2)b_2] \\ &+ S_0^2[n^2 - 3 - (n-1)^2 b_2]\}/n(n-2)^{(2)}S_0^2.\end{aligned} \qquad (1.42)$$

When the weights are binary, $S_0 = 2A$, and $S_1 = 4A$.

1.5.2 Interpretation of the results

Since we refer to I as a coefficient of spatial auto*correlation*, it is tempting to interpret the value of I as we would a correlation coefficient; that is, restricted to the range $[-1, +1]$ with values near ± 1 indicating a very strong relationship. However, it may be shown by the Cauchy–Schwartz inequality that

$$|I| \leqslant \frac{n}{S_0}\left[\frac{\text{var}(z_i^*)}{\text{var}(z_i)}\right]^{1/2}, \qquad (1.43)$$

where z_i^* is given by equation (1.21); note that the scaling (1.20) need not hold here. In general, the upper bound for $|I|$ will be less than unity, although it could exceed unity for an irregular pattern of weights if counties with extreme values of z_i are heavily weighted. When W is symmetric and n is large, the expected value of the upper bound *under the null hypothesis* is approximately

$$\frac{n}{S_0}\left[\frac{2nS_1 - S_2}{4n(n-1)}\right]^{1/2}.$$

Example For any lattice with symmetric binary weights, $S_1 = 2S_0$, and the bound approaches $(n/S_0)^{1/2}$. If the lattice is regular and mapped onto a torus, the expected value of the upper bound approaches $\tfrac{1}{2}$ for what is known (by analogy with chess) as the rook's case and $\frac{1}{\sqrt{8}}$ for the queen's case. For a time-series, **W** is asymmetric and $w_{ij} = 1$ only for $j = i-1$. The bound then approaches one.

Thus, for a qualitative interpretation of I, an initial scaling by a factor such as expression (1.43) is helpful.

To interpret spatial autocorrelation coefficients more generally, let us envisage that I (or some other statistic) has been evaluated at several levels of spatial separation, such as for first, second, third, ... order neighbouring cells. That is, we construct a correlogram; see chapter 5. Then Sokal (1979) provides the following summary of possibilities in the context of population densities, and it is possible to construct similar schemes for other spatial processes.

		Order of autocorrelation (spatial lag)	
		low	high
Sign of autocorrelation	positive	(1) dispersal from few sources (2) large favourable patches (3) gradient (trend)	(1) symmetrical surfaces (2) patchy arrangement
	negative	(1) heterogeneous study area (2) small patches	(1) gradient (trend)

In talking of patches, we must consider the relative magnitudes of distances between individuals in the same patch or, alternatively, patch diameter and the magnitude (diameter) of the cell or county observed. Thus, when patch diameter is greater than cell diameter, we can expect positive low-order correlations, but when cell diameter exceeds patch diameter, we may get negative low-order correlations and positive higher-order correlations, depending upon the degree of regularity in the occurrence of the patches. When the possibility of a gradient arises, it may be desirable to remove trends and then test the residuals; this topic will be pursued further in chapter 8.

1.6 Space-time interactions
1.6.1 Clustering in time and space
Certain events, such as cases of a notifiable disease, are recorded in terms both of their timing and of their location. Thus in our description of figure 1.1, the spread process of measles was examined by considering both the temporal and the spatial proximity of reported cases. As discussed in section 1.2.2, the features of interest are not just the grouping of cases in time alone, nor the grouping of cases in space alone, but the interaction effect in space and time. That is, we may accept the temporal and spatial patterns but search for some interdependence between the two.

An appropriate measure is given by I_{s-t}, such that

$$I_{s-t} = \sum_{(2)} w_{ij} y_{ij} \,, \tag{1.44}$$

where the $\{w_{ij}\}$ are spatial weights that may be used to define spatial contiguity as previously, whereas the $\{y_{ij}\}$ measure closeness in time. For example, if the ith event occurred at time t_i, we may define

$$y_{ij} \begin{cases} = 1\,, & \text{if } |t_i - t_j| \leqslant u\,, \\ = 0\,, & \text{otherwise}\,, \end{cases}$$

for a suitable choice of u.

The space-time interaction coefficient, I_{s-t}, given in expression (1.44) was first proposed by Knox (1964) and generalised by Mantel (1967). Comparisons with equations (1.5), (1.6), and (1.7) indicate that all the join-count statistics could be written in this form for suitable choices of the $\{y_{ij}\}$. If we relax the condition that the $\{y_{ij}\}$ be binary, we can think of I_{s-t} as a general autocorrelation coefficient (Hubert et al, 1981), where w_{ij} measures the 'closeness' of locations i and j in space, whereas y_{ij} measures their 'closeness' with respect to some other dimension (time, income, or whatever). We always take $y_{ii} = 0$ for all i. Hubert (1978) uses this approach to develop an I statistic based on ranks.

Under the null hypothesis of no association we may regard all the assignments of the $\{w_{ij}\}$ as fixed and then consider all possible permutations of the rows (and corresponding permutations of the columns) of the $n \times n$ matrix $\mathbf{Y} = \{y_{ij}\}$. That is, we are considering the randomisation hypothesis (R) of section 1.3.2. Using this approach, Mantel (1967) finds that

$$E(I_{s-t}) = \frac{S_0 T_0}{n(n-1)}\,, \tag{1.45}$$

where

$$T_0 = \sum_{(2)} y_{ij}\,.$$

To specify the variance, we set, by analogy with the definitions for the $\{w_{ij}\}$,

$$T_1 = \tfrac{1}{2} \sum_{(2)} (y_{ij} + y_{ji})^2\,, \quad \text{and} \quad T_2 = \sum_i (y_{i.} + y_{.i})^2\,,$$

where

$$y_{i.} = \sum_j y_{ij}\,, \quad \text{and} \quad y_{.i} = \sum_j y_{ji}\,.$$

Mantel's results for the variance were derived without restrictions upon the $\{y_{ij}\}$, but we shall assume that $y_{ij} = y_{ji}$ for all i and j. Then the variance may be written as

$$\text{var}(I_{s-t}) = \frac{S_1 T_1}{2n^{(2)}} + \frac{(S_2 - 2S_1)(T_2 - 2T_1)}{4n^{(3)}}$$

$$+ \frac{(S_0^2 + S_1 - S_2)(T_0^2 + T_1 - T_2)}{n^{(4)}} - [E(I_{s-t})]^2\,. \tag{1.46}$$

1.6.2 Tests of significance

Whenever $I_{s\text{-}t}$ reduces to any of the measures discussed in earlier sections, the corresponding normal approximations will apply. In general, we need more structure upon the $\{y_{ij}\}$ to be able to make statements about the form of the distribution, even for large n. Nevertheless, normality may often prove a reasonable approximation.

However, when the $\{w_{ij}\}$ and the $\{y_{ij}\}$ are both binary and the number of nonzero values for each is small (specifically, both S_0/n and T_0/n should stay finite as n goes to infinity), then the distribution of $\tfrac{1}{2}I_{s\text{-}t}$ is approximately Poisson with mean $S_0T_0/2n(n-1)$. The factor of 2 in the denominator is included to avoid double counting of the joins. This result is demonstrated in section 2.4.4.

1.7 Examples
1.7.1 Map pattern analysis

The BW join-count statistic given in equation (1.6) has been evaluated for the various distributions mapped in figures 1.3 and 1.4. Binary weights were used to test for spatial autocorrelation among contiguous areal units. The moments were evaluated under nonfree sampling [equations (1.33)-(1.34)]. The sampling distribution was assumed to be approximately normal, and the quantities, z, shown on the diagrams can be treated as normal deviates. On figure 1.3a, the high degree of spatial clustering of boroughs in standardised mortality ratio (SMR) category 1 is confirmed by the large deficit of BW joins. The map was reduced to two colours by giving a B coding to each borough in the SMR category (or categories) of interest, and to all others a W coding. The range of values for z when $n = 29$ is shown in figures 1.3(b)-1.3(d); a large negative z score is characteristic of strong positive autocorrelation with the BW statistic [figure 1.3(b)], a large positive value with a checkerboard pattern [figure 1.3(d)], and a value close to zero with a random pattern of B and W units [figure 1.3(c)]. Moderate spatial contagion is also detected for figure 1.4(b) (bronchitis deaths in Wales), but not for figure 1.4(a) (tuberculosis deaths). However, for Wales, n is only equal to 13 counties and, as we shall see in section 2.7, this makes the normal approximation to the sampling distribution of BW dubious.

1.7.2 Diffusion on graphs; measles in Cornwall

In a study of measles epidemics in Cornwall, 1966-70, Haggett (1976) reduced the map showing the twenty-seven local authority areas which report measles cases in Cornwall (see figure 1.7) to various binary graphs corresponding as closely as possible to given hypothetical diffusion processes. The graphs tried were

G-1 Local contagion assuming a spread of measles only between contiguous local authorities. Planar graph with 34 joins.

G-2 Wave contagion assuming spread by shortest paths from the Plymouth area, where measles is endemic. The disease is not endemic in any Cornish local authority. Planar graph with 28 joins.
G-3 Regional contagion assuming spread occurs only within two separate regional subsystems, one based on east Cornwall and one based on west Cornwall. Planar subgraphs with 32 joins.
G-4 Urban–rural contagion assuming spread within sets of urban and rural communities created as separate subgraphs. Nonplanar subgraphs with 181 joins.
G-5 Population size assuming spread down the population-size hierarchy from largest to smallest local authority. Nonplanar graph with 26 joins.
G-6 Population density assuming spread down the population-density hierarchy. Nonplanar graph with 26 joins.
G-7 Journey-to-work contagion assuming that (a) these flows provide a surrogate for spatial interaction between areas, and (b) measles spread followed high interaction links. Nonplanar graph with 58 joins.

To assess the relationship of each graph to the spread of measles epidemics, the 222 weekly maps, 1966–70, were translated into outbreak/no outbreak terms. Vertices on each of the graphs were coded B for outbreak or W for no outbreak; and the BW statistic discussed above was

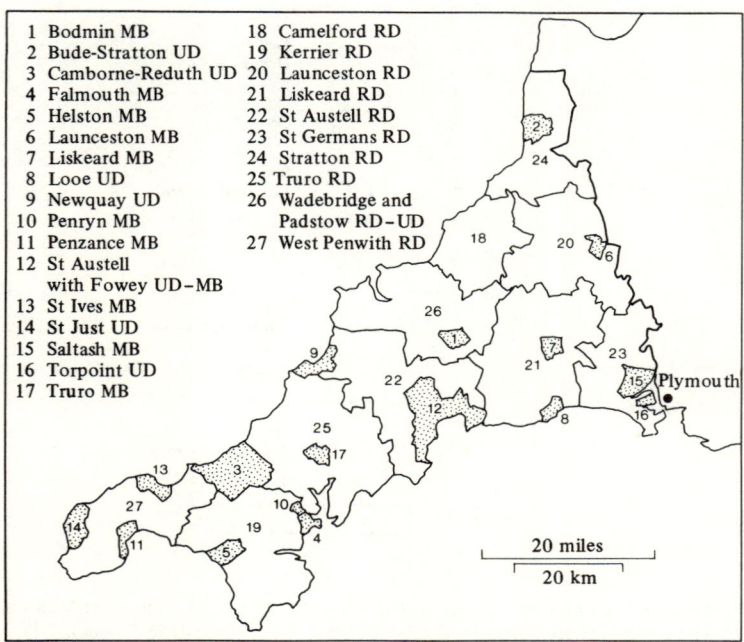

Figure 1.7. Cornish local authorities or General Register Office districts used to report disease data. Urban and metropolitan districts are stippled.

evaluated under nonfree sampling to measure the degree of autocorrelation (contagion) present in the graph. The greater the degree of correspondence between any given graph and the transmission path followed by the diffusion wave, the larger (negative) should become the z-score for BW joins.

Figure 1.8 shows the results for a forty-week epidemic starting in 1966, week 40. The epidemic peak is shown by the vertical pecked lines on the graphs. The population-size/density hierarchies are extremely important during the early build-up phase of the epidemic, whereas after the epidemic

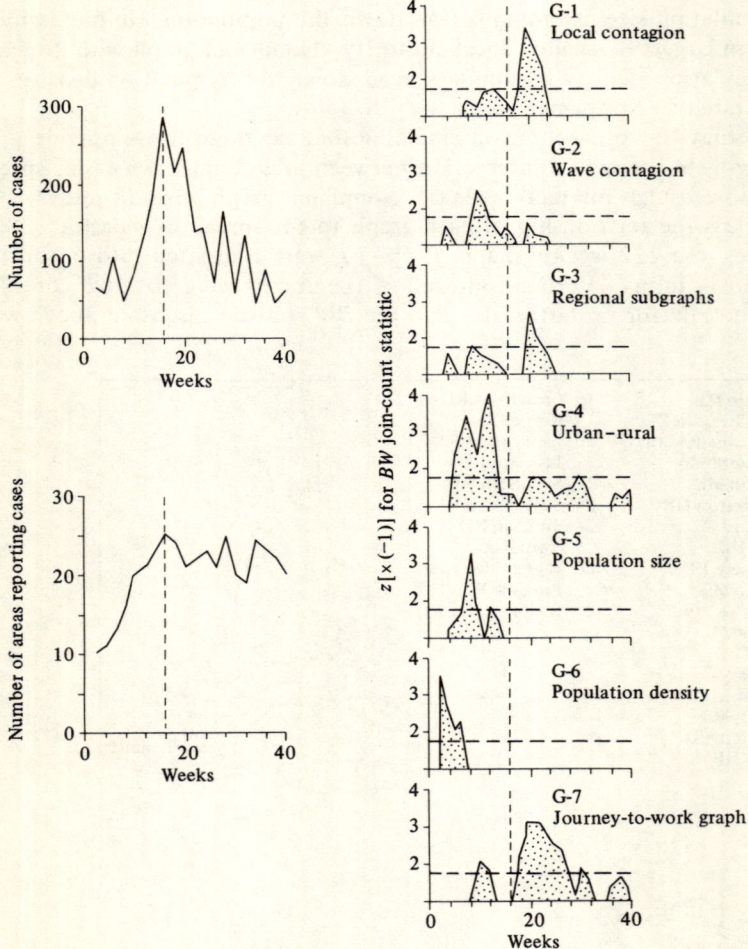

Figure 1.8. Diffusion on graphs. Values of the BW statistic [$\times (-1)$ and in standardised score form] for a measles epidemic in Cornwall, 1966 (week 40)–1967 (week 27). For definitions of graphs G-1 to G-7, see text. Vertical pecked lines indicate epidemic peak; horizontal pecked lines correspond with the $\alpha = 0\cdot 05$ significance level in a one-tailed test of significance. Source: Haggett (1976, page 141).

peak there is a marked switch to the locally contagious spatial spread graphs G-1, G-3, and G-7. The results suggest a simple diffusion model dominated by central place (size) effects in the early stages and localised spatial spread once the epidemic is fully established. This links back to the discussion of figure 1.1 earlier in the chapter.

1.7.3 Space-time interactions; cholera in the metropolis

The weekly death rate from cholera in each of the London boroughs is given in table 1.2. Their locations and identity numbers appear in figure 1.9. Using these data, we tested for spatial autocorrelation amongst contiguous boroughs in each week of the epidemic. The coefficient, I, defined in equation (1.15) was evaluated with $w_{ij} = 1$ if boroughs i and j shared a common boundary, and $w_{ij} = 0$ otherwise. The significance of I was tested using the randomisation assumption [equations (1.37) and (1.39)]. The results are shown in figure 1.10(a). The maps illustrate the spatial pattern of deaths for four weeks of the epidemic. Weeks 8 and 23 are associated with the build-up and fade-out phases of the epidemic, which was at its peak in week 20. Reference back to figure 1.2(b) indicates that the source in week 8 is in the area served by the Southwark and Vauxhall Water Company. That company's area continued to suffer extensively throughout the epidemic, which eventually faded out in the East End.

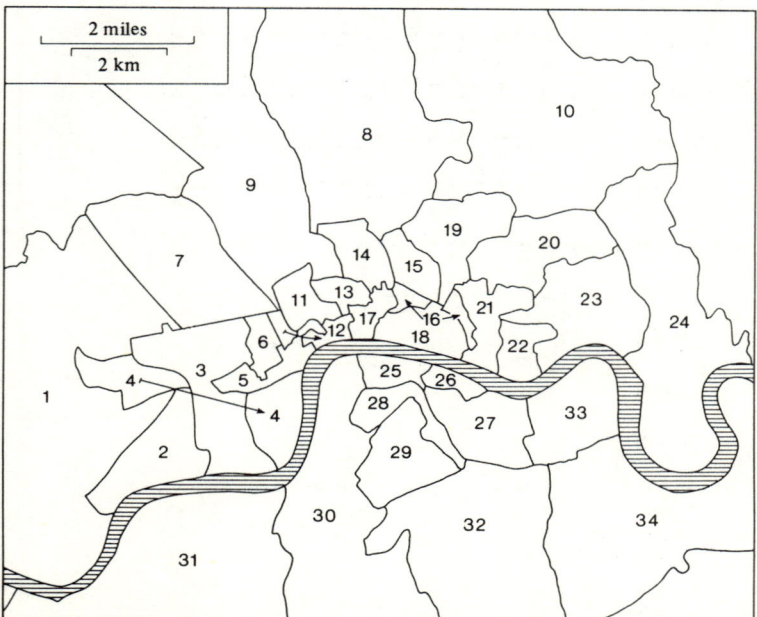

Figure 1.9. Locations and identity numbers of the London boroughs given in table 1.2.

Table 1.2. Weekly deaths from cholera in thirty-four districts of metropolitan London, April 28–November 24, 1849. Source: PP 1850, XXI, page 221.

District	Apr 28	May 5	May 12	May 19	May 26	Jun 2	Jun 9	Jun 16	Jun 23	Jun 30	Jul 7	Jul 14	Jul 21	Jul 28	Aug 4	Aug 11	Aug 18	Aug 25	Sep 1	Sep 8	Sep 15	Sep 22	Sep 29	Oct 6	Oct 13	Oct 20	Oct 27	Nov 3	Nov 10	Nov 17	Nov 24	Total deaths	1841 population	Deaths per 10000
1						1	1			1	1	2	4	4	5	11	12	13	23	35	40	11	12	12	6							193	61326	31
2					1	1	1				1	4	3	14	12	5	20	28	29	48	39	12	14	5	2	1						238	40179	59
3			1						3	13		8	4		13	9	6	6	10	18	21	5	7	8	2							128	66552	19
4								1	1	3	4	10	30	20	17	17	34	36	75	56	42	25	27	10	3			1				428	56712	75
5									2			2	3	1		6	7	8	10	9	13	8	5	2	2		1					90	25091	36
6								1	2	1		2	2		8	6	4	6	11	12	6	5	4	1	1		2					55	37398	15
7							1		2	1	3	11	4	6	10	7	24	34	41	51	28	14	10	5	8	1						258	138164	19
8			1						2	3	2	3	2	6	4	6	25	22	15	32	15	33	35	19	1		1	1				189	55690	34
9					2			2	2	3	1		1	10	8	14	20	45	54	56	45	36	13	9	8	3	2	1				327	129763	25
10												2		3	4	19	7	13	10	16	7	6	8	11			2					103	37121	27
11						1		1		1	3	5	13	6	15	3	28	28	30	27	45	8	2	1	5	1						248	54292	46
12							1			1				13	7	35	22	19	15	21	17	10	6	4	2	1						158	43598	36
13						1	3	1	1	2	4	12	4	9	8	8	12	21	9	20	15	2	2	3								138	44461	31
14									2	2	1	2	9	5	4	6	16	10	18	15	20	10	10	3	1	1		1				121	56708	21
15										2		3	8	5	11	8	13	14	22	28	33	9	6	2	3		2					177	49829	36
16												5	3	9	13	14	18	20	22	28	17	14	9	8	1	2	1					183	39655	46
17						1	3	1	4	22	19	20	18	35	32	18	40	35	45	42	43	21	19	4	2	2	2					437	29142	150
18		1						1		8	5	7	15	17	3	7	18	26	38	20	21	12	3		5	3						200	55920	36
19	1				2					1	1	4	6	17	17	21	98	121	139	109	91	52	31	29	12	1						759	83432	91

Table 1.2 (continued)

District	April 28	May 5	May 12	May 19	May 26	June 2	June 9	June 16	June 23	June 30	July 7	July 14	July 21	July 28	August 4	August 11	August 18	August 25	September 1	September 8	September 15	September 22	September 29	October 6	October 13	October 20	October 27	November 3	November 10	November 17	November 24	Total deaths	1841 population	Deaths per 10000
20							2		1	4	1	5	5	10	16	35	125	127	128	96	91	36	28	15	7	5	3					734	74088	99
21	1							2	3	4	2	7	14	20	30	28	45	55	74	58	48	39	17	20	4	2	2	1				478	71765	67
22								1	3	4		11	7	9	16	16	17	10	15	27	18	6	5	8	1		4					178	41350	43
23								1	4	12	9	11	35	22	31	24	55	58	64	59	49	30	17	11	3	4		1				500	90687	55
24									2	2	1	8	43	26	17	8	17	22	27	41	33	20	15	8	5	1	1					298	31122	96
25									3	4	9	14	14	32	46	47	57	52	65	75	49	29	15	6	6		2				1	526	32975	160
26								6	2	3	5	4	23	26	38	28	28	25	41	44	33	16	10	6	2	1						340	19837	171
27						3	3	11	7	6	13	23	64	64	67	56	53	53	70	101	72	32	9	10	6			1				773	34947	221
28							1	1	1	4	3	16	51	70	112	62	73	57	77	109	108	58	8	8	7	3						831	46644	178
29						1	1	1	1	6	5	13	53	66	86	55	65	65	93	157	137	66	15	15	2	1	1				1	861	54606	158
30							2	2	1	10	18	49	106	111	143	84	96	89	179	278	234	117	50	24	7	2		1		1		1565	112927	139
31						1				1	1	4	13	19	24	48	32	18	29	45	28	10	3	5	7	1						288	26337	109
32							1		5	3	2	12	37	43	43	32	33	35	56	79	53	16	9	7		3	1		1			468	37964	123
33								1		5	24	37	37	31	14	25	27	21	20	40	32	18	7	4		2						346	13917	249
34	1					2	1	6	1	2	8	12	28	34	37	29	72	51	75	50	73	44	26	17	8	1						578	55212	105

Key to districts: 1 Kensington, 2 Chelsea, 3 St George, Hanover Square, 4 Westminster, 5 St Martin-in-the-Fields, 6 St James, Westminster, 7 Marylebone, 8 Islington, 9 Pancras, 10 Hackney, 11 St Giles, 12 Strand, 13 Holborn, 14 Clerkenwell, 15 St Luke, 16 East London, 17 West London, 18 London City, 19 Shoreditch, 20 Bethnal Green, 21 Whitechapel, 22 St George-in-the-East, 23 Stepney, 24 Poplar, 25 St Saviour, 26 St Olave, Southwark, 27 Bermondsey, 28 St George, Southwark, 29 Newington, 30 Lambeth, 31 Wandsworth, 32 Camberwell, 33 Rotherhithe, 34 Greenwich.

The left-hand graph shows the number of deaths recorded in each week of the epidemic, and the right-hand graph plots the value of the I statistic in standard-deviate form. The pecked line is the $\alpha = 0 \cdot 05$ significance level in a one-tailed test of positive spatial autocorrelation. The degree of spatial contagion waxes and wanes in remarkable sympathy with the epidemic curve, and is indicative of substantial areal concentration of high-risk and low-risk zones when the epidemic was in full swing.

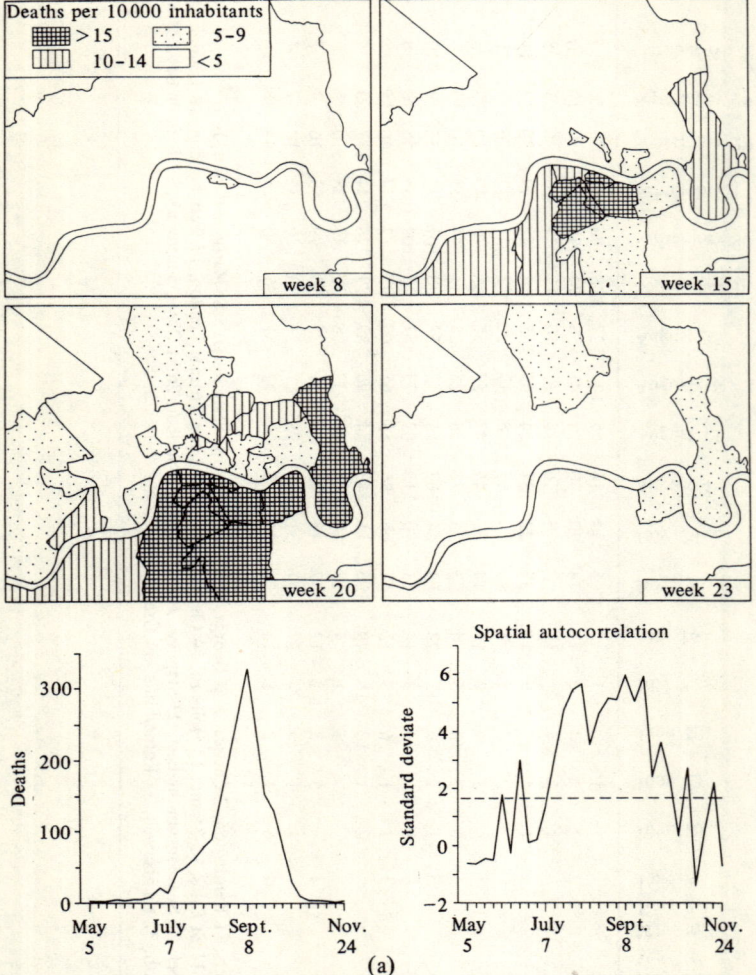

Figure 1.10. Deaths from cholera in London, 1849, and results of tests for spatial autocorrelation. (b) Cumulative deaths and results of space-time interaction analysis The area served by the Southwark and Vauxhall Water Company is shown on figure 1.2(b).

To try to determine whether the epidemic was dominated by the *in situ* rise and fall of death rates in the areas served by different water companies, or by a spatial diffusion process, we evaluated the space-time index, $I_{s\text{-}t}$, given in equation (1.44). The $\{w_{ij}\}$ were defined as above and we set $y_{ij} = 1$ if borough i was above the mean death rate per 10000 people at time t and area j, $j \neq i$, was also above the mean at $t-1$; $y_{ij} = 0$ otherwise. $I_{s\text{-}t}$ was computed and tested for significance using equations (1.45)-(1.46) for all pairs of maps during the epidemic, and we are thus searching for a relationship between high death rates in area i at t, and high death rates in boroughs contiguous to i at $t-1$; that is, a first order relationship for a space-time spatial diffusion process. The results are illustrated in figure 1.10(b). The set of maps shows a three-week sequence. The bonds show the space-time relationships examined for the cross-hatched area on the week-13 map, and may be compared with figure 1.5. The left-hand

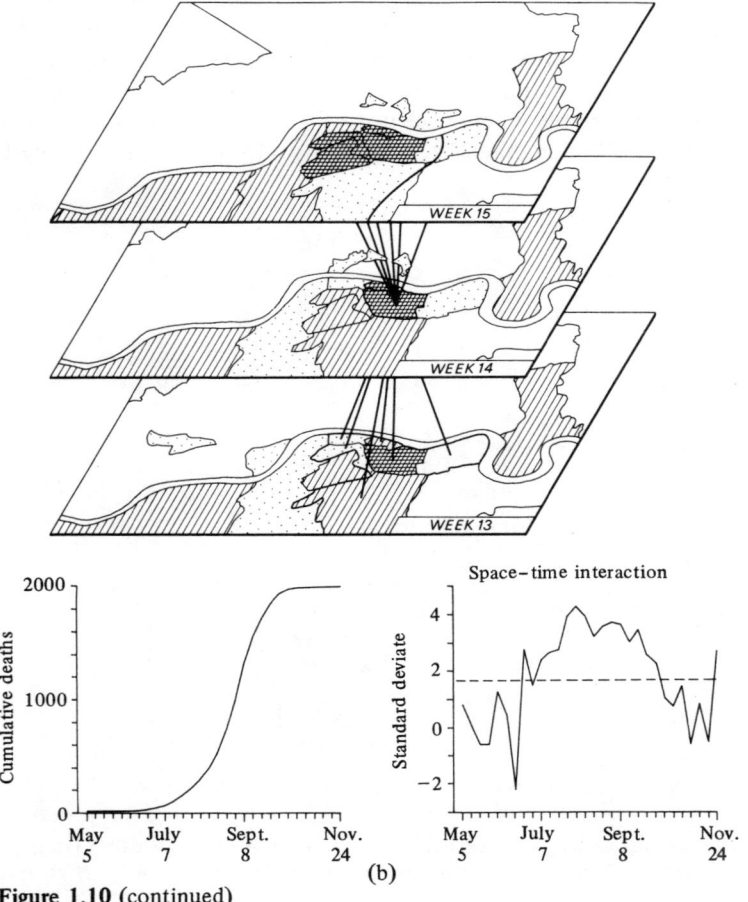

Figure 1.10 (continued)

graph indicates that the cumulative number of deaths with time has the logistic shape characteristic of most diffusion and epidemic processes, and the right-hand graph is a plot of $I_{\text{s-t}}$ in standard-deviate form (vertical axis) against time. $I_{\text{s-t}}$ is significant during the epidemic peak, which shows that *in situ* processes and diffusion are both important during the peak, whereas the strength of the diffusion processes weakens when the number of cases is low.

To check this interpretation, we constructed the following model. Let $x_{i,t}$ be the death rate per 10 000 people from cholera in borough i at time t. Then we postulate

$$X_{i,t} = a + b_1 x_{i,t-1} + b_2 \sum_{j \in J} w_{ij} x_{j,t-1} + \epsilon_{i,t} \tag{1.46}$$

in the manner of figure 1.5. The summation is over the j boroughs contiguous to i. Although there are problems posed by zero death rates in some boroughs during the build-up and fade-out phases of the epidemic, we fitted the model by ordinary least squares using a stepwise regression procedure. Interest centres upon the order in which variables were entered and the levels of explanation provided. If *in situ* growth is more important, we would expect variable 1 (coefficient b_1) to be entered first, whereas if a diffusion process is dominant, we would expect variable 2 (coefficient b_2) to appear first.

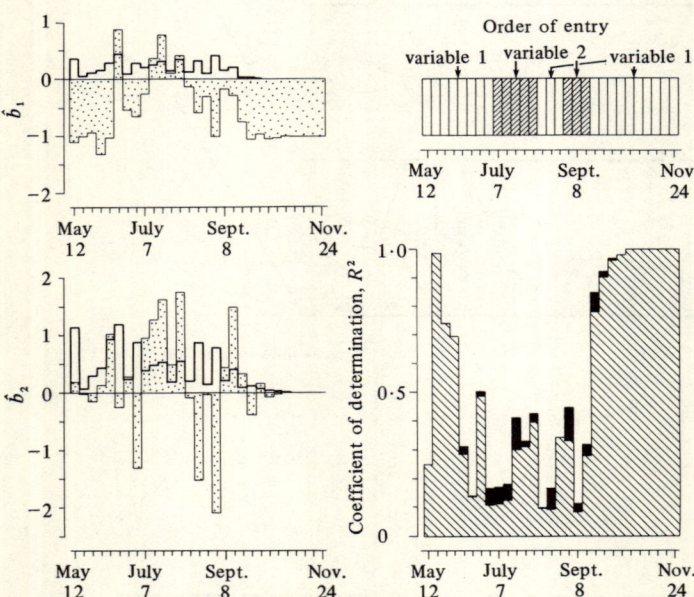

Figure 1.11. Results of stepwise regression analysis of 1849 cholera data. Graphs give the values of the coefficients of the variables, their standard errors, order of entry of variables, and values of R^2.

The results are shown in figure 1.11 for all pairs of maps during the epidemic. The left-hand diagrams give the estimated values of b_1 and b_2 (stipple) and their standard errors (solid line). The top right-hand diagram shows whether variable 1 or variable 2 was entered first for each week. The bottom right-hand diagram gives the values of R^2 (coefficient of determination) after one variable has been entered (cross-hatched) and after the second step (black). High levels of 'explanation' are achieved at the start and end of the epidemic, where the process is dominated by the purely temporal autoregressive component (variable 1). This is largely because of the zero death rates reported by several boroughs during these phases of the epidemic. Where continuous data records are available (beginning of June to early September), levels of R^2 for the model are not high. Space-time interaction (variable 2) is more important in the immediately preepidemic and postepidemic peak periods (July and late August), and temporal autocorrelation is dominant at the epidemic peak.

Given that the quality of the water supply is similar in neighbouring areas and the infectious nature of the disease, these patterns of spread seem reasonable. Indeed, the *in situ* growth may be interpreted as the local spread (within the recording unit) which is likely to predominate when the overall numbers of new cases are low and the disease occurs in isolated pockets.

1.8 Conclusions

In this chapter, the concepts of spatial, temporal, and space-time autocorrelation have been outlined, and various measures of autocorrelation in these domains have been defined. We have indicated that the sampling distributions of the measures approach normality as the size of the study area increases, and tests of significance for the coefficients have been proposed based upon this fact. Formulae for the location and scale parameters of the coefficients have been given. To illustrate the use of the methods, we have analysed the map patterns formed by the incidence of various diseases—measles in Cornwall, 1969-70; cholera in London, 1849; and tuberculosis and bronchitis in Wales, 1959-63—and have tried to indicate how interpreting the maps in the light of the statistical evidence can give insights into the underlying processes shaping the geographical patterns.

As noted above, normal distribution theory is essential to tests of hypotheses using the coefficients. Accordingly we shall turn, in the next chapter, to consider the circumstances under which this theory holds.

Distribution theory for the join-count, I, and c statistics

2.1 Introduction
In chapter 1 we suggested that the join-count, I, and c statistics could be tested for significance as standard normal deviates. Computational formulae for the first two moments of the various coefficients were given, and these we indicated should be used to specify the location, μ, and scale, σ^2, parameters of the normal curve. In this chapter we shall show how the moments given in chapter 1 were derived. Colour lattices are considered in section 2, and the Moran and Geary statistics are examined in section 3; some higher order moments for the various coefficients are also outlined. In section 4, we shall prove that the join-count, I, and c statistics are all asymptotically normally distributed under the null hypothesis of no spatial autocorrelation, which provides the basis for the test procedure outlined in chapter 1. However, in sections 5–7 we shall show that the sampling distributions may be badly nonnormal in small lattices, and alternative approximations will therefore be explored.

2.1.1 General expressions for the moments
In section 1.6.1 we introduced the general statistic

$$I_{\text{s-t}} = \sum_{(2)} w_{ij} y_{ij} \,. \tag{2.1}$$

We let Y_{ij} denote the random variable corresponding to y_{ij}. Then, given the set of weights, the expected value of $I_{\text{s-t}}$ is

$$\mu = E(I_{\text{s-t}}) = \sum_{(2)} w_{ij} E(Y_{ij}) \,, \tag{2.2}$$

where the expectation is taken with respect to the distribution of Y_{ij} or to the set of random permutations, as appropriate. Likewise, the variance is

$$\text{var}(I_{\text{s-t}}) = E(I_{\text{s-t}}^2) - \mu^2 \,, \tag{2.3}$$

where

$$E(I_{\text{s-t}}^2) = \sum_{(2)} w_{ij}(w_{ij} + w_{ji}) E(Y_{ij}^2) + \sum_{(3)} (w_{ij} + w_{ji})(w_{ik} + w_{ki}) E(Y_{ij} Y_{ik}) + \sum_{(4)} w_{ij} w_{kl} E(Y_{ij} Y_{kl}) \,, \tag{2.4}$$

where

$$\sum_{(2)} = \sum_{\substack{i=1 \\ i \neq j}}^{n} \sum_{j=1}^{n}, \qquad \sum_{(3)} = \sum_{\substack{i=1 \\ i \neq j \neq k}}^{n} \sum_{j=1}^{n} \sum_{k=1}^{n},$$

and

$$\sum_{(4)} = \sum_{\substack{i=1 \\ i \neq j \neq k \neq l}}^{n} \sum_{j=1}^{n} \sum_{k=1}^{n} \sum_{l=1}^{n} .$$

Since $w_{ii} = 0$ for all i, there are $n(n-1)$ terms in the expression (2.1) for I_{s-t}, although many of these will vanish in particular cases. Hence there are $[n(n-1)]^2$ possible terms in the expression on the right-hand side of equation (2.4).

The different types of terms and their frequencies are shown in figure 2.1 and it may be verified that the total number of terms is indeed $[n(n-1)]^2$. The weights mentioned in the figure are by way of example only; thus the first pattern also has weights $w_{ij}w_{ji}$, and the second has $w_{ij}w_{ki}$, $w_{ji}w_{ki}$, and $w_{ji}w_{ik}$ in addition to the one given. These permutations of the indices are reflected in expansion (2.4).

When the join pattern separates, as for the term $w_{ij}w_{kl}$ in figure 2.1, this reflects the fact that the terms in the expression may be partitioned into disjoint subsets; here, (i, j) and (k, l). Then, if the variates Y_{ij} and Y_{kl} are independent, they will make no contribution to var(I_{s-t}). To exploit this possibility we may rewrite var(I_{s-t}) as

$$\text{var}(I_{s-t}) = \sum_{(2)} w_{ij}(w_{ij} + w_{ji})\text{var}(Y_{ij}) + \sum_{(3)} (w_{ij} + w_{ji})(w_{ik} + w_{ki})\text{cov}(Y_{ij}, Y_{ik})$$
$$+ \sum_{(4)} w_{ij}w_{kl}\text{cov}(Y_{ij}, Y_{kl}) \,. \tag{2.5}$$

Whenever Y_{ij} and Y_{kl} are independent (for all $i \neq j \neq k \neq l$), the third term in expression (2.5) will be zero. In the remainder of this chapter we shall work with whichever of equations (2.4) and (2.5) seems more convenient.

Further simplification of these expressions is possible when we suppose that 'E(Y_{ij}) and cov(Y_{ij}, Y_{kl}) for any $i \neq j$ and $k \neq l$ are *not* dependent on the subscripts'.

Under the null hypothesis of no spatial autocorrelation, we may make these assumptions either in a specific distributional sense or by virtue of assuming that all permutations are equally likely. Then it follows from equation (2.2) that

$$\mu = \text{E}(Y_{ij})\sum_{(2)} w_{ij} = S_0 \text{E}(Y_{ij}) \,, \tag{2.6}$$

where $S_0 = \sum_{(2)} w_{ij}$, and E(Y_{ij}) is the same for any pair of distinct subscripts.

Likewise, from equation (2.5) we have

$$\text{var}(I_{s-t}) = \text{var}(Y_{ij})\sum_{(2)} w_{ij}(w_{ij} + w_{ji})$$
$$+ \text{cov}(Y_{ij}, Y_{ik})\sum_{(3)}(w_{ij} + w_{ji})(w_{ik} + w_{ki})$$
$$+ \text{cov}(Y_{ij}, Y_{kl})\sum_{(4)} w_{ij}w_{kl} \,, \tag{2.7}$$

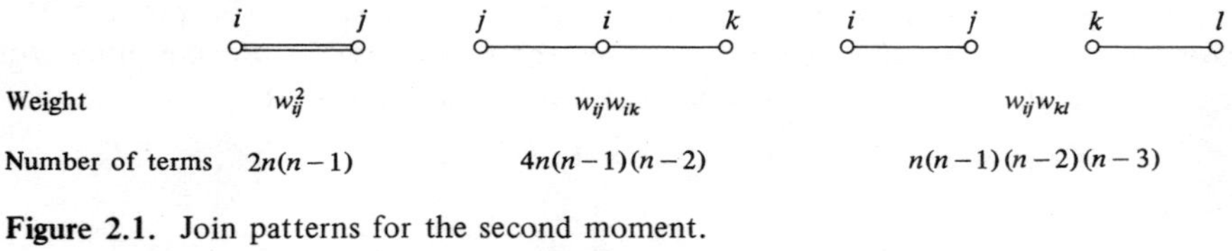

	$i \quad\quad j$	$j \quad\quad i \quad\quad k$	$i \quad\quad j \quad\quad k \quad\quad l$
Weight	w_{ij}^2	$w_{ij}w_{ik}$	$w_{ij}w_{kl}$
Number of terms	$2n(n-1)$	$4n(n-1)(n-2)$	$n(n-1)(n-2)(n-3)$

Figure 2.1. Join patterns for the second moment.

with a similar result for equation (2.4). Again the variance and covariances are the same for each set of distinct subscripts. We can now operate upon the sums of weights in expression (2.7) to obtain a simpler form. First of all, we remind the reader of the definitions:

$$S_1 = \tfrac{1}{2} \sum_{(2)} (w_{ij} + w_{ji})^2 , \quad \text{and} \quad S_2 = \sum_i (w_{i.} + w_{.i})^2 ,$$

where $w_{i.} = \sum_j w_{ij}$, and $w_{.i} = \sum_j w_{ji}$. We find that

$$\sum_{(3)} (w_{ij} + w_{ji})(w_{ik} + w_{ki}) = \sum_{(2)} (w_{ij} + w_{ji})(w_{i.} + w_{.j} - 2w_{ij})$$
$$= \sum_i (w_{i.} + w_{.i})^2 - \sum_{(2)} (w_{ij} + w_{ji})^2$$
$$= S_2 - 2S_1 ; \qquad (2.8)$$

further,

$$\sum_{(4)} w_{ij} w_{kl} = S_0^2 - \sum_i (w_{i.} + w_{.i})^2 + \sum_{(2)} w_{ij}(w_{ij} + w_{ji}) = S_0^2 - S_2 + S_1 . \quad (2.9)$$

Using these results, we can reduce expression (2.7) to

$$\text{var}(I_{\text{s-t}}) = S_1 \text{var}(Y_{ij}) + (S_2 - 2S_1)\text{cov}(Y_{ij}, Y_{ik}) + (S_0^2 - S_2 + S_1)\text{cov}(Y_{ij}, Y_{kl}) . \qquad (2.10)$$

A similar result is obtained for equation (2.4) as

$$E(I_{\text{s-t}}^2) = S_1 E(Y_{ij}^2) + (S_2 - 2S_1)E(Y_{ij}Y_{ik}) + (S_0^2 - S_2 + S_1)E(Y_{ij}Y_{kl}) . \quad (2.11)$$

We shall now proceed to use these results to develop the moments of the statistics considered in the previous chapter.

2.2 The join-count statistics

Following the method of section 1.3.1, we consider a lattice, regular or irregular, of n cells and allow the ith cell to be coloured black, B, or white, W. The binary variate, X_i for cell i, is defined as

$$X_i = \begin{cases} 1, & \text{if the } i\text{th cell is } B, \\ 0, & \text{if the } i\text{th cell is } W. \end{cases}$$

Consider the null hypothesis, H_0, that the colour for cell j is selected independently of that for cell i, $i \neq j$, or that

$$\text{prob}(X_i = 1, X_j = 1) = \text{prob}(X_i = 1)\text{prob}(X_j = 1) , \qquad (2.12)$$

with similar statements for X_i and/or X_j equal to zero. The process of choosing a particular colour is equivalent to success/failure sampling from a population. If the additional assumption is made that

$$\text{prob}(X_i = 1) = p_i = p , \qquad i = 1, 2, ..., n ,$$

we have either
(1) binomial sampling, if there is no restriction on the sampling process (known as the free sampling model); or
(2) hypergeometric sampling, if the total numbers of black and white cells are specified *a priori* (nonfree sampling).

Free sampling corresponds to the situation where p is known and so sampling takes place as if from an infinite population. More commonly, however, p has to be estimated from the data as, say,

$$p = \frac{\text{number of black cells}}{\text{total number of cells}} = \frac{n_1}{n}. \tag{2.13}$$

Although the moments will then only be estimates, the free sampling model may still be used. As stated in section 1.3.1, the investigator must decide the appropriate reference set for his hypothesis test. Thus the two models are not competitors but depend upon whether or not we condition upon exactly n_1 black cells.

The test statistic based on black–black (BB) joins is written as (see section 1.4.1)

$$BB = \tfrac{1}{2} \sum_{(2)} w_{ij} x_i x_j, \tag{2.14}$$

while for black–white (BW) joins we have

$$BW = \tfrac{1}{2} \sum_{(2)} w_{ij}(x_i - x_j)^2, \tag{2.15}$$

where x_i is the value observed for the variate X_i. We do not consider white–white (WW) joins separately since

$$WW = \tfrac{1}{2} \sum_{(2)} w_{ij}(1 - x_i)(1 - x_j),$$

where $1 - x_i = 1$ if cell i is W, and

$$BB + BW + WW = \tfrac{1}{2} \sum_{(2)} w_{ij}[x_i x_j + (x_i - x_j)^2 + (1 - x_i)(1 - x_j)]$$
$$= \tfrac{1}{2} \sum_{(2)} w_{ij}, \tag{2.16}$$

a constant. Result (2.16) follows because

$$x_i^r = x_i, \tag{2.17}$$

for all integers $r \geq 1$, since x_i can only be zero or one. Thus WW is a linear function of BB and BW and does not supply any additional information.

From equation (2.1) we note that BB and BW may both be written in the form

$$\tfrac{1}{2}(I_{\text{s-t}}) = \tfrac{1}{2} \sum_{(2)} w_{ij} y_{ij}, \tag{2.18}$$

where $y_{ij} = x_i x_j$ for BB, and $y_{ij} = (x_i - x_j)^2$ for BW. We shall now proceed to use the results of section 2.1.1 to derive the moments of these statistics.

2.2.1 Mean and variance of BB and BW

Let us first look at the free sampling model, where each drawing has a binomial distribution, and then examine nonfree sampling.

(1) *The free sampling model*

Each observation is an independent drawing with $\text{prob}(X_i = 1) = p$ for all i, so that

$$E(X_i) = 0 \times (1-p) + 1 \times p = p, \qquad (2.19)$$

and $E(X_i^r) = p$ also. Because each drawing is independent, it follows that $E(X_i X_j) = E(X_i)E(X_j) = p^2$ and so on. Thus for the BB statistic, $Y_{ij} = X_i X_j$ and we have

$$E(Y_{ij}) = p^2 ;$$

from expressions (2.2) and (2.18) this yields

$$E(BB) = \tfrac{1}{2} S_0 p^2 . \qquad (2.20)$$

Similarly

$$\text{var}(Y_{ij}) = E(X_i^2 X_j^2) - [E(X_i X_j)]^2$$
$$= E(X_i^2)E(X_j^2) - [E(X_i)E(X_j)]^2 = p^2 - p^4 ,$$

because of independence and the result that $E(X_i^2) = p$. Since $E(Y_{ij} Y_{ik}) = E(X_i^2 X_j X_k) = p^3$, $i \neq j \neq k$, and $E(Y_{ij} Y_{kl}) = E(X_i X_j X_k X_l) = p^4$, $i \neq j \neq k \neq l$, it follows that $\text{cov}(Y_{ij}, Y_{ik}) = p^3 - p^4$, and $\text{cov}(Y_{ij}, Y_{kl}) = 0$. Thus, equations (2.10) and (2.18) yield

$$4\,\text{var}(BB) = S_1(p^2 - p^4) + (S_2 - 2S_1)(p^3 - p^4)$$
$$= p^2(1-p)[S_1(1-p) + S_2 p] , \qquad (2.21)$$

corresponding to equation (1.26) when $k = 2$.

A similar argument for BW yields

$$E(Y_{ij}) = 2pq, \qquad \text{var}(Y_{ij}) = 2pq(1 - 2pq) ,$$
$$\text{cov}(Y_{ij}, Y_{ik}) = pq(1 - 4pq), \quad \text{and} \quad \text{cov}(Y_{ij}, Y_{kl}) = 0 ,$$

where $q = 1 - p$, and all the subscripts are different. Hence

$$E(BW) = S_0 pq , \qquad (2.22)$$

and

$$4\,\text{var}(BW) = 4S_1 pq + S_2 pq(1 - 4pq) , \qquad (2.23)$$

corresponding to equations (1.27) and (1.28) when $k = 2$. The results for $k > 2$ follow by similar arguments.

(2) The nonfree sampling model

The observations are no longer independent since they must satisfy the constraint

$$\sum_{i=1}^{n} X_i = n_1.$$

We must now consider sampling without replacement so that for any single cell

$$\text{prob}(X_i = 1) = \frac{n_1}{n},$$

whereas for any pair of cells,

$$\text{prob}(X_i = 1, X_j = 1) = \frac{n_1(n_1 - 1)}{n(n-1)} \tag{2.24}$$

and

$$\text{prob}(X_i = 1, X_j = 1, ..., X_m = 1) = \frac{n_1^{(b)}}{n^{(b)}}, \tag{2.25}$$

where $n^{(b)} = n(n-1) ... (n-b+1)$, and b distinct subscripts are represented in equation (2.25).

Once again we examine the BB statistic and we see from expression (2.24) that

$$E(Y_{ij}) = E(X_i X_j) = \frac{n_1^{(2)}}{n^{(2)}},$$

leading to

$$E(BB) = \frac{S_0}{2} \frac{n_1^{(2)}}{n^{(2)}}. \tag{2.26}$$

It is now more convenient to use expression (2.4) and we find, from equation (2.25), that

$$E(Y_{ij}^2) = \frac{n_1^{(2)}}{n^{(2)}}, \qquad E(Y_{ij} Y_{ik}) = \frac{n_1^{(3)}}{n^{(3)}}, \quad \text{and} \quad E(Y_{ij} Y_{kl}) = \frac{n_1^{(4)}}{n^{(4)}},$$

which yields

$$4\text{var}(BB) = S_1 \left[\frac{n_1^{(2)}}{n^{(2)}} - \frac{2n_1^{(3)}}{n^{(3)}} + \frac{n_1^{(4)}}{n^{(4)}} \right] + S_2 \left[\frac{n_1^{(3)}}{n^{(3)}} - \frac{n_1^{(4)}}{n^{(4)}} \right] + \frac{S_0^2 n_1^{(4)}}{n^{(4)}} - \left[\frac{S_0 n_1^{(2)}}{n^{(2)}} \right]^2. \tag{2.27}$$

As $n \to \infty$, $n_1^{(j)}/n^{(j)} \to p^j$, but result (2.27) does *not* approach that given in equation (2.21) because the last two terms do not vanish. Instead, they approach $-4S_0^2 p^3 q / n$. Thus, in the limit, for the nonfree sampling case, we have

$$4\text{var}(BB) = p^2 q \left(S_1 q + S_2 p - \frac{4 S_0^2 p}{n} \right).$$

Var(BW), given in equation (1.34), can be found by similar methods.

As with BB, $\text{var}(BW)$ for nonfree sampling does not approach $\text{var}(BW)$ for free sampling as $n \to \infty$. Rather, the limiting form is

$$4\,\text{var}(BW) = 4S_1 p^2 q^2 + pq(1 - 4pq)\left(S_2 - \frac{4S_0^2}{n}\right).$$

2.2.2 Higher moments

Derivation of the third and higher order moments follows exactly the same lines as for the first two moments. The third and fourth moments for the two and k-colour join counts are given in the literature as follows:

(1) *Free sampling*
$k = 2$, rook's case (Moran, 1948; Krishna Iyer, 1949),
queen's case (BB only, Dacey, 1965),
queen's, bishop's, row-only, and column-only cases (Cliff, 1969).
$k > 2$, all cases (Cliff, 1969);

(2) *Nonfree sampling*
all cases (Cliff, 1969).

The types of terms which occur in the third moments, and their weights, are shown in figure 2.2, and those for the fourth moment appear in figure 2.3. Terms marked with an asterisk have two or more separate links and, in the free sampling case, do not appear in the final forms of the moments.

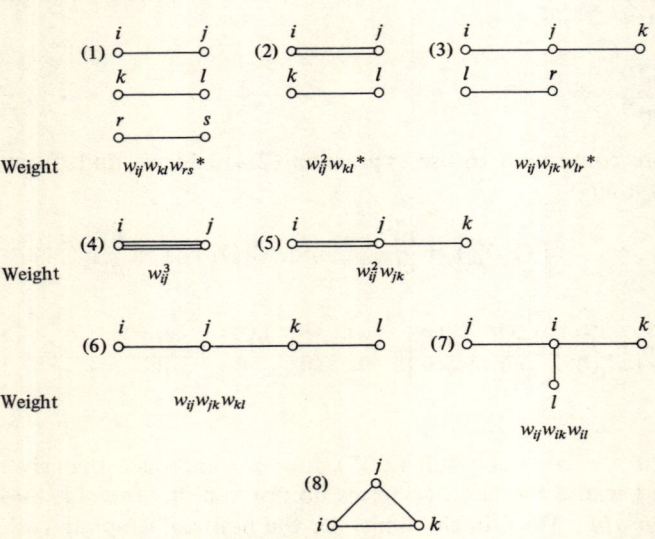

Figure 2.2. Join patterns for the third moment. Numbers are as allocated by Moran (1948).

Distribution theory for the join-count, I, and c statistics

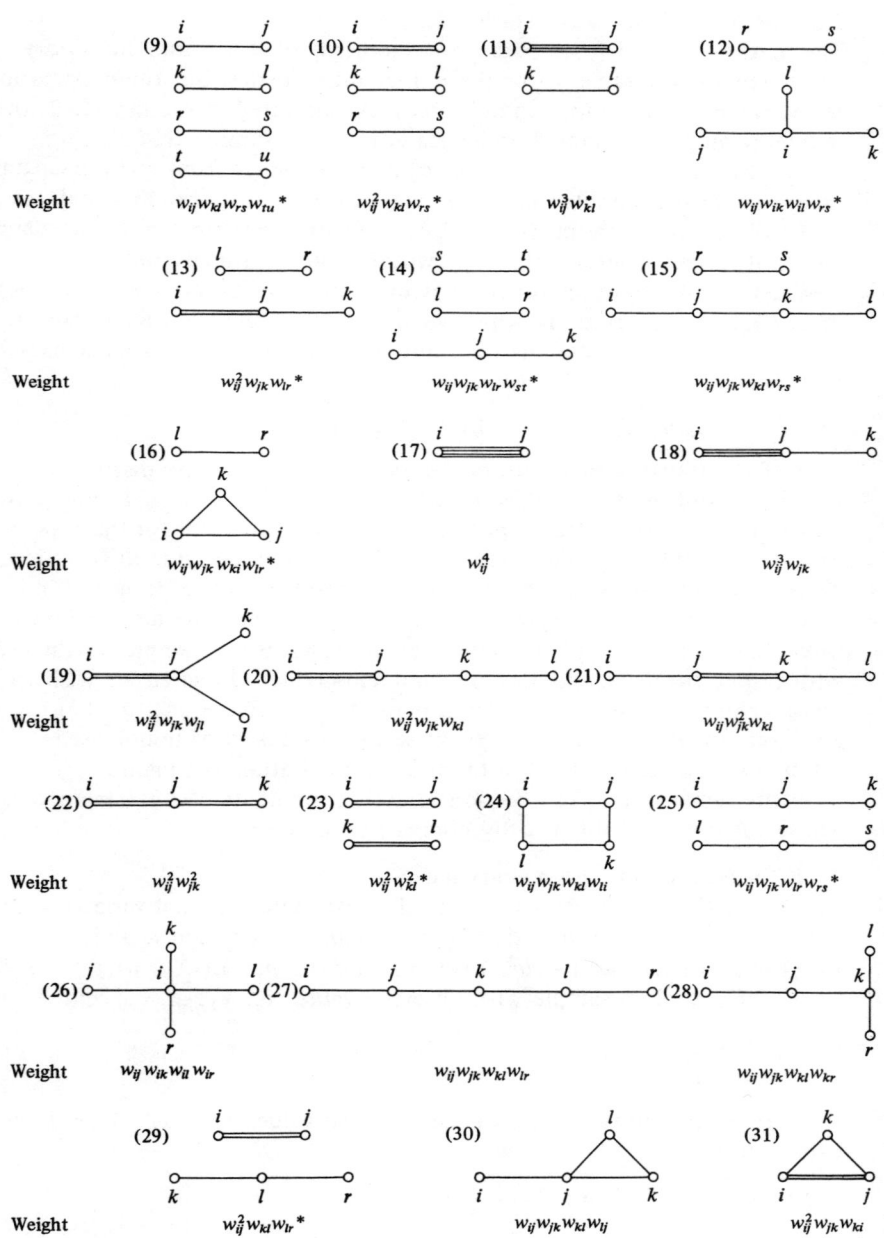

Figure 2.3. Join patterns for the fourth moment.

2.3 The Moran and Geary statistics

The method by which we obtain the moments of the Moran and Geary coefficients is the same as for the join-count statistics, but the expectation terms are somewhat more complicated. As indicated in section 1.3.2, the moments may be evaluated under H_0 either by assuming that:
1. the observations are random independent drawings from one (or separate identical) normal population(s)—assumption N; or by assuming that:
2. the observations are random independent drawings from one (or separate identical) population(s) with unknown distribution function(s).

When the distribution function is unknown, we may consider the set of $n!$ random permutations, which we refer to as assumption R. However, Ord (1981) demonstrates that, for any distribution whose variance exists, the expectations under H_0 are

$$E(I) = -1/(n-1) \quad \text{and} \quad E(I^2) = E[E_R(I^2)] .$$

Even if the fourth moment does not exist, $b_2 \leq n$ for any distribution, so that $E(I^2)$ still exists. That is, $E_R(I^2)$, the expected value of I^2 under the set of random permutations provides an unbiased estimator for the expected value of I^2 (and hence the variance of I) for any underlying distribution. Thus, when we come to approximate the sampling distribution of I in section 2.4, we may use $E_R(I^2)$ to estimate $E(I^2)$ when the distribution is unknown. The practical importance of this result is that we may evaluate I with respect to its 'randomisation' moments either (1) when we wish to consider only the set of $n!$ random permutations of the data, or (2) for any nonnormal population, when these expressions provide unbiased estimators for the moments without a randomisation argument.

In the remainder of this section, only the moments of I are evaluated. Those for c can be obtained by comparable operations.

2.3.1 The moments of I, given assumption N

Let $X_1, ..., X_n$ be independent identically distributed normal variates with mean, μ, and variance, σ^2; that is, X_i is $N(\mu, \sigma^2)$ for each i, and $E(X_i) = \mu$, $E[(X_i - \mu)^2] = \sigma^2$, $E[(X_i - \mu)^3] = 0$, and $E[(X_i - \mu)^4] = 3\sigma^4$. Thus, for a random sample with observed values $x_1, x_2, ..., x_n$, and

$$\bar{x} = \frac{1}{n} \sum_{i=1}^{n} x_i ,$$

the variates Z_i corresponding to the observed values $z_i = x_i - \bar{x}$ have expectations

$$\left. \begin{array}{ll} E(Z_i) = 0 , & E(Z_i^2) = \left(1 - \frac{1}{n}\right)\sigma^2 , \\[6pt] E(Z_i Z_j) = -\dfrac{\sigma^2}{n} , & E(Z_i^2 Z_j^2) = \dfrac{(n^2 - 2n + 3)\sigma^4}{n^2} , \\[10pt] E(Z_i^2 Z_j Z_k) = -\dfrac{(n-3)\sigma^4}{n^2} , & E(Z_i Z_j Z_k Z_l) = \dfrac{3\sigma^4}{n^2} . \end{array} \right\} \quad (2.28)$$

The key theorem which allows us to evaluate the moments of I under assumption N is due to Pitman (1937) and Koopmans (1942).

Theorem 1: If $X_1, ..., X_m$ are independent identically distributed normal variates, say $N(0, 1)$, then any scale-free function of them, $h(X_1, ..., X_m)$, is distributed independently of $Q = \sum_{i=1}^{m} X_i^2$.

Proof: Consider the moment generating function

$$M(t_1, t_2) = E[\exp(t_1 h + t_2 Q)] .$$

Then

$$M(t_1, t_2) \propto \int_{(m)} \exp(t_1 h)\exp[-\tfrac{1}{2}(1 - 2t_2)x_i^2] \prod_{i=1}^{m} dx_i ,$$

and therefore

$$M(t_1, t_2) = (1 - 2t_2)^{-m/2} E[\exp(t_1 h)] ,$$

so that h and Q are independent, and Q is distributed as chi-squared with m degrees of freedom.

If the moment generating function of h does not exist, the theorem follows on using the characteristic function instead.

In the present context, if all the x_i (and thus z_i) are replaced by λx_i in I (or c), the value of the statistic is unchanged. Thus I is truly a scale-free function of the x_i. Apart from a constant, the denominator of I is $\sum_{i=1}^{n} z_i^2$. Although the Z_i are not independent, they can be transformed into $(n-1)$ independent identical normal variates, so that ΣZ_i^2 is distributed as chi-squared with $(n-1)$ degrees of freedom.

Thus, from the Pitman–Koopmans theorem, I and Q ($= \Sigma Z_i^2$) are independent so that

$$E(I^p Q^p) = E(I^p)E(Q^p) , \quad \text{for any positive } p. \tag{2.29}$$

In Dacey's coefficient, I', given in equation (1.11), the denominator term is $\Sigma \alpha_i z_i^2$, which cannot be transformed into the form of Q, as is required by the theorem.

If we return now to I we find that, since

$$IQ = \frac{n}{S_0} \sum_{(2)} w_{ij} z_i z_j ,$$

it follows that

$$E(I^p) = \frac{E(I^p Q^p)}{E(Q^p)} = \left(\frac{n}{S_0}\right)^p \frac{E[(\sum_{(2)} w_{ij} Z_i Z_j)^p]}{E(Q^p)}. \tag{2.30}$$

Equation (2.30) allows us to evaluate the crude moments of I. Thus

$$E(I) = \frac{n}{S_0} \frac{E(\sum_{(2)} w_{ij} Z_i Z_j)}{E(\sum Z_i^2)}. \tag{2.31}$$

From equation (2.28), the denominator term has expectation $(n-1)\sigma^2$, which also follows directly from the fact that Q is distributed as chi-squared with $(n-1)$ degrees of freedom. For the numerator,

$$E(\sum_{(2)} w_{ij} Z_i Z_j) = \sum_{(2)} w_{ij} E(Z_i Z_j) = -\frac{\sigma^2 S_0}{n}.$$

Finally,

$$E(I) = \frac{n}{S_0}\left(-\frac{\sigma^2 S_0}{n}\right) \bigg/ (n-1)\sigma^2 = -\frac{1}{n-1}. \tag{2.32}$$

Evaluation of the second crude moment proceeds in the same way, with

$$E(I^2) = \frac{n^2 E[(\sum_{(2)} w_{ij} Z_i Z_j)^2]}{S_0^2 E[(\sum Z_i^2)^2]}. \tag{2.33}$$

From the chi-squared distribution the denominator term in equation (2.33) has expectation $(n-1)(n+1)\sigma^4$. By use of equation (2.4) the numerator term becomes

$$E[(\sum_{(2)} w_{ij} Z_i Z_j)^2] = S_1 E(Z_i^2 Z_j^2) + (S_2 - 2S_1) E(Z_i^2 Z_j Z_k)$$
$$+ (S_0^2 - S_2 + S_1) E(Z_i Z_j Z_k Z_l). \tag{2.34}$$

If we substitute the expectations (2.28) into equation (2.34), we obtain the expression

$$\frac{\sigma^4}{n^2}[(n^2 - 2n + 3)S_1 - (n-3)(S_2 - 2S_1) + 3(S_0^2 + S_1 - S_2)],$$

which reduces to

$$\frac{\sigma^4}{n^2}(n^2 S_1 - n S_2 + 3 S_0^2).$$

Finally

$$E(I^2) = \frac{1}{(n-1)(n+1)S_0^2}(n^2 S_1 - n S_2 + 3 S_0^2). \tag{2.35}$$

Higher order moments may be evaluated in the same way. However, the results increase rapidly in complexity as the following equation for $E(I^3)$ shows:

$$E(I^3) = -\frac{3(n-1)(n-5)S_1 + 2n^3 S_3 - 3n^2(S_0 S_1 + S_4) + 3S_0(3n S_2 - 5 S_0^2)}{2(n-1)(n+1)(n+3) S_0^3}, \tag{2.36}$$

where

$$S_3 = \sum_{(3)} (w_{ij} + w_{ji})(w_{ik} + w_{ki})(w_{kj} + w_{jk})$$

and

$$S_4 = \sum_{(2)} (w_{ij} + w_{ji})(w_{i.} + w_{.i})(w_{j.} + w_{.j}).$$

The third and fourth moments for c, under assumption N, are given in Geary (1954).

2.3.2 The moments of I under assumption R

If the set of $n!$ equally likely random permutations of the county sample values is considered, the moments of I and c may be evaluated under the null hypothesis of no spatial autocorrelation, *conditional* upon the n sample values observed. Thus

$$\text{prob(county } k \text{ has observation } z_i) = \frac{1}{n},$$

for all i and k, so that, for each county, the average Z_i observed is

$$\frac{1}{n}z_1 + \frac{1}{n}z_2 + \ldots + \frac{1}{n}z_n = \frac{1}{n}\sum z_i = \bar{z},$$

which is zero by definition. We write this as $E_R(Z_i) = 0$, with the understanding that the expectation is being taken over the $n!$ random permutations.

Likewise

$$E_R(Z_i^2) = \frac{1}{n}\sum z_i^2 = m_2, \quad \text{say.}$$

Given that a particular cell has value z_i, a neighbouring cell can take any z_j, other than z_i, with probability $1/(n-1)$ so that

$$E(Z_i Z_j) = \frac{1}{n-1} E_R[Z_i(Z_1 + \ldots + Z_{i-1} + Z_{i+1} + \ldots + Z_n)]$$

$$= \frac{1}{n-1} E_R\left[Z_i\left(\sum_{j=1}^{n} z_j - Z_i\right)\right]$$

$$= -\frac{1}{n-1} E_R(Z_i^2) = -\frac{m_2}{n-1}, \tag{2.37}$$

since $\sum_{i=1}^{n} z_i = 0$ by definition. Because the denominator of I is now fixed,

$$I = \frac{1}{m_2 S_0} \sum_{(2)} w_{ij} Y_{ij}, \quad \text{where} \quad Y_{ij} = Z_i Z_j.$$

It follows from equations (2.2) and (2.37) that

$$E_R(I) = \frac{S_0 E(Y_{ij})}{m_2 S_0} = -\frac{1}{n-1}, \tag{2.38}$$

the same value as under assumption N. Higher moments of Z_i may be evaluated in the same way, and we find that

$$\left.\begin{aligned} E(Z_i^2 Z_j^2) &= \frac{nm_2^2 - m_4}{n-1}, \\ E(Z_i^2 Z_j Z_k) &= \frac{2m_4 - nm_2^2}{(n-1)^{(2)}}, \\ E(Z_i Z_j Z_k Z_l) &= \frac{3nm_2^2 - 6m_4}{(n-1)^{(3)}}, \end{aligned}\right\} \qquad (2.39)$$

where

$$n^{(b)} = n(n-1) \ldots (n-b+1), \quad \text{and} \quad nm_4 = \sum_{i=1}^{n} z_i^4.$$

The variance of I may now be derived using equation (2.11). After some rearrangement of terms, and setting $b_2 = m_4/m_2^2$, we obtain

$$E(I^2) = \frac{n[(n^2 - 3n + 3)S_1 - nS_2 + 3S_0^2] - b_2[(n^2 - n)S_1 - 2nS_2 + 6S_0^2]}{(n-1)^{(3)} S_0^2}.$$
(2.40)

When the x_i are ranks (1, 2, ..., n), the I statistic has mean $-1/(n-1)$ and its variance may be derived from expression (2.40) on setting $b_2 = 3(3n^2 - 7)/5(n^2 - 1)$. This expression for b_2 rapidly approaches its limiting value of $1 \cdot 8$ as n increases; thus $b_2 = 1 \cdot 776$ when $n = 10$.

As noted at the beginning of section 2.3, $E_R(I^2)$ provides an unbiased estimator for $E(I^2)$ for any distribution. In particular, when the underlying distribution is normal, the *expected value* of $E_R(I^2)$ given in equation (2.40) reduces to $E(I^2)$ in equation (2.35).

Again, if we consider the Dacey coefficient I', we note that the denominator term $\Sigma \alpha_i z_i$ is not invariant under random permutations, so that the moments of I' cannot be expressed in a usable form.

The higher moments of I under assumption R are much more complicated than under assumption N because $E(Z_i^3) \neq 0$. For the third moment of I, see Cliff and Ord (1969, pages 37-38).

When the Z_i are 0 or 1, and n_1/n is near to $\frac{1}{2}$, I corresponds approximately to the BW join-count statistic under nonfree sampling, since

$$I = -\frac{2n(BW)}{S_0 n_2} + \frac{2n(n_2 - n_1)(BB)}{S_0 n_1 n_2} + \frac{S_0 n_1}{n_2}.$$

2.4 The distributions of the test statistics

If we are to use the statistics discussed above for hypothesis testing, we need to know the form of the sampling distribution under H_0. Except for very small lattices, exact evaluation of the distribution function is impractical and approximations must be found. In section 2.4.1 we shall show that, given assumption N, the distributions of the I and c statistics approach normality, as n increases, under fairly mild conditions. The large sample distributions of the join-count statistics under free sampling are discussed in section 2.4.2; following this discussion asymptotic distributions under randomisation are examined in sections 2.4.3 and 2.4.4. However, the reader willing to accept these proofs should move on to section 2.5.

2.4.1 The I and c statistics

Once again we shall concentrate upon I, since the results for c can be obtained in similar fashion.

In order to establish the asymptotic normality of the statistics we use the following theorem due to Hoeffding (1952).

Theorem 2: Let $F(y)$ be a distribution function uniquely determined by its moments, μ_k, and let $\{F_n(y)\}$, $n = 1, 2, ...$, be a sequence of distribution functions with moments μ_{nk}. If $\mu_{nk} \to \mu_k$ as $n \to \infty$ for $k = 1, 2, ...$, then $F_n(y) \to F(y)$ in probability at every continuity point of $F(y)$.

In the present context, let $F_n(y)$ correspond to the distribution of any of the statistics, given n observations, and let $F(y)$ correspond to the normal distribution. Then the statistic will have a distribution that approaches the normal if its moments converge to those of the normal.

To obtain the results in general terms which allow ready extension to the regression residuals case (see chapter 8), we use a matrix formulation.

Let $x^T = (x_1, ..., x_n)$, $z^T = (z_1, ..., z_n)$. Then

$$z = Mx ,$$

where

$$M = I_n - \frac{1}{n} 11^T . \qquad (2.41)$$

I_n is the identity matrix of order n, and $1^T = (1, 1, ..., 1)$, a $(1 \times n)$ row vector. The test statistic I may be written as

$$I = \frac{z^T W z}{z^T z} , \qquad (2.42)$$

where $W = \{w_{ij}\}$, the matrix of weights, and the weights have been scaled so that $\sum_{(2)} w_{ij} = n$. By using equation (2.41), equation (2.42) may be transformed into

$$I = \frac{x^T T x}{x^T M x^T} , \qquad (2.43)$$

where $T = M^T W M$, and $M^T M = M^2 = M$, since M is idempotent. We assume, without loss of generality, that the X variates have zero means and unit variances. Then, under assumption N, the moment generating function of $x^T T x$ is

$$E[\exp(\theta x^T T x)] = (2\pi)^{-n/2} \int_{(n)} \exp(\theta x^T T x - \tfrac{1}{2} x^T x) dx_1 ... dx_n$$

$$= |I_n - 2\theta T|^{-\frac{1}{2}} . \qquad (2.44)$$

It follows from equation (2.44) that the cumulants of $x^T T x$ are

$$\kappa_j = 2^{j-1}(j-1)! \sum_{i=1}^{n} \nu_i^j , \tag{2.45}$$

where $\nu_1, ..., \nu_n$ are the eigenvalues (latent roots) of the matrix T. For a more detailed discussion see Durbin and Watson (1951, pages 418-421). From equations (2.30) and (2.45) the cumulants of I are

$$E(I) = \frac{n\bar{\nu}}{n-1} , \tag{2.46}$$

where

$$n\bar{\nu} = \sum_{i=1}^{n} \nu_i ,$$

$$\kappa_2(I) = \text{var}(I) = \mu_2(I) = \frac{2}{n^2-1} \sum_{i=1}^{n} (\nu_i - \bar{\nu})^2 = \frac{2n}{n^2-1} \sigma_\nu^2 , \tag{2.47}$$

$$\kappa_3(I) = \mu_3(I) = \frac{8 \sum (\nu_i - \bar{\nu})^3}{(n-1)(n+1)(n+3)} , \tag{2.48}$$

and

$$\kappa_4(I) = \mu_4(I) - 3[\mu_2(I)]^2$$
$$= \frac{48 \sum (\nu_i - \bar{\nu})^4}{(n-1)(n+1)(n+3)(n+5)} + \frac{96(n+2)[\sum (\nu_i - \bar{\nu})^2]^2}{(n^2-1)^2(n+3)(n+5)} , \tag{2.49}$$

where $\mu_j(I)$ denotes the jth moment of I about its mean. Then from theorem 2, the set of necessary and sufficient conditions for the distribution of I to approach normality as n increases is that, for $j \geq 3$,

$$\frac{\kappa_j(I)}{[\kappa_2(I)]^{j/2}} \text{ be } o(1) , \tag{2.50}$$

where o means 'of smaller order than'. That is, the ratios given in expression (2.50) should go to zero as $n \to \infty$. For $j = 3, 4$ it can be seen that this condition reduces to the requirement that

$$\sum_{i=1}^{n} (\nu_i - \bar{\nu})^j \Big/ \left[\sum_{i=1}^{n} (\nu_i - \bar{\nu})^2 \right]^{j/2} \text{ be } o(1) . \tag{2.51}$$

Indeed the same requirement holds for all $j \geq 3$.

If the eigenvalues of W are $\lambda_1, ..., \lambda_n$, then this condition becomes

$$\sum_{i=1}^{n} \lambda_i^j \Big/ \left(\sum_{i=1}^{n} \lambda_i^2 \right)^{j/2} \text{ be } o(1) . \tag{2.52}$$

For any matrix, $A = \{a_{ij}\}$, we define the trace of A to be $\text{tr}(A) = \sum a_{ii}$. Then the term $\bar{\lambda}$, where

$$\bar{\lambda} = \frac{1}{n} \sum_{i=1}^{n} \lambda_i ,$$

is zero since the trace of **W** is zero. Further, since the trace of $\mathbf{W}^j$ is $\sum_{i=1}^{n} \lambda_i^j$ for all j, and

$$\text{tr}(\mathbf{W}^2) = \sum_{(2)} w_{ij} w_{ji} = \tfrac{1}{2} S_1$$

for symmetric **W**, the condition (2.52) becomes

$$\frac{\text{tr}(\mathbf{W}^j)}{S_1^{j/2}} \text{ be } o(1) \text{ for all } j \geq 3 . \tag{2.53}$$

Simpler necessary and sufficient conditions than (2.53) are not known, but two sets of sufficient conditions are available. In Cliff and Ord (1973), the author derived a condition which may be relaxed to the requirement that

$$\max_i \frac{\lambda_i^2}{S_1} \text{ be } O(n^{-1}) , \tag{2.54}$$

where O denotes 'of the same order as'. The sufficiency of this condition is readily checked by putting $\lambda_i^2 \leq c_1^2 S_1/n$ in (2.53), where c_1^2 is an arbitrary positive constant.

Recalling that **W** is symmetric and $S_0 = n$, Sen (1976) showed that if we impose the condition

$$\max_i w_{i.} \leq c_1 < \infty , \tag{2.55a}$$

and

$$0 < \lim_{n \to \infty} \frac{S_1}{n} = \gamma^2 < \infty , \tag{2.55b}$$

then condition (2.50) is satisfied and $n^{\frac{1}{2}} I$ is asymptotically normal with mean zero and variance γ^2. It may be shown (Gantmacher, 1959, page 57) that condition (2.55a) implies that

$$\max_i \lambda_i \leq c_1 ,$$

since **W** is a matrix of nonnegative elements. Thus conditions (2.55) imply condition (2.54). The reverse is not true as the following example demonstrates.

Example 1 Suppose that the study area is partitioned into two subareas containing m_n and $n - m_n$ counties respectively. Further, we assume that each of the counties within the subarea of size m_n is totally connected with weights $w_{ij} = a_n$ if ii and ij are in this subarea. The remaining counties are taken to have finite weights with a finite number of other counties. Let m_n be of order n^α, and a_n be of order $n^{-\beta}$, such that $2\beta = \alpha$ and $\alpha \leq \tfrac{2}{3}$. Then it follows that S_0 and S_1 are both of order n (we can make $S_0 = n$ by a suitable choice of scaling factor), $\max|\lambda_i|$ is $O(1)$, and $\max w_{i.}$ is of order $\tfrac{1}{2}\alpha$ in n. Thus condition (2.54) holds but conditions (2.55) do not. This example is artificial, but it shows that *I may still be normal when there is a subsystem whose counties have a large number of weak connections, provided that the subsystem is not too large.* Varying α and β, it is easy to find cases which satisfy condition (2.50) but do not meet condition (2.54).

Notwithstanding these situations, Sen's condition is easier to understand and will prove a good yardstick in most practical situations. It is never easy to relate limit theorems to approximations for small or moderate sized samples, but conditions (2.55) are satisfied whenever each county has a clearly defined number of nonzero links with finite weights—which will often be the case; that is, the number of links is not a function of lattice size, beyond boundary effects.

To show that the conditions are not always satisfied, let us consider a simple counterexample.

Example 2 Consider the star lattice with a single articulation point, as shown in figure 2.4. Label the centre 1 and the other points 2, 3, ..., n. Assign binary weights to each link so that the matrix of weights is

$$W = \begin{bmatrix} 0 & 1 & \cdots & 1 \\ \hline 1 & & & \\ \vdots & & 0 & \\ 1 & & & \end{bmatrix}.$$

The eigenvalues of W are $(n-2)$ zeros and $\pm(n-1)^{1/2}$. Thus

$$\bar{\lambda} = 0, \quad \text{as always}, \quad \sum \lambda_i^2 = 2(n-1), \quad \sum \lambda_i^3 = 0,$$

and

$$\sum \lambda_i^4 = 2(n-1)^2; \quad \text{thus} \quad \frac{\sum \lambda_i^4}{(\sum \lambda_i^2)^2} = \tfrac{1}{2}.$$

Hence the distribution of I is not asymptotically normal for this lattice. Cliff and Ord (1972) show that a beta distribution is appropriate for a star lattice.

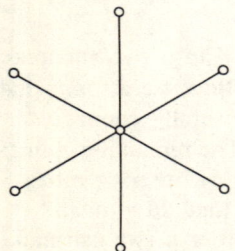

Figure 2.4. Star lattice with single articulation point.

So far in this section we have assumed that the random variables are normally distributed. However, Sen (1976) has shown that[3] if $X_1, ..., X_n$ are independent and identically distributed variates with $E(X_i^4) < \infty$, and conditions (2.55) hold, then $n^{1/2}I$ is asymptotically normal with mean zero and variance γ^2. Thus, when $E(b_2)$ is finite, we may use a normal approximation for the distribution of I even when the data are nonnormal. The Monte Carlo results of section 2.5 also suggest that the normal approximation is reasonable for moderate sample sizes.

2.4.2 The join counts under free sampling

The proof of asymptotic normality for BB and BW under free sampling follows from a theorem of Noether (1970), which may be stated as follows:

Theorem 3: Consider the statistic $T = \Sigma_{(2)} w_{ij} y_{ij}$, where y_{ij} is a uniformly bounded function of the observed values x_i and x_j, and y_{ij} and y_{kl} are independent unless they have at least one common subscript. Also, put $S_0 = \Sigma_{(2)} w_{ij}$. Then as the sample size, n, tends to infinity, T will be asymptotically normally distributed if $S_0^{-2} \text{var}(T)$ is exactly of order n^{-1}.

Proof: Carried out by showing that the moments of T approach those of the normal distribution as $n \to \infty$. For details, the reader should consult Noether (1970).

To see how this theorem can be applied to the statistics given in chapter 1, consider the BW join count and let

$$y_{ij} = \begin{cases} 1, & \text{if the join is } BW, \\ 0, & \text{otherwise.} \end{cases}$$

For regular lattices, S_1/S_0^2 and S_2/S_0^2 are both of order n^{-1}. Provided that $0 < p < 1$, it follows that $S_0^{-2} \text{var}(T)$ is exactly of order n^{-1}, and the statistic is asymptotically normally distributed.

For irregular lattices, given that S_0 is of order n, conditions (2.55) are sufficient to ensure that S_1/S_0^2 and S_2/S_0^2 are both of order n^{-1} since S_1/n approaches γ^2, and $4n \leq S_2 \leq 4nc_1^2$.

As a counterexample, consider the star lattice described earlier. Here, $S_0 = 2(n-1)$, $S_1 = 4(n-1)$, and $S_2 = 4n(n-1)$, so that S_1/S_0^2 is of order n^{-1}. However, S_2/S_0^2 is of order 1, so that $S_0^{-2}\text{var}(T)$ is of order 1 and the theorem does not hold. Since the theorem expresses only a sufficient condition, this does not demonstrate that the asymptotic distribution *is* nonnormal, but warns that this may well be the case (nonnormality may be established by showing that the Pearson β_1 coefficient does not approach zero as $n \to \infty$).

Asymptotic normality for these statistics may also be demonstrated directly, as in Moran (1948) for example.

---

[3] This statement differs slightly from that of Sen, but it is asymptotically equivalent.

2.4.3 Asymptotic normality under assumption R

To establish the asymptotic normality of statistics under assumption R, we again make use of theorem 2 in section 2.4.1. As under assumption N, it is necessary to show that conditions (2.50) hold for $j \geq 3$. Hoeffding (1952) demonstrates that the asymptotic distributions under N and R are the same under fairly general conditions.

The details in any particular case are rather tedious but it may be shown that the distribution is asymptotically normal provided that
(a) conditions (2.55) hold, and

(b) $\dfrac{\sum z_i^j}{(\sum z_i^2)^{j/2}}$ is $O(n^{1-j/2})$ for $j = 3, 4, \ldots$, \hfill (2.56)

where O means 'terms of this order in n'. As before, this is a sufficient rather than a necessary and sufficient condition, but should suffice for most practical purposes. For further discussion, see Strauss (1977). Conditions (2.55) will be violated by configurations like the star lattice, and condition (b) precludes data sets such as

$z_1 = \ldots = z_k = -a$,

$z_{k+1} = \ldots = z_{2k} = a$,

$z_{2k+1} = \ldots = z_n = 0$,

for k finite, since for $j = 2m$ the ratios in condition (2.56) reduce to $(2k)^{1-m}$, $m = 2, 3, \ldots$, which are $O(1)$. For binary data, the requirement given in (2.56) reduces to

$0 < \lim\limits_{n \to \infty} \dfrac{n_1}{n} < 1$.

2.4.4 The Poisson limit for BB joins

Knox (1964) was the first to use the I_{s-t} statistic defined in section 1.6.1. He used binary weights, both for spatial and for temporal variables ($w_{ij} = 0$ or 1, and $y_{ij} = 0$ or 1) in equation (1.44). For the applications he had in mind, the numbers of pairs of counties for which $w_{ij} = 1$, or $y_{ij} = 1$, were small and Knox conjectured that the BB join-count statistic had a distribution which was approximately Poisson. This conjecture relates only to nonfree sampling since we are considering the set of all possible permutations of fixed sets of $\{w_{ij}\}$ and $\{y_{ij}\}$.

Barton and David (1966) verified this conjecture and we now outline their proof, adapted to our own situation. From section 2.2.1 we see that, for BB under nonfree sampling and when n is large,

$E(BB) = \tfrac{1}{2} S_0 p^2$,

$\mathrm{var}(BB) = \tfrac{1}{4} p^2 q \left[S_1 q + S_2 p - \dfrac{4 S_0^2 p}{n} \right]$.

Since the weights are binary, we know that $S_1 = 2S_0 = 4A$, and if we set $\lambda = \frac{1}{2}S_0 p^2$ then

$$E(BB) = \lambda ,$$

$$\text{var}(BB) = \lambda(1-p)^2 + \frac{\lambda pq}{2nS_0}(nS_2 - 4S_0^2) . \tag{2.57}$$

To keep the numbers of nonzero w_{ij} and y_{ij} small we require that $S_0 = c_1 n$, and $np^2 = c_2$, for some constants $0 < c_1 < \infty$, $0 < c_2 < \infty$. This has the effect of ensuring that the expected number of BB joins, λ, $\lambda = \frac{1}{2}c_1 c_2$, stays finite as n increases. Then, provided that S_2 is $O(n)$, it follows that

$$\text{var}(BB) = \lambda[1 + O(n^{-\frac{1}{2}})] .$$

The equality of the mean and variance is a feature of the Poisson model so that the Poisson distribution is likely to be a reasonable approximation in this case. It should be noted that Barton and David impose an additional constraint,

$$nS_2 - 4nS_0 - 4S_0^2 \text{ be } O(n^{\frac{1}{2}}) ,$$

to obtain the stronger result, $\text{var}(BB) = \lambda[1 + O(n^{-1})]$. This is a natural constraint for the space-time interaction problem, although it looks rather artificial in the context of join counts.

As we shall see in section 2.6, it is the BB join-count statistic under nonfree sampling which is least susceptible to a normal approximation for small p (as the results of this section could lead us to expect). Therefore, the Poisson approximation is a useful alternative for small values of p and nicely complements the normal approximation obtained in section 2.4.3.

2.5 Evaluation of the distribution functions

The proofs of normality given above are asymptotic only, and may not provide a reasonable approximation for the small lattices met in applications (typically n is less than 50). Empirical investigations (Cliff and Ord, 1971) suggest that this is indeed the case, and alternative approximations to the distribution functions (DFs) of the coefficients are required in such cases if we are sensibly to test the calculated values of the coefficients. One way of refining results would be to use the third (and possibly fourth) moments of the statistics to define a Gram-Charlier expansion. This has not been pursued to date because of the practical difficulties of evaluating the higher order moments numerically, even when the theoretical forms are known. However, for regular lattices and under assumption N, this approach deserves further consideration. We recall that the third and fourth moments are available as listed in sections 2.2.2 and 2.3.2.

A second approach is to use Monte Carlo methods to examine the form of the DFs of the coefficients. There are four factors which will affect

the shape of the DFs. These are:
1. the shape of the county system and the average number of joins per county (defined as A/n);
2. the weights used $\{w_{ij}\}$;
3. the distribution of the variate, X;
4. the size of the county system, n.

In 1971, the authors carried out an extensive Monte Carlo study of the DFs of all the measures of autocorrelation described in chapter 1. The findings for the I coefficient were presented in Cliff and Ord (1971), and these are summarised in section 2.5.1. We then go on, in sections 2.5.2 and 2.6, to describe the findings for the c and join-count statistics.

2.5.1 Monte Carlo study for I

Various data sets were taken and the sampling frame comprised the $n!$ possible arrangements of the data values among the n counties in a given county system. In any one set of runs of the simulation program, three of the four factors '1'-'4' above were held constant, and the fourth was allowed to vary, so that its effect on the DF of I could be explored. Since, for testing purposes, we need to evaluate the DF of I only in the tails, the centre of the distribution was largely ignored, and interest was focussed on the tail regions for conventional probability levels of up to 10% in a one-tailed test.

Each simulation run consisted of generating m random permutations of the data set. In each run, m was usually 300, and anything from one to six such runs were carried out for a particular combination of the factors. I was computed for each permutation. The empirical distribution of I was constructed, and the observed percentage points of this distribution were compared with the cutoff points obtained from various theoretical approximations. The 100α percentage point of I, I_α, in these approximations was usually of the form,

$$y_\alpha = \frac{I_\alpha + k_\alpha(n-1)^{-1}}{\sigma(I)},$$

or

$$I_\alpha = y_\alpha \sigma(I) - k_\alpha(n-1)^{-1}, \tag{2.58}$$

where y_α is treated as a standardised normal deviate. $\sigma(I)$ is the standard deviation of I, and k_α is a constant depending only on α. The straightforward use of the normal distribution corresponds to taking $k_\alpha = 1$ for all α. It is clear that, whatever k_α is chosen, equation (2.58) approaches the normal approximation as n increases.

A total of 36 different factor combinations was considered. Binary weights were most commonly used, as it soon became evident that the main effect of general weights was exerted through increases in the standard deviation.

As a result of this study, the following procedure was proposed for evaluating the percentage points of I. First, for any given lattice of size n, consider a notional regular lattice of size $t \times t = n$. We define A/n ratios as

$$R = \frac{2(t-1)}{t} < 2, \qquad (2.59)$$

in the rook's case, and

$$Q = \frac{2(t-1)(2t-1)}{t^2} < 4 \qquad (2.60)$$

in the queen's case. The upper limits occur only if the lattice is mapped onto a torus. The choice of approximation is then made as follows.

Step 1 Calculate A/n for the lattice under consideration (a necessary step anyway).

Step 2 If $R < A/n \leqslant Q$, use equation (2.58) with $k_\alpha = (10\alpha)^{\frac{1}{2}}$. If $A/n \leqslant R$ or $A/n > Q$, use equation (2.58) with $k_\alpha = 1$ for all α. As previously stated, these approximations apply only to the tail regions ($\alpha \leqslant 0.10$). Note that α is taken as the one-tail probability. For a two-tailed test of size α^*, put $\alpha = \frac{1}{2}\alpha^*$ in k_α.

Example 3 Consider a 25 county system with $A = 55$. Take $t = n^{\frac{1}{2}} = 5$. Then $R = 1 \cdot 6$ and $Q = 2 \cdot 88$. As $A/n = 2 \cdot 2$, the approximation should be used with $k_\alpha = (10\alpha)^{\frac{1}{2}}$. Thus, when $\alpha = 0 \cdot 05$, $k_\alpha = 0 \cdot 71$ and when $\alpha = 0 \cdot 025$, $k_\alpha = 0 \cdot 5$.

The justification for the proposed rules is that, in a wide variety of situations, the resulting value for I_α did not generally differ significantly from the conventional empirical cutoff levels when tested using chi-squared. For fuller details see Cliff and Ord (1971).

The approximations are not satisfactory when:
(1) binary data are used, except when the underlying population is roughly symmetric and the lattice size is fairly large; see section 2.6 for further discussion.
(2) one county, or a small group of counties, in the system figures in a high percentage of joins (see the conditions for asymptotic normality in section 2.4).
(3) n is small ($\leqslant 10$, say). A complete enumeration of the distribution function is often feasible in such cases, particularly when the lattice is regular and binary weights are used; see section 2.7.

The rules suggested are clearly *ad hoc* and should not be applied blindly; for example, if R is near A/n, look at both approximations. However, used with care, we believe that the resulting inference will not be seriously in error. Finally we stress again that the suggested transformations do not seek to approximate the entire distribution of I by the normal, but only the tail regions.

2.5.2 Monte Carlo study for c

In view of the extended discussion of the form of the sampling distribution of I in section 2.5.1 and in Cliff and Ord (1971), the discussion for c can be fairly brief.

Eight different runs were performed on three distinct lattices for $n = 25$ with m between 200 and 600. The data sets were the same as those for the study of I (Cliff and Ord, 1971). The 100α percentage points for the tail areas of the distribution of c, using the normal approximation, are given by

$$c_\alpha = y_\alpha \sigma(c) + 1 , \qquad (2.61)$$

where y_α is the standard normal deviate, $\sigma(c)$ is the standard deviation of c, and $E(c) = 1$. This approximation works well for the lower tail of c (corresponding to positive spatial autocorrelation), but not so well for the upper tail (negative spatial autocorrelation), where the approximation is markedly conservative; that is, the nominal probability that $c > c_\alpha$ in the upper tail may be considerably greater than the true probability. For the upper tail, the revised formula,

$$c_\alpha = 1 + y_\alpha \sigma(c) - k_\alpha (n-1)^{-1} , \qquad (2.62)$$

is recommended, where $k_\alpha = (10\alpha)^{\frac{1}{2}}$, as for I.

2.6 Monte Carlo studies for the join-count statistics, $k = 2$

The binary form of the data means that the progress to normality of the join-count statistics can be very slow, since the probability measure is concentrated at a few points in an irregular manner. We refer to such distributions as being 'lumpy'. Because of the difficulties caused by this lumpiness, we feel it worthwhile to give a fairly full account of the simulation work done on these distributions, so that the extent of the empirical backing for the approximation suggested can be assessed.

2.6.1 The experimental procedure

As in the simulation studies of the sampling distribution of the I and c statistics, a variety of lattices were used together with different proportions (nonfree sampling) or probabilities (free sampling) for black and white cells. Since nonbinary weights make the distribution less lumpy and somewhat easier to approximate, we stoically used binary weights throughout the study. However, the final results may reasonably be applied to the nonbinary weighted forms of the statistics.

To indicate the scope of the sampling experiments, we shall give details for the study of the *BB* join count under nonfree sampling. The lattices used were as follows:
(1) regular lattices, queen's case ($\delta_{ij} = 1$ if, and only if, two cells had a common edge or vertex) for squares of side 5(1)10. The square of side 6 was also mapped onto a torus, so that all counties had eight joins;

(2) the circle comprising 156 cells (equivalent to a circular time series of this length);
(3) the county systems of England and Wales, and of Eire, excluding County Dublin.
In cases (2) and (3), $\delta_{ij} = 1$ if, and only if, the ith and jth counties had a common boundary. The total number of counties, n, and the total number of joins, A, and A/n, are given in table 2.1. For each lattice, the number of black cells, n_1, was taken as the nearest integer to np, $p = 0 \cdot 1(0 \cdot 1)0 \cdot 9$. For each lattice and value of p, $m(= 600)$ random permutations were generated.

The normal and chi-squared distributions were considered to be plausible approximations to the sampling distribution of BB. Both approximations were fitted using the first two moments of BB given in equations (1.31) and (1.32). Let μ and σ represent, respectively, the mean and standard deviation of the statistic BB. Then, for the normal approximation, N, the 100α percentage point for BB, BB_α say, is given by

$$BB_\alpha = y_\alpha \sigma + \mu, \qquad (2.63)$$

where y_α is the 100α percentage point of the standard normal curve. For the chi-squared approximation,

$$BB_\alpha = \gamma x_\alpha, \qquad (2.64)$$

where $\gamma = \sigma^2/2\mu$ and x_{α_2} is the 100α percentage point of chi-squared with $\nu = 2\mu^2/\sigma^2$ degrees of freedom. Values for noninteger ν were obtained by linear interpolation.

For a given lattice and value of n_1, the upper and lower 10, 5, 2·5, 1, and 0·5 percentage points of N and χ^2 were evaluated and used to partition the generated sampling distribution of BB into eleven classes with boundaries:

$BB < L(0 \cdot 005)$, $L(0 \cdot 005) \leq BB < L(0 \cdot 01)$, ...,

$L(0 \cdot 05) \leq BB < L(0 \cdot 10)$, $L(0 \cdot 10) \leq BB < U(0 \cdot 10)$,

$U(0 \cdot 10) \leq BB < U(0 \cdot 05)$, ..., $U(0 \cdot 005) < BB$.

Here $L(\alpha)$ and $U(\alpha)$ represent, respectively, the lower and upper 100α percentage points, and BB represents the number of the m permutations in each class. Thus if $L(0 \cdot 005) = 4 \cdot 43$, the permutations which yielded 0, 1, 2, 3, or 4 BB joins were recorded in the first class. The goodness-of-fit between the observed frequencies so obtained and the corresponding expected frequencies was examined using the X^2 test with ten degrees of freedom (for the eleven classes). This was done for every lattice and value of n_1. The results of the X^2 test for all cases are given in table 2.1.

Table 2.1. Results of X^2 test of significance for BB joins, nonfree sampling.

Lattice	n	A	$\frac{A}{n}$	Value of p															
				0·1		0·2		0·3		0·4		0·5		0·6		0·7		0·8		0·9	
				N	χ^2	N	χ^2	N	χ^2	N	χ^2	N	χ^2	N	χ^2	N	χ^2	N	χ^2	N	χ^2
				approximation																		
10 × 10 queen's case	100	342	3·42	—		12·6	18·8	18·0	11·7	10·0	11·7	9·9	14·8	6·1	8·2	9·2	9·2	12·1	12·1	10·3	10·3
9 × 9 queen's case	81	272	3·36	—		15·3	58·1	19·4	7·5	20·2	8·3	7·7	9·7	11·3	11·3	12·5	12·5	13·2	7·1	7·8	6·4
8 × 8 queen's case	64	210	3·28	—		—		19·0	10·6	18·7	12·7	16·3	7·4	6·0	10·0	11·0	12·3	14·8	14·4	10·2	10·2
7 × 7 queen's case	49	156	3·18	—		—		11·8	22·0	15·3	6·9	21·7	13·8	18·1	13·4	20·3	15·7	25·4	13·2	+	
156 cell circle	156	156	1·00	—		—		12·3	27·8	10·2	9·6	9·9	19·7	14·6	14·6	14·3	14·3	+		+	
English and Welsh counties	53	119	2·25	—		—		—		19·1	12·9	37·2	7·4	20·5	9·0	16·1	5·5	9·3	7·2	13·5	12·2
6 × 6 queen's case	36	110	3·06	—		—		—		—		8·3	30·5	23·1	11·1	21·2	6·5	27·5	14·6	+	
5 × 5 queen's case	25	72	2·88	—		—		—		—		17·0	9·4	13·9	12·1	7·9	3·9	17·4	9·4	+	
Eire (excluding County Dublin)	25	55	2·20	—		—		—		—		—		—		7·6	20·5	11·0	11·0	+	
6 × 6 queen's case on a torus	36	144	4·00	—		—		—		15·0	31·9	6·0	7·3	+		+		+		+	

Significance levels (one-tail) $X^2 = 18·3$ (5%), $X^2 = 23·2$ (1%).
− indicates that the distribution is too lumpy in the negative tail to make two-tailed test for spatial autocorrelation sensible.
+ indicates that the distribution is too lumpy in the positive tail to make two-tailed test for spatial autocorrelation sensible.

2.6.2 Analysis of the results

In table 2.1, some X^2 values have been omitted because the lumpy nature of the tail areas made it impossible sensibly to fit any continuous approximation. Fortunately, when p is small, the Poisson approximation is often useful as we shall see in section 2.6.3. Also, it is noticeable that greater lumpiness occurs in the lower tail when p is small, as shown by columns 1-3 of the example in table 2.2. It is clear that no reasonable test of *negative* spatial autocorrelation is feasible in such cases.

If we restrict attention to $p \geq 0.25$, and $p \leq 0.8$, or thereabouts, then the following conclusions may be drawn:
1. The normal and the chi-squared approximations are both fairly successful for lattice sizes commonly encountered in practice. However, the χ^2 form appears to have the edge.
2. For values of p at equivalent distances from $p = 0.0$ and $p = 1.0$, the approximations perform much better for the larger value of p. Thus for $p = 0.7$ and 0.8 the approximations are better than for $p = 0.3$ and 0.2 respectively. *Therefore the join count(s) for the more common colour(s) should be used in this range.* Nothing is lost by using one colour rather than another provided the BW count is also examined; see equation (2.16).
3. From asymptotic power considerations (see chapter 6), counties should be coloured so that the probability of a county being B is near 0.5.

Before making a final recommendation about the form of approximation, it is instructive to look at a singularly ill-behaved case, the 9×9 regular lattice, queen's case, with $n_1 = 16$. The results appear in table 2.3, and give the cumulative observed frequencies for the normal and chi-squared approximations, along with the expected frequencies required to achieve exactly the probability levels specified in the first column. The X^2 test values based on the noncumulative counts are given at the foot of each column.

The 'square root' (SR) transformation referred to in table 2.3 is the transformation commonly used to approximate the chi-squared by the

Table 2.2. Sampling results for the 5×5 regular lattice (queen's case) with $n_1 = 3$ (nonfree), and $p = 0.1$ (free), for $m = 600$ random permutations.

| Number of BB joins | Observed frequency | | Approximation | |
	nonfree	free	Poisson $\lambda = 0.72$	binomial $N = 6, \theta = 0.12$
0	239	350	292	279
1	293	146	210	228
2	50	58	76	78
3	18	25	18	14
4	0	8	3	$\Big\{$ 1.5
≥ 5	0	13	1	

Note: $E(BB) = 0.72$ in both these cases.

normal when the number of degrees of freedom ν is large. Typically in using the χ^2 approximation we found that the value for ν was large, so that little is lost by using the square root transformation. The 100α percentage point of BB is given by

$$BB_\alpha = \sigma^2 \left[y_\alpha + \left(\frac{4\mu^2 - \sigma^2}{\sigma^2} \right)^{\!\!1/2} \right]^2 \!\! \bigg/ 4\mu \, , \qquad (2.65)$$

where y_α is the 100α percentage point of the normal distribution.

If the X^2 figures in table 2.3 are taken at their face value, the case for using the normal approximation appears overwhelming. However, if the cumulative figures in table 2.3 are taken into account, the chi-squared approximation would seem to give a somewhat better indication of 'where to draw the line'. The reason for this seemingly paradoxical behaviour is the lumpy nature of the distribution. A slight difference between the approximations can empty a whole class, as happens with the 26 observations for $\alpha = 0 \cdot 10/0 \cdot 05$ in the lower tail. This reinforces the conclusion that the join count(s) for the more common colour(s) should be used.

The results are analysed in a different way in table 2.4. If the approximations are correct, then for $m = 600$ we would expect 15 observations to fall below the lower 2½ percentage points, and the same number above the upper 2½ percentage points. Consider the 52 data/ lattice combinations in table 2.1 for which X^2 values are recorded. Of the 90 combinations examined, these 52 were the ones for which the N and χ^2 approximations could sensibly be used to provide a two-tailed test for spatial autocorrelation. For these 52, the average numbers of observations

Table 2.3. An analysis of cumulative frequencies for the BB statistic, nonfree sampling ($n_1 = 16$), for 9×9 regular lattice, queen's case.

Nominal probability levels	Cumulative expected frequencies	Cumulative observed frequencies using approximation		
		normal (N)	chi-squared (χ^2)	square root (SR)
Upper tail				
0·10	60	77	77	77
0·05	30	38	38	38
0·025	15	22	11	22
0·01	6	11	5	5
0·005	3	5	3	3
Lower tail				
0·10	60	43	43	43
0·05	30	17	43	17
0·025	15	6	17	17
0·01	6	1	6	6
0·005	3	1	6	6
X^2		15·3	58·1	33·1

which in fact fell above and below the upper and lower 2½ percentage points, and the standard deviations, are given for each approximation. In addition the number of times that each approximation was liberal (observed number of BB joins exceeded the required number) is given. From table 2.4 we can see that the chi-squared approximation, although a little conservative, is fairly well balanced in both tails. By contrast the normal approximation is slightly 'heavy' in the upper tail, and decidedly 'light' in the lower tail. The same analysis is performed for the $0 \cdot 5\%$ points in each tail. Here the heaviness in the upper tail, and lightness in the lower tail, of the normal approximation is even more marked.

Finally, on the basis of the results presented in this section, we conclude that the normal approximation is not bad, but that the chi-squared (or square root, which is virtually the same thing) is somewhat better. Either equation (2.64) or (2.65) can be readily applied.

Table 2.4. An analysis of the number of observations for the BB 5% points (two tails).

Item	Approximation			
	normal (N)		chi-squared (χ^2)	
	upper tail	lower tail	upper tail	lower tail
Required number of observations (out of 600)	15	15	15	15
Average observed number (52 runs)	15·9	11·0	12·9	13·0
Standard deviation	5·2	3·5	3·8	3·9
Range	5–30	4–21	5–21	4–23
Number of times observed > required	27	4	10	13

2.6.3 The Poisson approximation

When p^2 and S_0/n are small and the $\{w_{ij}\}$ are binary, we may use the Poisson approximation developed in section 2.4.4. All that is necessary is to equate the Poisson parameter, λ, to the mean for the BB statistic given in equation (1.31).

We give the Poisson approximation for the 5×5 regular lattice queen's case ($n_1 = 3$) in table 2.2. Although the fit in the lower tail ($BB = 0, 1, 2$) is poor, the upper-tail fit is good, yielding prob($BB \geq 3$) = $0 \cdot 037$ against the observed value of $0 \cdot 03$. Since BB cannot exceed 3 in this case, the Poisson approximation works fairly well.

It is evident that any test of fit of the Poisson approximation to the entire observed BB distribution will decisively reject the null hypothesis that the data follow the Poisson law. However, when we restrict attention to the three classes $\leq 2, 3, \geq 4$, the test statistic is $X^2 = 4 \cdot 2$ on two degrees of freedom, indicating a reasonable accord. *It should be stressed that we are concerned with the performance of the Poisson approximation in the upper tail only as a test of positive autocorrelation.*

Besag and Diggle (1977) have carried out several tests with the $I_{\text{s-t}}$ statistic by generating a set of random permutations with which to compare the observed value (this procedure is discussed in section 2.7). Their results suggest that the Poisson approximation may be too heavy in the upper tail. Indeed, the variance of $I_{\text{s-t}}$ is often less than the mean, a mean, as suggested by the form of equation (2.57). An alternative approximation is given by the binomial distribution with index N set equal to the lesser of $\frac{1}{2}S_0$ and $n_1^{(2)}$, and proportion θ set equal to $E(BB)/N$. This approximation seems to work better than the Poisson for Besag and Diggle's examples and for the example given in table 2.2, but it remains to be tested systematically.

2.6.4 The other join-count statistics
Similar extensive analyses have been carried out for BB joins under free sampling, and for BW joins both under free and under nonfree sampling. As a result of these studies the following approximations are recommended:
1. BB joins (free sampling). The third central moment for BB, $\mu_3(BB)$, may be evaluated without too much difficulty using the following equation,

$$\mu_3(BB) = \tfrac{1}{2}j_4 p^2(1-p)^2(1-2p)^2 + 3j_5 p^3(1-p)^2(1-2p) + 3j_6 p^4(1-p)^2$$
$$+ j_7 p^4(1-p)(1-2p) + j_8 p^3(1-p)^2(1+2p) \,, \qquad (2.66)$$

where j_4-j_8 are given by

$$8j_4 = \sum_{(2)} (w_{ij} + w_{ji})^3 \,,$$

$$8j_5 = \sum_{(2)} (w_{ij} + w_{ji})^2 (w_{i.} + w_{.i}) \,,$$

$$8j_6 = \sum_{(2)} (w_{ij} + w_{ji})(w_{i.} + w_{.i})(w_{j.} + w_{.j}) \,,$$

$$8j_7 = \sum_{i=1}^{n} (w_{i.} + w_{.i})^3 \,,$$

$$8j_8 = \sum_{(3)} (w_{ij} + w_{ji})(w_{ik} + w_{ki})(w_{jk} + w_{kj}) \,.$$

Note that j_4-j_8 are simplified coefficients and do not correspond directly to the join patterns in figure 2.2.

The best approximation is then the beta distribution, β, with variable $x = BB/R$, range [0, 1], and a and b are parameters, where

$$a = \frac{\mu(2\mu_3 \mu + \mu^2 \sigma^2 - \sigma^4)}{2\sigma^4 \mu - \mu_3 \mu^2 + \mu_3 \sigma^2} \,, \qquad b = \frac{a(a+1)\sigma^2}{\mu^2 - a\sigma^2} \,, \qquad \text{and} \quad R = \frac{\mu(a+b)}{a} \,.$$

See Ord (1972). The test is most easily carried out using Snedecor's F. For the upper tail, put $F = x/(1-x)$ with $2a$ and $2b$ degrees of freedom in the standard tables. For the lower tail, use $F^* = (1-x)/x$ with $2b$ and $2a$ degrees of freedom.

If the beta approximation is not used, the square root or chi-squared approximation suggested in the previous section is somewhat better than the normal. Again it is recommended that the more common colours be used, noting that the negative tail is very lumpy when $p \leq 0.3$ in small lattices ($n \leq 50$).

2. *BW* joins. The results for *BW* joins proved to be rather similar to those for the *c* statistic. Let μ and σ denote, respectively, the mean and variance of the *BW* statistic. Then the cutoff point for the 100α percentage tail area for *BW*, BW_α say, is given by

$$BW_\alpha = y_\alpha \sigma + \mu - k_\alpha, \qquad (2.67)$$

where, in the upper tail, $k_\alpha = 1 - (10\alpha)^{1/2}$ both for free and for nonfree sampling; and, in the lower tail, $k_\alpha = 2[1 - (10\alpha)^{1/2}]$ for nonfree sampling, and $k_\alpha = 3[1 - (10\alpha)^{1/2}]$ for free sampling. The excessive lumpiness noted in the *lower* tail for *BB* joins under free sampling also occurs in the *upper* tail for *BW* joins under free sampling, although not for nonfree sampling. Fortunately, we are usually interested in testing for the presence of positive spatial autocorrelation, and these lumpy tails correspond to negative spatial autocorrelation. Where we do wish to test for negative spatial autocorrelation with *BW* joins, the approximation suggested in equation (2.67) works tolerably well. It is unlikely that any worthwhile improvement can be made except by complete enumeration of the probabilities in the tail area.

2.7 Monte Carlo tests

When the exact distribution function is intractable, and approximations to it are of doubtful accuracy, a procedure devised by Barnard and developed in Hope (1968) may be used.

If m values of the statistic concerned are generated under H_0, we should reject H_0 at the $100[(j+1)/(m+1)]$ percent level if the observed value of the statistic exceeds at least the $(m-j)$ smallest generated values (assuming the critical region to be the upper tail). This is known as a Monte Carlo test.

Table 2.5. Welsh counties: the percentage changes in population, 1961 and 1951.

County	Code	% change	Rank	Contiguous counties
Anglesey	A	2·05	5	C
Brecon	B	−1·7	8	D, E, H, J, M
Carnaervon	C	−2·4	9	A, F, I
Cardigan	D	0·5	7	B, E, I, K, L, M
Carmarthen	E	−2·5	10	B, D, H, L
Denbigh	F	1·8	6	C, E, G, I
Flint	G	3·2	3	F
Glamorgan	H	2·1	4	B, E, J
Merioneth	I	−5·9	12	C, D, F, K
Monmouth	J	4·4	1	B, H
Montgomery	K	−3·8	11	D, F, I, M
Pembroke	L	3·4	2	D, E
Radnor	M	−7·8	13	B, D, K

$n = 13$, $A = 21$, $S_2 = 332$.

As an example we take data on the percentage change in population, 1951–1961, in the counties of Wales, including Monmouth. These data are drawn from the Preliminary Census Report (General Register Office, 1961), and are reproduced here in table 2.5. Against the null hypothesis of no spatial autocorrelation in population changes by county we range the alternative of positive spatial autocorrelation, looking for a drift from the rural northern counties of Wales to the more industrialised south. Ranks were used instead of the raw data so that comparisons between the Hope procedure and our approximation would not be obscured by particular features of the data (rank tests are reasonably efficient; see section 6.4.4). The coefficient I was evaluated using binary weights: $\delta_{ij} = 1$ if the counties i and j had a common boundary, and $\delta_{ij} = 0$ otherwise.

Under H_0, using randomisation, $E(I) = -0.0833$ and $\sigma(I) = 0.1808$. The observed value of I for the rank data is $I_{obs} = 0.1106$, yielding the standard deviate $z = 1.07$. From tables of the normal distribution (that is using the normal approximation) $\text{prob}(I > I_{obs}) = 0.1423$, and there is not sufficient evidence to reject H_0. Inspection of more detailed data reveals a complex pattern of migration to the towns and city suburbs, away from the rural and mining areas, rather than the simple southward shift suggested by H_1.

Table 2.6. The results of repeated Monte Carlo tests.

Number (m) of generated values of I used in the test	Number of $I > I_{obs}$	Frequency	prob($I > I_{obs}$)
$m = 19$	0	1	0.05
	1	6	0.10
	2	8	0.15
	3	7	0.20
	4	4	0.25
	5	5	0.30
	total	31	
$m = 49$	4	1	0.10
	5	1	0.12
	6	4	0.14
	7	1	0.16
	8	2	0.18
	9	1	0.20
	10	2	0.22
	total	12	
$m = 99$	13	2	0.14
	14	1	0.15
	15	2	0.16
	16	0	0.17
	17	1	0.18
	total	6	

We now examine the value of I using the Hope procedure. We generate 19 dummy values of I and reject H_0 at the 5% level if I_{obs} exceeds all these, or at the 10% level if I_{obs} is largest or next to largest. The same procedure has been carried out for 49 and 99 dummy values of I. The results appear in table 2.6 and show the findings for 31 replications of the 19-dummy test, and 12 and 6 replications respectively for the 49-dummy and 99-dummy tests. Using only 19 dummy values, it can be seen that the results vary considerably. For the 49-dummy test the null hypothesis would be rejected (just) only once in twelve at the 10% level. By the time 99 dummy values are used, the probability of a wrong inference for these data is negligible, even at the 10% level.

On the basis of this and other experiments (Cliff and Ord, 1971), we conclude that the Monte Carlo test procedure is adequate, provided that $m = 49$ or 99 rather than 19. Of course, other values of m are feasible. For further discussion on the choice of m, see Marriott (1979).

Finally, we compare the results of the straightforward normal approximation, the Hope procedure (with $m = 599$), and the transformation suggested in section 2.5.1. By the Hope procedure, the $\alpha = 0.05$, and $\alpha = 0.10$, percentage points of I (upper tail only) are $I(0.05) = 0.2415$, and $I(0.10) = 0.1735$. The value of R given by equation (2.59) is 1.44 ($t = \sqrt{13} \approx 3.60$), and $Q = 2.49$. Since $A/n = 1.62$, the $k_\alpha = (10\alpha)^{\frac{1}{2}}$ approximation for $\alpha = 0.05$ is used. The results are compared in table 2.7. The improvement rendered by the revised approximation for $\alpha = 0.05$ is clear; the original approximation tends to overstate the significance of the results.

Table 2.7. Approximations to the percentage points of I for the Welsh rank data [a].

Nominal percentage point	Values of I: Hope procedure ($m = 599$)	Normal approximations with:	
		$k_\alpha = 1$	$k_\alpha = (10\alpha)^{\frac{1}{2}}$
0·05	0·2415 (0·05)	0·2140 (0·07)	0·2385 (0·05)
0·10	0·1735 (0·10)	0·1483 (0·12)	0·1483 (0·12)

[a] The numbers in brackets below the values for I indicate the proportion of generated I values out of 600 (599+1 observed) which exceed the approximate percentage point.

2.8 Conclusions

In this chapter, we have examined the asymptotic and small-sample sampling distributions of the various measures of spatial autocorrelation defined in chapter 1. The conditions under which asymptotic normality holds for the measures have been stated, and a series of small-sample approximations to the sampling distributions has been explored. Finally, a Monte Carlo test procedure has been presented and illustrated.

Map comparison with application to diffusion processes

3.1 Introduction
In this chapter, we shall show how the join-count and I statistics may be used to evaluate the spatial goodness-of-fit between the observed and theoretical maps of some stochastic process; the rationale behind the proposed testing procedure is outlined in section 3.2.1. If the procedures are applied to diffusion processes, particular problems arise and these are discussed in section 3.2.2. The use of the methods is illustrated in section 3.3, where we evaluate the spatial goodness-of-fit between some observed and theoretical maps of the diffusion of an innovation, described in Hägerstrand (1953). Finally, other approaches are considered in section 3.4.

3.2 Map comparison
A common product of geographical research is a stochastic model of some spatial process, which yields, as its end product, a map whose degree of spatial correspondence to the real-world pattern modelled we wish to evaluate. For example, it might be required to determine the spatial agreement between some theoretical arrangement of central places and the observed location of cities within Europe. This kind of map comparison is usually undertaken as part of an attempt to determine whether the model, which produced the theoretical map, is an acceptable description of reality. There is some discussion in the literature [see, for example, Tobler (1965), Gale (1971), Lankford (1974), Yapa (1976), and Cummings et al (1973)] of methods which can be used to compute the degree of spatial correspondence between theoretically derived maps and observed maps. In this chapter we again look at the problem in some detail, and illustrate how the join-count statistics and a variant of the coefficient I may be used to provide an appropriate test.

3.2.1 The test
Suppose that it is desired to evaluate the degree of spatial correspondence between, for example, a theoretically derived, or *expected*, map showing the production of wheat in kilos per hectare in the English counties, and the corresponding map of *observed* production figures. From these two maps a third map, showing the county-by-county *differences* between the expected and observed maps, can be constructed by calculating the quantity (observed − expected) county values. This is analogous to constructing a map of residuals from regression. Zero residuals represent counties in which the theoretical and observed values are identical; positive residuals represent areas of underestimation by the model; negative residuals represent areas of overestimation of the actual county values by the model. Interest is centred upon the spatial arrangement in the differences map of the nonzero residuals. If the observed map is a possible realisation of the stochastic

process underlying the theoretical map, then the nonzero residuals should be randomly located in the differences map; in other words, the observed and expected maps should differ only by chance. A systematic spatial arrangement of the nonzero residuals (for example, clustering of positive and negative residuals) suggests that the observed map cannot be regarded as an outcome of the process underlying the theoretical map, and that the model is not an acceptable description of reality.

Clearly, for this kind of analysis to work, we require the theoretical map to be the 'expected' outcome of the process postulated. If a model of the process is specified, but not the map itself, we can derive an expected map by obtaining several independent realisations for the model and then fixing the expected value of the ith county ($i = 1, 2, ..., n$) as the average of the values recorded in the ith county by the several solutions. If the expected map is so obtained, and a nonrandom spatial arrangement of the nonzero residuals in the differences map is then detected (that is, H_0 is rejected), the implication is that either:

(1) a correct inferential decision has been made—the theoretical and observed maps cannot be regarded as outcomes of the same process; or
(2) a type I error has been made. It might be that the observed map is in fact a true realisation of the process postulated, but that it is one of those outcomes which occur rarely by chance (five times in a hundred for $\alpha = 0.05$). That is, the observed map is a true but atypical outcome of the process.

By similar arguments, if H_0 is accepted we cannot with certainty conclude that our model is successful since the possibilities of a type II error and equifinality must be considered.

We examine two ways of testing for a nonrandom spatial arrangement of the nonzero residuals in a differences map. The first is to create a three-colour map by coding negative residuals black, B, positive residuals grey, G, and zero residuals white, W. We then test the distributions of BB, GG, and BG joins for departures from randomness by using equations (1.12)-(1.13), and (1.31)-(1.34), or by using I and c given in equations (1.15)-(1.16). This is a sign test. In the second method, if the mean values of the random variable on the observed and expected maps are allowed to differ, we can use a variant of I which preserves the magnitude of the differences as well as the signs. The required variant, I, is given by

$$I = \frac{n \sum_{(2)} w_{ij} x_i x_j}{S_0 \sum_{(1)} x_i^2} \, , \tag{3.1}$$

where x_i is the value of the residual in county i and $\sum_{(1)} = \sum_{i=1}^{n}$. Note that in equation (3.1) the x values are not taken about their mean as in equation (1.15). The residuals *may* sum to zero, but this is not a precondition for the test to be applied. As a result, when $x_i = 0$ (that is, for counties in which the observed and expected values are the same), no contribution is made either to the numerator or to the denominator of equation (3.1), and the coefficient therefore measures only the spatial autocorrelation in the nonzero residuals. We define the jth crude moment of the $\{x\}$ as

$$m_j = n^{-1} \sum_{(1)} x_i^j , \qquad (3.2)$$

and

$$a_1 = \frac{nm_1^2}{m_2} , \qquad a_2 = \frac{m_4}{m_2^2} , \qquad a_3 = \frac{m_3 m_1}{m_2^2} . \qquad (3.3)$$

Then, by using randomisation, it can be shown that

$$E_R(I) = (n-1)^{-1}(a_1 - 1) ,$$

and

$$E_R(I^2) = \frac{1}{(n-1)^{(3)} S_0^2} \{ n [(n^2 - 3n + 3)S_1 - nS_2 + 3S_0^2]$$
$$- a_2 [(n^2 - n)S_1 - 2nS_2 + 6S_0^2] - na_1 [2nS_1 - (n-3)S_2 + 6S_0^2]$$
$$+ na_3 [4(n-1)S_1 - 2(n-1)S_2 + 8S_0^2] + na_1^2 (S_1 - S_2 + S_0^2) \} .$$
$$(3.4)$$

Apart from its particular use here, this model is of general interest, since it does not assume $\sum_{(1)} z_i = 0$, as in equations (1.37)–(1.39), to obtain the moments for I.

In the above discussion we have only considered methods for examining the degree of *spatial* correspondence between maps. In any real problem, the researcher would also have to check the aspatial goodness-of-fit, for example by comparing the frequency distribution of the observed number of counties with 0, 1, 2, ..., k objects in them with the expected frequency distribution using the chi-squared test. Care is needed in the interpretation of the results of such tests. As is shown by the example in section 3.3, a good spatial fit does not guarantee that the aspatial goodness-of-fit test will produce a satisfactory result or vice versa.

3.2.2 Application to diffusion processes

Suppose that the actual pattern is from a *diffusion* process, and that we wish to evaluate the goodness-of-fit between the observed maps and model maps of the process. We then face certain difficulties in applying the testing procedures outlined above. These have been pointed out by Brown and Moore (1969).

"Thus a viable comparative test of empirical and simulated diffusion patterns must have two important characteristics; it must preserve the property of relative location of individual cell values (i.e. the appropriate distance–decay characteristics), and it must be independent of specific directional bias.

Relatively little progress has been made towards the establishment of such a test. The contiguity ratio [that is, the tests of spatial autocorrelation] which has been suggested as a method of evaluating differences between two spatial patterns, does not appear appropriate to the spatial diffusion situation. This is because random differences in the directional biases of observed and simulated patterns will produce systematic variation in the spatial distribution of individual cell differences."

The requirements of an appropriate test as specified by Brown and Moore may be summarised as follows:
(1) relative location of cell values must be preserved;
(2) the test must allow for systematic directional biases which occur by chance in the observed and simulated map patterns.
An aspatial statistic such as the chi-squared test fails on both counts, whereas the autocorrelation tests are affected by requirement (2). Let us now consider this problem of directional bias in the observed and expected maps in more detail. The problem appears to be specific to diffusion processes and, although the methods given in section 3.2.1 yield an appropriate test where this problem does not exist, a modified procedure is required to handle diffusion phenomena.

The following example serves to illustrate Brown and Moore's point (2). Suppose that the county system is as shown in figure 3.1. The middle county contains a single adopter of an innovation, and every other county contains a single potential adopter. The observed number of adopters of the innovation in each county at the end of the diffusion process might be like that shown in figure 3.1(a). We can model the process leading to this observed map in the following way:
(*a*) As on the observed map, the initial adopter is located in the middle county of the model plane.
(*b*) Each potential adopter accepts the innovation as soon as he is contacted by an adopter.
(*c*) In generation 1 of the model, the initial adopter contacts the potential adopter in the contiguous county to his right with probability $0 \cdot 95$ and the potential adopter in the contiguous county to his left with probability $0 \cdot 05$.
(*d*) In generations 2, 3, and 4, only the adopter who accepted in the previous time period may make a new contact, and he must contact the potential adopter in the county contiguous to the one in which he is located.

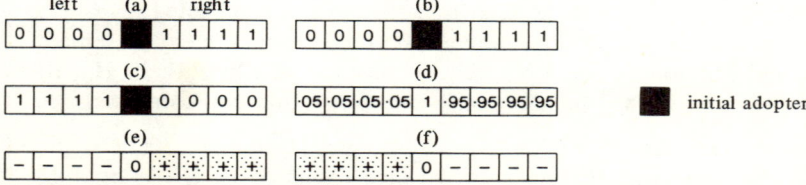

Figure 3.1. An example to illustrate Brown and Moore's point (2). For definitions, see text.

This model will produce a map either like that shown in figure 3.1(b) with probability equal to 0·95, or like that shown in figure 3.1(c) with probability equal to 0·05, yielding the map of expected values shown in figure 3.1(d). Then the possible differences (signs only) will be like figure 3.1(e) or 3.1(f) according to whether the observed map is like (b) or (c). A conventional join-count test, based on the assumption of independence between neighbouring cells, at $\alpha = 0\cdot05$ would reject H_0, given the pattern of signs shown in map (e), or (f), as

prob(no plus–minus joins) = $\frac{1}{35}$ = 0·029

whatever the outcome of the observed process. A valid test at $\alpha = 0\cdot05$ would reject H_0 whenever the observed process moved to the left.

It therefore appears that, if we apply the spatial autocorrelation tests to a differences map for a diffusion process and reject H_0, we must consider two possible explanations for this result:
1. the observed map is not a realisation of the process underlying the theoretical map;
2. we have committed a type I error, because the systematic variation in individual cell differences results from chance differences in the directional orientation of the observed and modelled patterns of adopters. Moreover, the true probability of a type I error may be considerably greater than the 'nominal' level used in the test because of the directional bias which may arise by chance.

The problem of directional bias appears to be peculiar to diffusion because the spread of phenomena such as innovations, diseases in human populations, and urban blight frequently displays a strong contagious element. Thus once the process moves off in a particular direction, often purely by chance, it will tend to retain that orientation because of the operation of the so-called 'neighbourhood effect' (Hägerstrand, 1953).

We now describe a procedure which deals with this problem, based upon the Hope procedure described in section 2.7.

Step 1: generate m independent realisations of the diffusion model and, from these *and the observed map*, compute an expected map by simply averaging over the $(m+1)$ realisations. See Marriott (1979) for some guidelines about the size of m.

Step 2: for each model map and for the observed map, compute a goodness-of-fit statistic between that map and the expected map. For example, Pearson's product moment correlation coefficient or sums of squared differences could be used. Under the null hypothesis H_0, these $(m+1)$ statistics will be identically distributed and equicorrelated.

Step 3: rank the $(m+1)$ statistics and reject H_0 at the $100(j+1)/(m+1)\%$ level if the statistic for the observed map has rank $(m-j+1)$ or worse (one-tailed test). Rules for two-tailed tests may be formulated in a similar manner. In ranking the test statistics, if Pearson's r is used, call the highest positive value rank 1; if the sum of squared differences is used, call the smallest sum rank 1. In this procedure, we reject H_0 only if the difference in directional bias between the observed and expected maps is judged so severe that it cannot be regarded as a chance occurrence at the stated level of significance.

We shall now illustrate the various goodness-of-fit methods discussed above for some maps given in Hägerstrand (1953).

3.3 Empirical example
3.3.1 The Hägerstrand model
Hägerstrand selected the Asby district of Sweden to study the spatial pattern of acceptance of a subsidy which the Swedish government granted from 1928 onwards to farmers of small units (less than 8 hectares of tilled land) if they enclosed woodland on their farms and converted it to pasture. The study area was divided into 125, 5×5 km², cells. The total number of farms in each cell which had accepted the subsidy by the ends of 1929, 1930, 1931, and 1932 respectively was recorded. Hägerstrand then developed a Monte Carlo model to simulate the recorded numbers of adopters in each cell up to the end of the three observation years, 1930-2. The model is described in detail in Hägerstrand's paper, but some discussion is required here to make interpretation of the results given later meaningful.

Hägerstrand assumed that the decision of any potential adopter to accept the subsidy was based solely upon information received orally at face to face meetings between the potential adopter and adopters or carriers. He further assumed that the probability of a potential adopter being paired with a carrier was inversely related to geographical distance between the teller and the receiver. Founded on these two assumptions, Hägerstrand developed a Monte Carlo model with the following structure.
(1) The input number and spatial locations in the model of carriers and potential adopters (that is, all farms $\leqslant 8$ hectares of tilled land which had not accepted the subsidy) were the actual configurations in 1929.
(2) A potential adopter accepted the subsidy as soon as he was contacted by a carrier.
(3) In each iteration of the model every carrier told one other person, carrier or noncarrier. The probability, P_i, that any given carrier would contact a receiver located in the ith cell of the model plane, is given for each cell in figure 3.2. These probabilities were determined from an analysis of migration and telephone traffic data (Hägerstrand, 1967, pages 165-246).

It was assumed that the carrier was located in the centre cell of this target or Mean Information Field (MIF). Outside the target, $P_i = 0$. As in the study lattice, each cell of the target was 5×5 km². The address of a carrier's contact in each generation was determined in two steps. First, a random number m, from a rectangular distribution, selected the ith cell if

$$\sum_{r=1}^{i-1} Q_r < m \leq \sum_{r=1}^{i} Q_r ,$$

where Q_i, the probability of a contact in cell i with population n_i, is given by

$$Q_i = \frac{P_i n_i}{\sum_{i=1}^{25} P_i n_i} .$$

A second random number from a rectangular distribution in the range 1 to n_i located the receiver in the cell. If he was identical with the teller, a new address was sampled.

(4) To take into account the reduction in interpersonal communication likely to be caused by physical features such as rivers and forests, two simplified types of barrier were introduced into the model plane, zero-contact, and half-contact barriers. When an address was directed over a zero-contact barrier, the telling was cancelled. When the address crossed a half-contact barrier, the telling was cancelled with probability 0·5. However, two half-contact barriers in combination were considered equal to one zero-contact barrier.

Using this model, Hägerstrand performed a series of computer runs to simulate the spatial pattern of acceptance of the improved pasture subsidy by farmers in the study area. The observed patterns and results from three runs of the model are given in Hägerstrand (1953, pages 23-25). The diagrams for one of these runs are redrawn here as figure 3.3. In figure 3.3, the ith iteration of the model is indicated by g_i; g_0 gives the initial configuration of adopters in 1929 under model rule (1), above.

0·0096	0·0140	0·0168	0·0140	0·0096
0·0140	0·0301	0·0547	0·0301	0·0140
0·0168	0·0547	0·4431	0·0547	0·0168
0·0140	0·0301	0·0547	0·0301	0·0140
0·0096	0·0140	0·0168	0·0140	0·0096

Figure 3.2. Hägerstrand's Mean Information Field.

Map comparison with application to diffusion processes 73

For generations $g_1 - g_4$, we colour coded cells B if cells with adopters in them in g_{i-1} gained more adopters in g_i, W if no change occurred, and G (stippled) if a cell received adopters for the first time. The numbers in squares at the lower right-hand corner of each map give the total number of adopters on each map. For g_5, we have recorded the actual number of

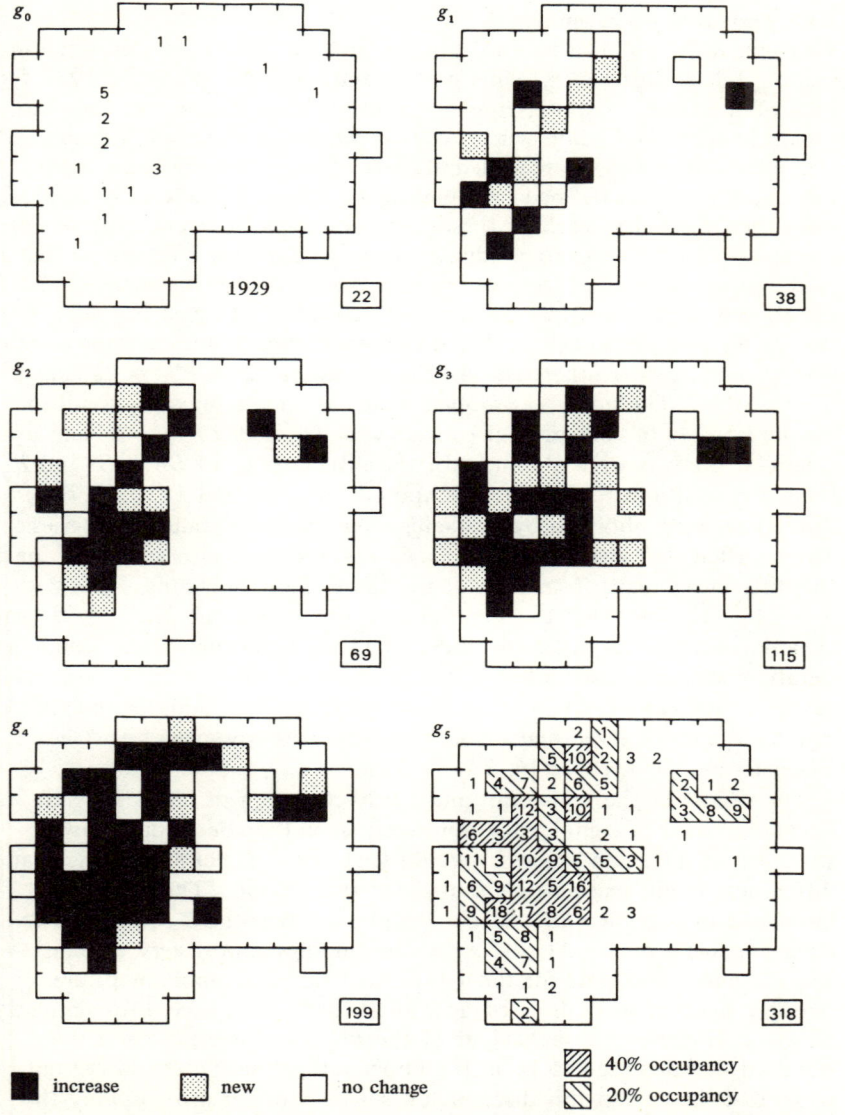

Figure 3.3. Pattern of adopters for one run of the Hägerstrand model. For definitions see text.

adopters in each cell, and shaded the cells according to the proportion of the total potential adopters in each cell who had been contacted by g_5. The striking feature of the map series is the strong contagious build-up of adopters in areas of existing adopters, a point to which we shall return later.

Next, we shall evaluate the spatial goodness-of-fit of the three simulation series given in Hägerstrand to the observed pattern of adoption.

3.3.2 Analysis of simulation results

To begin with, we need to illustrate the method of analysis described in section 3.2.1. Differences maps were constructed for 1931 and 1932 for each of the three computer runs reported by Hägerstrand, and these are shown in figure 3.4. For clarity, cells in which no observed and no simulated adopters were recorded are left blank. The nonfree sampling join-count statistics defined in equations (1.12)–(1.13) and (1.31)–(1.34) were then applied to each of the differences maps to test for spatial autocorrelation in the nonzero residuals. The n_1 cells with a negative residual in the differences maps were coded B (cross-hatched in figure 3.4), and the n_2 cells with a positive residual were coded G (stippled in figure 3.4). We set $w_{ij} = 1$ if two cells in the differences maps had a common edge or vertex, and $w_{ij} = 0$ otherwise. In the equations, $S_0 = \frac{1}{2}S_1 = 864$, and $S_2 = 25056$. The nonfree sampling model was used since we conditioned on the number of cells present of each type (n_1 cells of type B, and n_2 of type G). Results were obtained for the distributions of BB, GG, and BG joins. In addition, results were computed for 1931 and 1932 for the differences map obtained by averaging over the three simulation series. The coefficient I, given in equations (3.1)–(3.4), was also evaluated for the differences maps from each of the three simulation runs, again with $n = 125$. The fact that this coefficient does not assume $\Sigma_{(1)} z_i = 0$ has advantages here. As discussed under rule (3) of the simulation model, an iteration was completed when each carrier had contacted one other person, carrier or noncarrier. As a result the total number of adopters on the simulated and observed maps is not the same for any given year (see Hägerstrand, 1953, pages 16, 23–25), and so $\Sigma_{(1)} z_i \neq 0$.

The results of the join-count and I tests are given in tables 3.1 and 3.2. One-tailed tests of significance were used since the contagious growth structure of diffusion processes would lead us to expect systematic spatial differences exhibiting positive spatial autocorrelation. That is, we are interested in excessive BB and GG counts, a deficient BG count, and a large positive I value. At first sight, the findings appear very conclusive. The BB and GG counts for the 1931 and 1932 differences maps are strongly significant in all cases, as is the I coefficient, except for simulation-series 2. However, the behaviour of the BG count is erratic and the standard deviates tend to lie in the upper rather than in the lower tail.

To cast light upon this discrepancy, let us consider more closely the assumptions underlying the tests. The join-count statistics assume under H_0

that the probability of a cell being coded a particular colour is the same for all cells; but a careful inspection of the differences maps suggests that a cell in the southeast corner, far away from the main region of adopters, has a much higher chance of being coded W than a cell in the rest of the

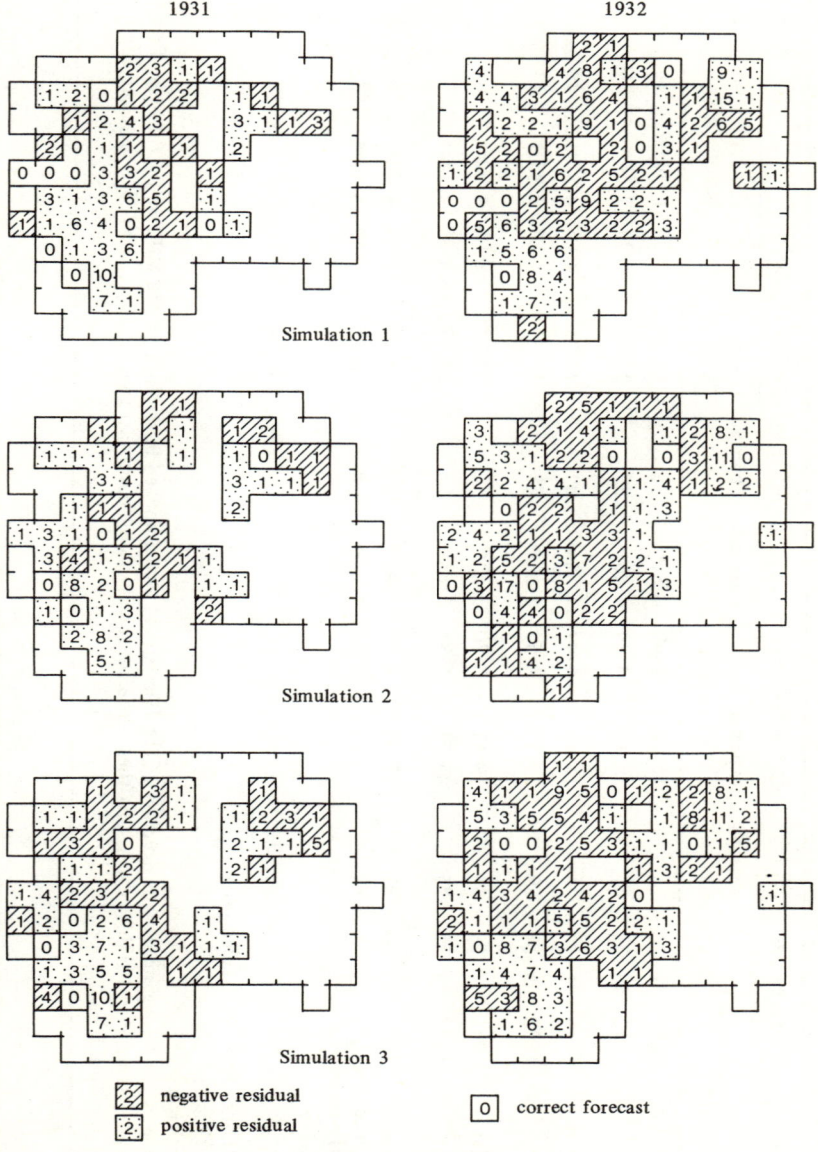

Figure 3.4. Differences maps for Hägerstrand's three simulation series.

Table 3.1. Values of the join-count statistics for the Hägerstrand differences maps.

Differences map	Year	n_1	n_2	BB joins (negative residuals)				GG joins (positive residuals)				BG joins			
				observed	expected	σ	standard deviate, z	observed	expected	σ	standard deviate, z	observed	expected	σ	standard deviate, z
Simulation series 1	1931	21	26	26	11.71	2.97	4.81**	38	18.11	2.95	6.74**	31	30.43	7.25	0.08
	1932	38	32	61	39.19	4.89	4.46**	39	27.65	4.27	2.66**	84	67.78	6.61	2.45
Simulation series 2	1931	21	30	23	11.71	2.97	3.80**	42	24.25	4.05	4.38**	43	35.12	5.02	1.57
	1932	35	38	59	33.17	4.58	5.63**	58	39.19	4.89	3.85**	73	74.14	6.86	−0.17
Simulation series 3	1931	27	29	33	19.57	3.70	3.63**	42	22.63	3.94	4.92**	61	43.65	5.53	3.14
	1932	41	35	75	45.71	5.81	5.85**	52	33.17	4.55	4.14**	76	79.99	7.06	−0.57
Average of differences over the three simulation series	1931	32	36	51	27.65	4.27	5.46**	55	35.12	4.68	4.24**	83	64.21	5.94	3.16
	1932	38	53	55	39.19	4.89	3.23**	108	76.81	6.12	5.10**	122	112.26	8.00	1.22

Significance levels (one-tailed test):
* 0·05 level
** 0·01 level

Because of the contagious growth structure of the Hägerstrand model implied in the form of the MIF, we hypothesise, if the observed and simulated maps do have systematic spatial differences, that these differences will exhibit positive spatial autocorrelation. We therefore apply one-tailed tests of significance to the standard deviates.

study area. A similar assumption is made in the case of I, namely that the probability of a cell in a differences map having a particular x_i value is the same for each cell.

In order to approximate more closely this assumption of equal probability in the join-count and I statistics, the analysis for these coefficients was reworked, redefining n as the number of nonblank cells in the differences maps and considering only joins between cells which were not blank. The effective number of observations and the map structure varied for each differences map and, although the analysis was worked through for all differences maps in the case of the join-count statistics, for I we examined only the average differences map. The results are given in table 3.3 and in table 3.4.

The picture now changes considerably. All the observed BB and GG counts continue to exceed expectation, but only five out of sixteen are significant at the $\alpha = 0.05$ level. On the other hand the BG join counts are now much further away from their expectations under the null hypothesis, and indicate fewer such links than would be expected. Negative standard deviates for BG joins imply clustering of areas of overestimation and underestimation on the differences maps (cf section 1.5.2). The more powerful I statistic (for a discussion of relative efficiencies, see chapter 6) now firmly rejects H_0 for 1931 and 1932. Criticisms can be made of both analyses, but we feel that the second is closer in spirit to the assumptions underlying the testing procedures.

Table 3.2. Values of I for the Hägerstrand differences maps.

Value of I	Differences map					
	simulation series 1		simulation series 2		simulation series 3	
	1931	1932	1931	1932	1931	1932
Expected	0·015	−0·008	0·034	−0·005	−0·001	−0·008
Observed	0·275	0·176	0·115	0·064	0·247	0·235
σ	0·046	0·046	0·054	0·044	0·044	0·046
Standard deviate	5·67**	4·03**	1·49	1·56	5·60**	5·26**

** Significant at $\alpha = 0.01$ level (one-tailed test).

Table 3.3. Values of I for the reduced n Hägerstrand average differences map.

Year	Statistic				
	n	I	$E(I)$	σ	standard deviate
1931	71	0·303	0·030	0·061	4·51**
1932	92	0·205	−0·011	0·056	3·89**

** Significant at $\alpha = 0.01$ level (one-tailed test).

Table 3.4. Values of the join-count statistics for the reduced n Hägerstrand differences maps.

Differences map	Year	n_1	n_2	n	BB joins				GG joins				BG joins			
					observed	expected	σ	standard deviate, z	observed	expected	σ	standard deviate, z	observed	expected	σ	standard deviate, z
Simulation series 1	1931	21	26	56	26	20·04	3·59	1·66*	38	31·02	4·55	1·53	31	52·12	5·43	−3·89**
	1932	38	32	79	61	54·53	5·38	1·20	39	38·48	4·75	0·11	84	94·33	6·98	−1·48
Simulation series 2	1931	21	30	56	23	19·08	3·48	1·13	42	39·55	4·71	0·52	43	57·27	5·49	−2·60**
	1932	35	38	82	59	44·07	5·15	2·90**	58	52·07	5·45	1·09	73	98·52	7·15	−3·57**
Simulation series 3	1931	27	29	60	33	32·32	4·34	0·16	42	37·39	4·51	1·02	61	72·11	5·91	−1·88*
	1932	41	35	80	75	64·09	5·52	1·98*	52	46·51	4·97	1·10	76	112·17	7·30	−4·95**
Average of differences over the three simulation series	1931	32	36	71	51	41·72	4·64	2·00*	55	52·59	4·95	0·41	83	96·89	6·78	−2·05*
	1932	38	53	92	55	49·21	5·15	1·12	108	96·45	6·19	1·87*	122	140·97	8·14	−2·33**

Significance levels (one-tailed test):
* $\alpha = 0·05$ level
** $\alpha = 0·01$ level.

Table 3.5. Results of tests of significance for Hägerstrand's differences maps using the approximations given in equations (2.65) and (2.67).

Simulation series differences map		Year	BB joins			GG joins			BG joins		
			observed	BB_α (upper tail)		observed	GG_α (upper tail)		observed	BG_α (lower tail)	
				$\alpha = 0{\cdot}05$	$\alpha = 0{\cdot}01$		$\alpha = 0{\cdot}05$	$\alpha = 0{\cdot}01$		$\alpha = 0{\cdot}05$	$\alpha = 0{\cdot}01$
Corresponding to table 3.1	1	1931	26**	16·86	19·41	38**	23·12	25·47	31	17·92	12·17
		1932	61**	47·45	51·24	39**	34·91	38·30	84	56·32	51·01
	2	1931	23**	16·86	19·41	42**	31·15	34·40	43	26·28	22·06
		1932	59**	40·93	44·52	58**	47·45	51·24	73	62·27	56·79
	3	1931	33**	26·31	29·35	42**	29·36	32·53	61	33·97	29·40
		1932	75**	55·53	60·04	52**	40·88	44·44	76	67·79	62·17
	average	1931	51**	34·91	38·30	55**	43·04	46·49	83	53·85	49·00
		1932	55**	47·45	51·24	108**	87·04	91·60	122	98·51	92·25
Corresponding to table 3.3	1	1931	26	26·18	29·08	38	38·74	42·33	31**	42·63	38·10
		1932	61	63·57	67·64	39	46·50	50·18	84	82·30	76·70
	2	1931	23	25·03	27·86	42	47·50	51·13	43**	47·68	43·11
		1932	59**	52·76	56·72	58	61·54	65·38	73**	86·21	80·49
	3	1931	33	39·67	43·05	42	45·00	48·48	61*	61·83	56·97
		1932	75*	73·34	77·47	52	54·87	58·66	76**	99·61	93·79
	average	1931	51*	49·54	53·09	55	60·90	64·63	83*	85·19	79·73
		1932	55	57·87	61·79	108*	106·76	111·31	122*	127·03	120·64

Significance levels (one-tailed test):
* $\alpha = 0{\cdot}05$ level
** $\alpha = 0{\cdot}01$ level.

In tables 3.1, 3.2, and 3.4 we have tested the various coefficients for significance using the straightforward normal approximation. In the case of the variant of I used in tables 3.2 and 3.3, we have not undertaken any small-sample studies to examine the suitability of this approximation in finite sized lattices. However, in the case of the join-count statistics used in tables 3.1 and 3.4, the approximation given in equations (2.65) and (2.67) can be applied. This yields the results shown in table 3.5, and supports the comments made in previous paragraphs.

It is, of course, possible that the fairly conclusive rejection of H_0 by the above analyses for the various differences maps is caused by Brown and Moore's point—that the autocorrelation tests are not independent of directional biases in the observed and simulated maps. We therefore look to the Hope-type procedure outlined in section 3.2.2 for further evidence.

The Hope-type procedure. Hägerstrand has given the results for three simulation series, and we have generated another 96 expected maps for 1931 and 1932 yielding, together with Hägerstrand's results, m (= 99) model maps for each of those years. The steps in the procedure were then

Step 1: from the 99 generated maps and the corresponding observed map (k = 100 maps in all) we computed, for 1931 and 1932, an 'average expected' map, where $\bar{x}_i$, the number of adopters in the ith cell of the average map, was given by

$$\bar{x}_i = \frac{1}{100} \sum_{k=1}^{100} x_{ik}, \qquad i = 1, 2, ..., 125 \ .$$

Step 2: the spatial goodness-of-fit of each of the k (= 100) maps to the corresponding average map was determined as follows. We converted each of the k maps into a differences map by calculating the quantities,

$$z_{ik} = x_{ik} - \bar{x}_i, \qquad i = 1, 2, ..., 125 \ ; \quad k = 1, 2, ..., 100 \ ,$$

where x_{ik} is the number of adopters in the ith cell of the kth map. We then computed for the kth differences map the test statistic, I_k, given in equation (3.1), using the $\{z_{ik}\}$. Binary weights were employed as before.

Step 3: the I_k (k = 1, 2, ..., 100) were ranked from smallest (rank 1) to largest (rank 100), and the sampling distribution thus obtained is shown in figure 3.5. Given the procedure outlined above, the simulated maps which correspond 'best' with the average map have low degrees of spatial autocorrelation in their differences maps.

Step 4: the test statistic between the observed map and the average map for 1931 was 5·36 and rank 100 (the only observation in the end class of the sampling distribution shown in figure 3.5), and that for 1932 was 4·26 and rank 98. As discussed in section 3.2.2, we can reject H_0 for 1931 at the α = 0·01 level, and for 1932 at the α = 0·03 level (one-tailed test).

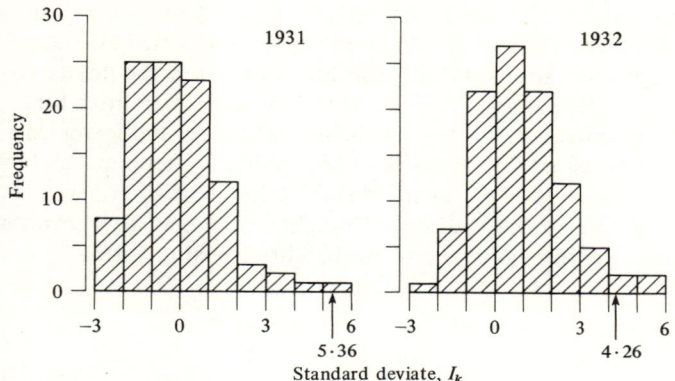

Figure 3.5. Empirically generated sampling distributions of I for a Hope-type analysis of Hägerstrand's spatial diffusion model. The I values between the observed and average expected maps are arrowed.

3.3.3 Interpretation of the results

If figures 3.3 and 3.4 are compared it will be noticed that the areas in which the simulation model overestimated numbers of adopters (negative residuals in figure 3.4) are clustered very closely along the line of the original adopters of the innovation in 1929. Conversely, areas in which the simulation produced fewer adopters than actually occurred (positive residuals) form a tyre-shaped ring around the hub of overestimation. This basic pattern is evident in all the differences maps of figure 3.4 and accounts for the significant negative values of the standard deviate for BG joins in table 3.4. If we consider the cells containing the original adopters in 1929, by 1932 the model had, for simulation series 1, overestimated numbers of adopters in eight of them, and underestimated in three (8:3). The corresponding proportions for series 2 and 3 are 9:3 and 9:2.

This tendency of the Hägerstrand model to produce overconcentration of adopters in core areas of a diffusion process has been considered by Tinline (1971) and Mollison (1975; 1977). Tinline's approach was to take the observed maps for two particular years, say 1930 and 1931, and to find the 5×5 matrix operator which transformed the 1930 map into the 1931 map with the minimum amount of error. His methods were the same as those employed by Tobler (1969a) to examine population and land-use changes in two US cities. Tinline defined operators of the form

$$\{Z_{pq}(t+1)\} = a + \sum_{j=-k}^{k} \sum_{i=-k}^{k} b_{ij} z_{p+i, q+j}(t) , \qquad (3.5)$$

where $Z_{pq}(t)$ denotes the random variable on a matrix map in cell (p, q) at time t, and a $(2k+1) \times (2k+1)$ operator is postulated. The $\{b_{ij}\}$ and a are coefficients which Tinline estimated using least squares (LS) and minimum absolute deviations (MAD). He computed results both allowing for, and ignoring, the susceptible population (potential adopters) in each cell in 1929. Figure 3.6 plots the average value of the $\{b_{ij}\}$ obtained (vertical axis of each graph) at various distances in cell units from the centre (0) of the operator (horizontal axis). The corresponding probabilities for Hägerstrand's MIF appear as the top left-hand graph. On the lower pair of graphs, the stippled area shows the range of $\{b_{ij}\}$ values. There is a tendency for the values of the coefficients in the distant cells of all the operators to rise after decaying away from the central weight. This may be contrasted with the negative exponential decay of probabilities postulated by

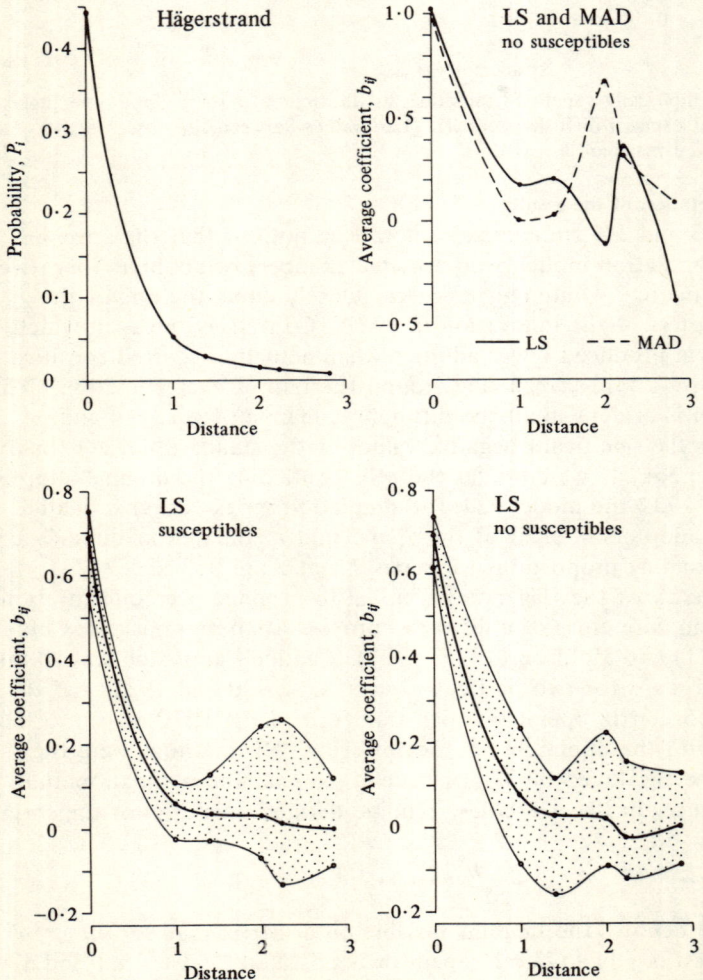

Figure 3.6. Coefficient values for Tinline's 5 × 5 matrix operators. Source: Tinline (1971, pages 85–89).

Hägerstrand in his MIF. Tinline and Mollison suggest that the rise in the peripheral coefficients of the operators reflects a build-up of the spread process at the operator edges, and that attempts to break out and diffuse widely are constrained, as in the MIF, by the fact that zero contact is enforced outside the 5 × 5 grid. The results imply that the problem of overestimation in areas of original adoption produced by Hägerstrand's

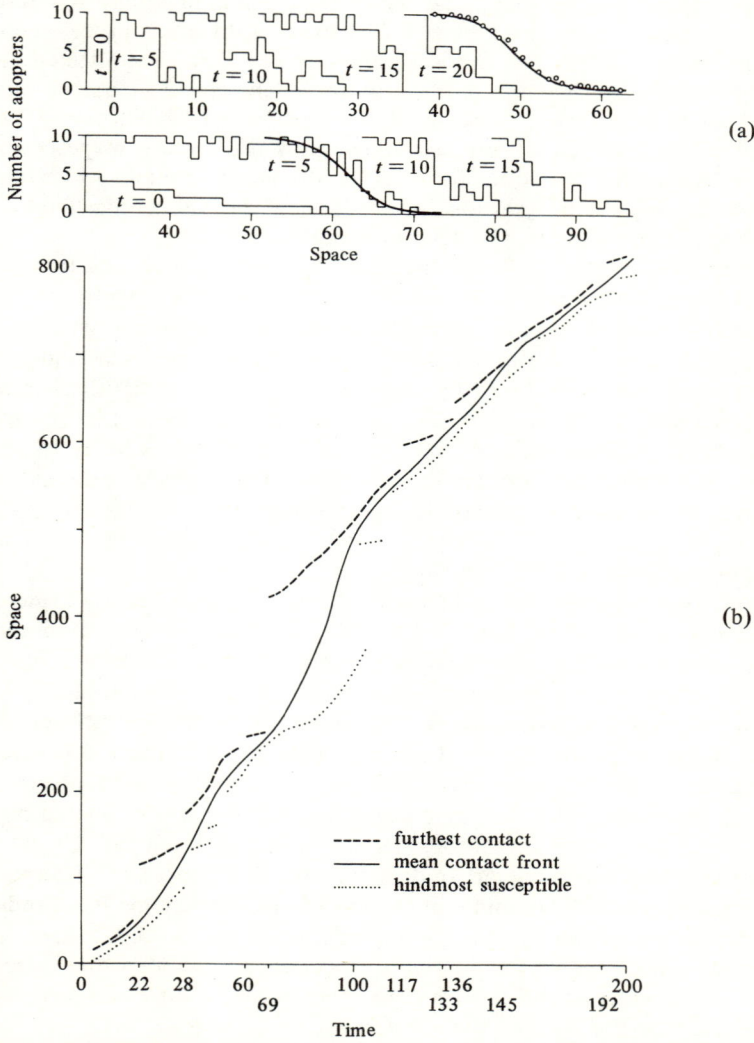

Figure 3.7. Form of diffusion waves produced by (a) exponentially bounded and (b) nonexponentially bounded mean information fields. Source: Mollison (1977, pages 307-308).

model could be mitigated by using a larger MIF and having a less rapidly decaying probability field.

The form of a diffusion wave when an MIF of the kind employed by Hägerstrand is used has been investigated by Mollison (1977). Mollison found that when the MIF has contact probabilities which fall off with distance between adopters and potential adopters at least as fast as exponentially (like Hägerstrand's), the diffusion wavefront moves out from the origin at a steady velocity; the waveform is maintained and few long distance contacts are made. Figure 3.7(a) shows a diffusion process along a line for such an MIF. The initial adopters ($t = 0$) are concentrated at the origin on the upper graph, but are more widely spread along the line on the bottom graph. The position of the wavefront is shown for subsequent time periods. The open circles show the average shape of the wavefront between $t = 105$ and $t = 200$; the heavy curve is a logistic. Such a steady logistic progression for a diffusion wave is familiar from many studies (for example, Casetti and Semple, 1969).

When an MIF which is not exponentially bounded is used, a different picture emerges [see figure 3.7(b)]. As time goes on, the wavefront disintegrates and long distance contacts (pecked line) are made well in advance of the mean wavefront (solid line). Mollison's evidence that a compact waveform is produced by exponentially bounded MIFs, whereas spatially irregular contacts are produced by MIFs which are not so bounded, supports further the conclusion that the Hägerstrand MIF is likely to lead to overconcentration of adopters in areas where the diffusion process originated, at the expense of more peripheral regions.

3.4 Other methods

The comparison of observed and simulated maps has also been considered in Cummings et al (1973), Lankford (1974), and Yapa (1976). Cummings et al tackle the problem of comparing two graphs (that is, two sets of nodes and edges). They evaluate the degree of spatial correspondence between the graphs by assuming that the distribution of edge differences between a pair of graphs can be described either by the binomial (cf free sampling) or by the hypergeometric (cf nonfree sampling) distributions. Lankford approaches the issue by comparing observed and simulated maps by the use of canonical correlation. His method was developed for data on origin-destination flows rather than for diffusion processes. However, the techniques both of Cummings et al and of Lankford assume independent observations. As shown by the counterexample in section 3.2.2, this assumption is not tenable for diffusion processes, and so we would argue that neither provides a valid test for map comparison for diffusion processes.

Yapa criticises the tests of this chapter on two grounds:
(1) The Hope procedure can be costly in computational time and effort.

Although it is a valuable procedure when the sampling distribution of a model is unknown, Yapa argues it can be dispensed with for Hägerstrand type models because the underlying probability structure can be identified as multinomial. The problem then resolves into determining whether the observed and simulated maps can be regarded as outcomes of the same multinomial distribution.
(2) Examination of cell-by-cell differences between observed and simulated maps may produce misleading results because of a confounding of localised nonrandom differences with the overall goodness-of-fit of the model.

Cliff and Ord (1980), however, have shown that Yapa's derivation of the multinomial is incorrect. Whereas for an aspatial test of fit, based upon X^2, an approximation based upon the multinomial is applicable when the proportion of adopters in the population is low (that is, in the early stages of spread), the spatial features of the model *do* affect the underlying distribution in the geographical case. For the present, there seems to be no viable alternative to a Monte Carlo approach. Thus, in agreement with Brown and Moore, we do not accept Yapa's assertion that "the issue of autocorrelation in diffusion maps becomes irrelevant to the problem of determining the statistical significance of an observed map vis-à-vis the Hägerstrand simulation". The scale issue raised by Yapa is important. Clearly all analysis is scale dependent. We would agree that significance at one spatial scale does not imply significance or otherwise at other spatial scales. Therefore, it is important to evaluate the spatial pattern at different scales, possibly using the methods described in chapter 5.

3.5 Conclusions

In this chapter, the problem of comparing observed and theoretical maps has been considered, and tests based upon the join-count and I statistics have been proposed which enable this to be done readily. However, it has been shown that special problems occur for diffusion processes, in that the spatial features of diffusion models affect the sampling distributions of the test statistics under the null hypothesis. A procedure based upon Monte Carlo methods has been outlined which overcomes these difficulties, and despite attempts, particularly by Yapa, to formulate a simple, direct test based upon the multinomial, the Monte Carlo approach seems at present to be the only valid alternative.

The use of the Monte Carlo method has been illustrated by an application to the maps produced by the classic Hägerstrand (1953) model of the adoption by farmers of a wood clearance scheme in Central Sweden in the 1920s and 1930s. The results of the test suggest that the mean information field employed by Hägerstrand incorporates too rapid a decay with distance in the probability of contact between adopters and potential adopters. This results in a pattern of spread which is slower and more spatially confined than that observed in reality, a conclusion which is supported by the findings of other workers whose research is reported in section 3.3.3.

4

The analysis of spatial point patterns

4.1 Introduction

In earlier chapters, we have considered the set of sites or counties to be fixed, and have focused attention upon some variable of interest recorded at those sites. Thus we might wish to examine incomes per head in different cities, or the size of trees in a forest. However, the *locations* of the cities or trees themselves are of direct importance in many studies. Thus Burghardt (1959, page 322), in an analysis of large river towns along the central Mississippi valley, argued that the towns "reveal an interesting uniformity of spacing along the rivers [in the central lowlands of the USA]". Again, in her classic text on mathematical ecology, Pielou (1977) emphasises the importance attaching to the relative locations of individuals in the study both of aggregation within a species and of the competition between species.

It is clear that, in all cases, the individual will occupy a finite area in the plane. Nevertheless in some processes individuals will occupy only a very small part of the plane, and it is then convenient to think of the individuals as being located at different *points* in the plane, or forming a *spatial point pattern*[4]. Conversely there are other processes which involve a partition of the plane, such as the ground cover of plant species or market areas of different retail outlets. In these instances an areal process must be defined. In this chapter we shall concentrate only upon point processes, although the study of areal patterns and mosaics is very important, well reviewed in Getis and Boots (1978, pages 121-163), Roach (1968), and Matérn (1979).

One approach to the analysis of point patterns is to sample small zones in the plane and to count the number of individuals occurring in each such zone. These sample zones are commonly known as quadrats (even when circular!), and we shall follow this (ab)usage. The quadrats may be *contiguous* (usually laid out on a regular grid) or may be located randomly in the study area. Section 4.2 is devoted to the analysis of data collected by quadrat sampling.

A risk underlying the use of quadrats is that any spatial pattern detected may be dependent upon the size of the quadrat (Pielou, 1977, pages 222-223), although this can be alleviated to some extent for contiguous quadrats (see section 5.3). To avoid this difficulty, we may study the distances from individuals to others nearby, known as *nearest-neighbour distances*; these measurements and their applications will be considered in section 4.3.

As we note in section 4.1.1 below, data obtained from random samples require a different perspective to data available as a map; section 4.4

---

[4] To avoid overworking the term *point*, we shall always refer to *individuals* located in the plane.

concentrates upon the particular problems relating to the analysis of mapped data.

The theory of spatial stochastic processes has developed in several disciplines; a major reference is Matérn (1960), and the comprehensive review paper by Whittle (1963) is still a valuable source of ideas. More recently Bartlett (1975, chapters 1 and 3) has provided an overview of current theory and gives several interesting examples.

4.1.1 Random or mapped data?

Many studies, particularly in ecology, use quadrats selected at random from the study area. This has the advantage that the observations may be taken as independent, thereby enabling the researcher to use statistical methods based upon this assumption. However, randomly located observations tell us very little about spatial pattern unless several different sizes of quadrat are used. Therefore, in section 4.2 we usually talk in terms of a set of contiguous quadrats (or a complete map of the study area), although some of the methods described could be used for randomly selected observations.

When we consider distance measurements, the picture changes; nearest-neighbour distances convey information on the spatial pattern, whether taken from randomly selected locations (or individuals) or considered for all individuals in a mapped area. When sampling at random, the observations are effectively independent provided that the number of observations taken is not too large relative to the total study area. In principle, search areas for the nearest events (neighbours) of different locations *may* overlap, but we may reasonably regard the observations as independent if twice the maximum observed nearest-neighbour distance is less than the minimum distance between sampling locations. We shall examine the effects of using mapped, rather than randomly selected, data in section 4.4.

4.2 The method of quadrat counts

If we superimpose a regular grid of quadrats over the map of a study area, then the count of the number of objects in each subarea enables us to reduce the areal distribution of objects displayed on that map to a frequency distribution. This method of data reduction has been discussed in detail by Greig-Smith (1964) and by Kershaw (1964). It has been applied to geographical problems by, among others, Birch (1967), Dacey (1964; 1966a; 1966b; 1968; 1969a), Getis (1964), Harvey (1966), Hudson (1967), McConnell (1966), Malm et al (1966), Olsson (1966), and Rogers (1965).

4.2.1 The Poisson process

In order to understand something of the spatial process generating the pattern of individual objects, we need to develop theoretical frequency distributions corresponding to hypothetical processes. The standard benchmark for spatial patterns is the *Poisson process* which may be described in the following way:

(a) Let a quadrat have unit area and let that quadrat be subdivided into N small subareas, each of area $1/N$.
(b) For each small subarea, define the probabilities,
   prob(no individuals in subarea) = $p_0 = 1 - \lambda/N + O(N^{-2})$,
   prob(1 individual in subarea) = $p_1 = \lambda/N + O(N^{-2})$,
   prob(2 or more individuals in subarea) = $O(N^{-2})$,
where $O(N^{-2})$ means terms of order N^{-2} or higher in $(1/N)$.
(c) As a consequence of (b) the number of individuals, X, in the quadrat follows the *binomial* distribution,

$$\text{prob}(X = x) = \binom{N}{x} p_1^x p_0^{N-x}, \qquad x = 0, 1, ..., N. \tag{4.1}$$

(d) Finally, if we allow $N \to \infty$ while keeping Np_1 fixed, $Np_1 = \lambda$, then in the limit the probability distribution becomes

$$\text{prob}(X = x) = \frac{\exp(-\lambda)\lambda^x}{x!}, \qquad x = 0, 1, ...; \tag{4.2}$$

this is the *Poisson* distribution.

The steps in this argument are standard; see Kendall and Stuart (1977, page 132). For any other area of size A, say, the Poisson distribution has parameter λA rather than λ. The mean of X, $E(X)$, is given by $E(X) = \lambda$, and we refer to λ as the *intensity* of the process or the expected number of individuals per unit area.

The important assumptions underlying the Poisson process are that:
first, there are no interactions between different subareas, whether inhibitory or attractional;
second, there is no possibility of multiple groupings of individuals within each subarea (no point clusters);
third, there is no tendency for neighbouring areas to display similar traits (for example, because of common proximity to an important resource).

All in all, the Poisson process looks uninteresting and unlikely ever to occur in reality! Its main role is as a 'no-dependence' null hypothesis against which data may be tested for departures from the three assumptions listed in the preceding paragraph. Two major departures from the simple Poisson scheme will be considered in this section.

(1) The first form is the true contagion or generalised Poisson process. Suppose that clusters of objects have been observed, such as plants in a field or houses in the study area, with each cluster containing one or more objects. The number of objects in each cluster follows a *generalising* distribution. Thus, if the generalising distribution is logarithmic, the generalised distribution is written as Poisson ∨ logarithmic, which is equivalent to the negative binomial, whereas Poisson ∨ Poisson yields the Neyman type A. Insofar as the existence of a cluster means that an object is 'more likely' to have other similar objects nearby, we say that these processes represent 'true contagion'.

(2) The second form is the apparent contagion or compound Poisson process. We consider the objects to be generated by a Poisson process as in form (1), but instead of identifying clusters and looking at the number of objects in each cluster, we assume that the number of objects in the ith quadrat is given by a simple Poisson process with mean λ_i, where λ_i may vary from quadrat to quadrat. That is, we assume that λ is itself a random variable and that its distribution may be specified. Thus the final distribution of the random variable X is the Poisson *compounded* with some other distribution. If λ follows a gamma distribution, the compound distribution is Poisson $\wedge$ gamma, which is the negative binomial. Likewise, Poisson $\wedge$ Poisson yields the Neyman type A.
If no contagion exists, or if $\lambda_i = \lambda$ for all i, the resulting distribution is the simple Poisson.

The most widely used compound and generalised Poisson distributions are summarised in table 4.1, where references to the original sources are given. Ord (1972, pages 135–147) and Rogers (1974, chapter 4) describe procedures for selecting and then fitting a particular distribution.

In addition to these distributions, which all exhibit clustering (real or apparent), we may consider the binomial distribution corresponding to the binomial process described earlier. If N is kept finite, this induces an

Table 4.1. Compound and generalised Poisson distributions.

Name of distribution (plus references)	Compound or generalised form ($P =$ Poisson)
Negative binomial (Pascal) (Fisher, 1941; Feller, 1943; Quenouille, 1949)	$P \vee$ logarithmic $P \wedge$ gamma
Polya–Aeppli (Polya, 1931)	$P \vee$ geometric Pascal $\wedge_k$ Poisson [a]
Neyman Type A (two parameter) (Neyman, 1939; Feller, 1943)	$P \vee P$ $P \wedge P$
Thomas (Thomas, 1949)	$P \vee P$ (one 'parent' plus random number of offspring)
Poisson–binomial (McGuire et al, 1957; Sprott, 1958)	$P \vee$ binomial Binomial $\wedge P$
Poisson–Pascal (Katti and Gurland, 1962; Shumway and Gurland, 1960)	$P \vee$ Pascal Pascal $\wedge P$
Short (Cresswell and Froggart, 1963; Kemp, 1967)	$P \vee$ (Poisson with zeros added)

[a] In the case of the Polya–Aeppli distribution the notation '$\wedge_k$' means that the compounding is done on the negative binomial parameter, k, defined in equation (4.3).
Source: Ord, 1972, page 126.

element of regularity into the spatial pattern, since we now have an upper bound on the number of individuals which may appear in any given area. Indeed, if we compute the population variance-to-mean ratio, which we designate φ and which is often known as the *index of dispersion*, the following ordering of the distributions is possible (Rogers, 1974, page 29).

Clustered point patterns	Negative binomial Poisson-Pascal Neyman type A Poisson binomial	$(\varphi > 1)$
Random point patterns	Simple Poisson	$(\varphi = 1)$
Regular point patterns	Binomial	$(\varphi < 1)$

decreasing clustering

However, the scheme must be used with caution, since many distributions are embedded in more general families which may include elements of both clustering and regularity.

4.2.2 True or apparent contagion?

If we return to table 4.1, it is important to note that both the true and the apparent contagion processes can yield the same final distribution, as is shown by the examples of the negative binomial and Neyman type A. When we obtain a 'good fit' of one of these distributions to an observed frequency array, we are faced with the problem of determining whether the generating process was of the true or apparent contagion form. Ord (1972, sections 6.6 and 7.8) has shown that, in the case of the negative binomial, if quadrat count data are available for two or more time periods, the index of the real contagion model will increase linearly with time, whereas for the apparent contagion model it remains constant. Bissell (1972a; 1972b; 1973) has extended this type of analysis by fitting the negative binomial to data records of different lengths, which would correspond to irregular areas in the present context.

However, we are often faced with only a single data set when investigating the generating process, so we now proceed to develop an approximate method which will enable us to distinguish true and apparent contagion in this case. The presentation is in terms of the negative binomial distribution but a parallel development would be possible for other cases.

Suppose that the study area has been partitioned into quadrats of a certain unit size, and that a 'good fit' of the negative binomial to the frequency array of objects in the quadrats has been obtained. Write the negative binomial density as

$$\text{prob}(X = x) = \binom{k+x-1}{x} p^k (1-p)^x \,, \qquad (4.3)$$

with parameters p and k. Consider a Poisson distribution with parameter λ, and a logarithmic distribution with parameter α. The Poisson $\vee$ logarithmic model (the 'true contagion' negative binomial) then has parameters

$$k = -\frac{\lambda}{\ln(1-\alpha)}, \quad \text{and} \quad p = \alpha.$$

We are considering point clusters and a Poisson distribution for numbers of clusters in an area, so that counts in neighbouring quadrats are independent. If s of the original quadrats are combined so that the Poisson parameter for cluster numbers becomes $s\lambda$, it follows that the number of individuals still has a negative binomial distribution but the parameters are now

$$k = -\frac{s\lambda}{\ln(1-\alpha)}, \quad \text{and} \quad p = \alpha.$$

For the Poisson $\wedge$ gamma distribution (the 'apparent contagion' negative binomial) the probabilities for a quadrat of unit area are given by

$$\text{prob}(X = x) = \int_0^\infty \exp(-\lambda) \frac{\lambda^x}{x!} g(\lambda) d\lambda, \tag{4.4}$$

where the gamma distribution has density function,

$$g(\lambda) = \frac{\lambda^{b-1} a^b \exp(-a\lambda)}{\Gamma(b)}, \quad \lambda > 0. \tag{4.5}$$

When the integral (4.4) is evaluated, we obtain the negative binomial with parameters $k = b$, and $p = a/(a+1)$. If we regarded neighbouring quadrats as independent, the combination of s quadrats would produce the same results as for the true contagion scheme (new $k = s \times$ old k, p unchanged). However, if we suppose that the apparent contagion is owing to the presence of some other resource *whose availability varies only slowly across space*, then the neighbouring quadrats will have similar λ values. Pushing this argument to the extreme, we shall assume that all s quadrats have the *same* value of λ. The compound distribution is then given as the sum of s independent, identical Poisson variates mixed, or compounded, with the gamma distribution given in equation (4.5). The s-fold sum has a Poisson distribution with parameter $s\lambda$, so that the resulting negative binomial has for its parameters,

$$k = b, \quad \text{and} \quad p = \frac{a}{a+s}.$$

Thus by calculating estimates for p and k for different sized lattices, we can see which of the models appears to be nearer the truth. This is only an approximate method of analysis because one cannot be certain that the quadrats combined to form larger quadrats have the same value of λ initially in the compound Poisson model. However, the procedure would seem to provide a reasonable check, provided it is recognised that we have made

the auxiliary practical assumptions that
> true contagion implies small 'clusters' (that is, much smaller than the quadrat size, so that the number of clusters overlapping a quadrat boundary is small);
> apparent contagion implies that the mean rate of occurrence, λ, varies slowly over space, so that the 'clusters' cover areas much larger than the typical quadrat.

Without these auxiliary assumptions, it is evident that the two distinct mechanisms cannot be distinguished from the data alone.

These auxiliary assumptions suggest an alternative approach, namely to use the spatial autocorrelation measures. If the generalised Poisson model holds, each quadrat will be an independent realisation of the negative binomial, provided that it is large enough to contain the entire cluster. Therefore we should find little or no spatial autocorrelation between adjacent quadrats. On the other hand, if the compound Poisson model is true and we postulate that λ varies slowly over space, we should expect to detect positive spatial autocorrelation between adjacent quadrats. Although we have used the negative binomial as a specific example to show how the spatial autocorrelation measures may be used to discriminate between real and apparent contagion, the same arguments would hold for other distributions. The set of possible conclusions from an autocorrelation analysis is summarised in table 4.2.

A recurring problem in studies of this kind is the choice of quadrat size, which is often rather arbitrary. Changing the quadrat size will usually affect the degree of spatial dependence between quadrats, and the larger the quadrat, the weaker the dependence will be. Thus by starting with very small quadrats and aggregating, as is done in the next section, we can monitor changes in the parameters of the selected Poisson distribution and examine the plausibility of the true or apparent contagion models.

Table 4.2. Selection of an appropriate theoretical Poisson model.

Item		Simple Poisson model	Generalised/compound Poisson model
Spatial autocorrelation	not detected	simple Poisson	true contagion
	detected	apparent contagion?	apparent contagion

4.2.3 A study of settlement patterns

In his classic study on settlement patterns, Matui (1932) selected two regions in the Tonami Plain, Japan, one in Hukuno Town and the other in Demati Town. He partitioned the former into 1200 quadrats, a single quadrat having each side measuring 100 metres, by superimposing over the area a regular 30 × 40 grid lattice. Demati Town was partitioned by a 25 × 30 grid lattice into 750 equal sized quadrats, each of side 130 metres.

For each lattice, Matui then counted the number of quadrats with 0, 1, 2, ...
isolated houses in them, and fitted a simple Poisson distribution to the two
resulting frequency arrays. Dacey (1968) repeated this sort of analysis for
twenty-one maps showing the location of isolated houses in Puerto Rico,
with the exception that he fitted a negative binomial, rather than a simple
Poisson, to the resulting frequency arrays. The results obtained by Matui
and Dacey are compared in Dacey (1969a), whose conclusions are consistent
with those given later in this section. Matui's Hukuno Town data are
reproduced in figure 4.1, and a map of the town and a housing-density
map are given in figure 4.2.

From Matui's original lattice several new lattices have been created by
combining adjacent quadrats in various ways. Let t be the number of
adjacent quadrats on a column, and s be the number of adjacent quadrats
on a row of the original lattice which were combined to form a new quadrat.
The combinations shown in table 4.3 were tried. Lattice 1 refers to
figure 4.1. To show how monitoring changes in the parameters of a
Poisson distribution with changes in quadrat size enables us to distinguish
real from apparent contagion models, we fitted the negative binomial by
maximum likelihood to the observed frequency distributions of quadrat

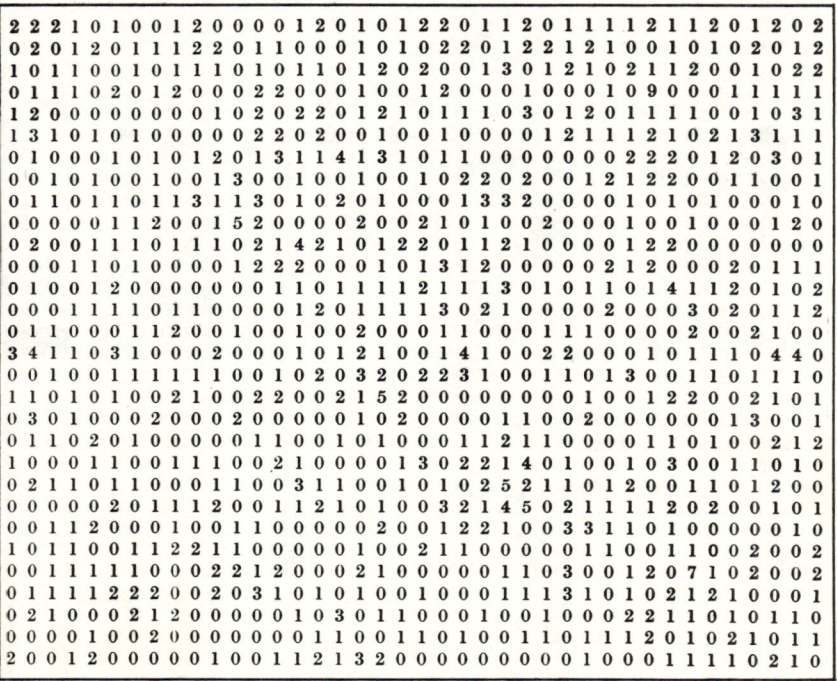

Figure 4.1. Quadrat counts of houses in Hukuno Town, Tonami Plain, Japan. Source: Matui (1932, page 254).

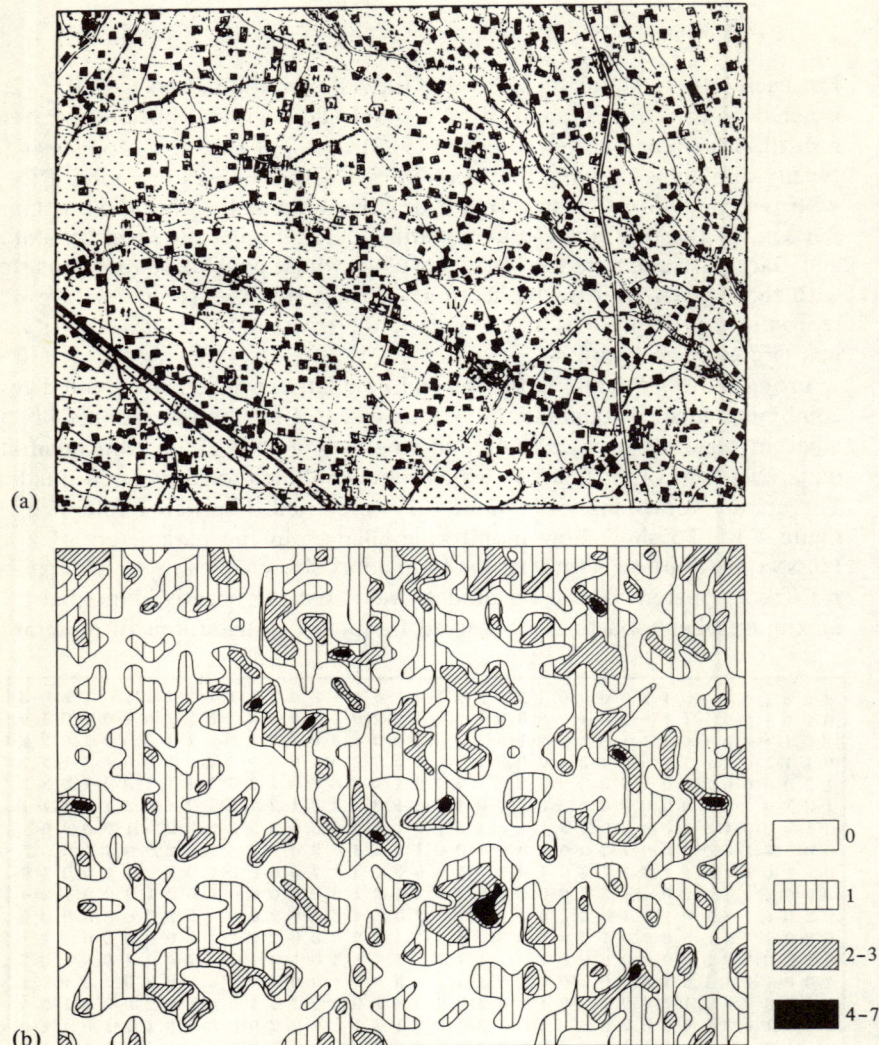

Figure 4.2. (a) Section of topographic map (1:20000 Military Land Survey) showing the western part of Hukuno Town. (b) Density of houses per 100 m² in the same area.

Table 4.3. Lattices derived from Matui's data.

Identity number	t	s	Lattice dimensions	Identity number	t	s	Lattice dimensions
1	1	1	30 × 40	6	3	2	10 × 20
2	1	2	30 × 20	7	1	4	30 × 10
3	2	1	15 × 40	8	2	4	15 × 10
4	2	2	15 × 20	9	3	4	10 × 10
5	3	1	10 × 40				

Table 4.4. Fitting the negative binomial model to lattices 1–9.

Class value	Lattice number																	
	1		2		3		4		5		6		7		8		9	
	O[a]	E[a]	O	E	O	E	O	E	O	E	O	E	O	E	O	E	O	E
0	584	590·2	151	153·4	137	145·9	14	21·7	49	52·3	2	4·2	20	21·0	0	1·1	0	0·1
1	398	392·3	189	188·4	207	191·9	56	50·0	97	95·4	13	14·0	46	49·2	2	4·5	0	0·5
2	168	154·7	138	133·8	139	139·3	72	63·4	98	96·6	31	25·2	67	63·2	17	9·8	0	1·5
3	35	47·1	72	71·9	69	73·7	54	58·6	77	71·6	31	32·5	63	59·0	10	15·3	5	3·2
4	9	12·2	37	32·4	33	31·8	43	44·1	44	43·3	34	33·4	48	44·6	18	19·1	3	5·2
5	4	2·8	5	12·9	8	11·8	27	28·7	16	22·7	34	29·2	23	29·1	27	20·5	8	7·3
6	0	0·6	2	4·7	4	3·9	18	16·6	13	10·6	13	22·5	18	16·9	22	19·5	11	9·1
7	1	0·1	4	1·6	1	1·2	7	8·8	2	4·6	13	15·7	4	9·0	15	16·8	16	10·2
8	0	0·0	0	0·5	1	0·3	4	4·4	1	1·8	12	10·1	4	4·4	10	13·5	11	10·5
9	1	0·0	2	0·2	0	0·1	2	2·0	1	0·7	11	6·0	4	2·1	6	10·1	8	10·1
10					1	0·0	1	0·9	1	0·3	3	3·4	2	0·9	9	7·1	5	9·2
11							0	0·4	0	0·1	0	1·8	1	0·4	6	4·8	7	7·9
12							2	0·2	1	0·0	1	0·9	0	0·2	2	3·1	6	6·5
13											0	0·5			1	1·9	6	5·2
14											0	0·2			1	1·2	2	4·0
15											1	0·1			4	0·7	5	2·9
16																	3	2·1
17																	1	1·5
18																	0	1·0
19																	2	0·7
20+																	1	1·5
X^2	6·84		5·81		3·65		6·27		3·49		13·12		8·39		17·12		9·95	
df	3		4		4		7		5		9		7		10		12	

] = cells combined for purposes of X^2 test. [a] O = observed; E = expected.
Note: expected frequencies < minimum listed are put in the end group.

counts of houses for lattices 1 to 9. The goodness-of-fit between the observed and expected frequency arrays was evaluated using X^2. The results are given in table 4.4, and indicate that the negative binomial is a good fit to the data for lattices. None of the X^2 values exceeds the *nominal* 5% value, although the nine tests are highly interdependent.

More recently, Haggett et al (1977, page 430) have fitted several other Poisson mixture distributions to the data for lattice 1, and it was found that several of these (notably the Neyman type A, Thomas, and Short in table 4.1) fitted equally well. Thus, there is no real evidence to prefer the negative binomial on this lattice, although it would be interesting to see how the different schemes fare as the quadrat size increases.

Given this rather weak vote for the negative binomial, let us now examine the results in greater detail. The maximum likelihood estimates for the parameters k and p for the nine lattices are shown in figure 4.3 together with the values we should 'expect' for lattices 2-9 from the generalised and compound schemes, given the results for lattice 1. If we compare the 'expected' parameter values with the maximum likelihood estimates, it appears that, with the exception of lattice 3, the compound model holds better than the generalised model.

It is now possible to illustrate how to use the spatial autocorrelation measures to provide a further check. The degree of spatial autocorrelation between adjacent quadrats in lattices 1-9 was examined using I as given in

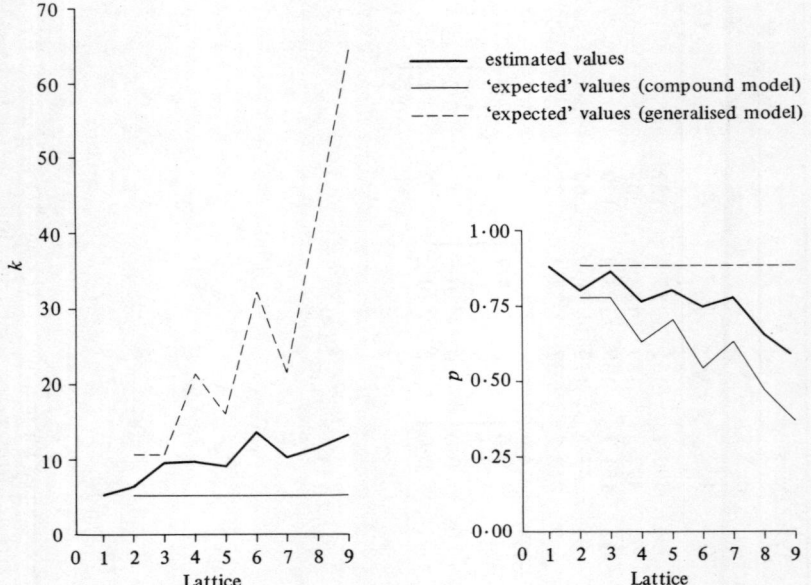

Figure 4.3. Parameter values for the negative binomial distribution applied to Matui's data. Lattices 1-9 are as defined in table 4.3. Source: Haggett et al (1977, page 434).

equation (1.15), The moments of I were evaluated under randomisation (R)—see equations (1.37)-(1.39); the differences in the standard deviates between assumptions R and N never exceeded 0·02. Two forms for **W** were tried:
(1) $w_{ij} = 1$ if the ith and jth quadrats had an edge in common, and $w_{ij} = 0$ otherwise (the rook's case);
(2) $w_{ij} = 1$ if the ith and jth quadrats had a common edge or vertex, and $w_{ij} = 0$ otherwise (the queen's case).
Matui (1932) and Ginsberg (1958) have suggested that the settlement pattern in this part of Japan is one of nucleated villages. We would thus expect individual houses to be clustered, but clusters to be fairly independent. However, the compound model implies similar densities in neighbouring areas. We therefore postulate, under H_1, positive spatial autocorrelation between quadrats and use a one-tailed test of significance. The results of the test are given in table 4.5. For all the lattices there is fairly clear evidence of positive spatial autocorrelation. Reference to table 4.2 shows the test results confirm the evidence of the parameter changes discussed above, namely that the compound model is more plausible than the generalised form.

The reader may recall from earlier in the section that Matui originally fitted the simple Poisson model to lattice 1. We repeated this exercise and also fitted the Poisson distribution to the observed frequency arrays for lattices 2-9. The goodness-of-fit between the observed and the expected frequency arrays was again evaluated using the X^2 statistic, and the results are given in table 4.6. It is evident from these results that the closest fit of the simple Poisson model to the observed frequency distribution of quadrat counts of houses occurs for lattices 2, 3, 4, and 7. As table 4.5 shows, for these four lattices there is a significant degree of spatial auto-correlation both in the rook's and in the queen's cases, although it is less marked in the queen's case for lattice 7. From table 4.2 these results suggest that we are again dealing with a compound Poisson process, although in the case of lattice 7 the apparent contagion is rather weak.

To test the strength of the departure from the simple Poisson model, we used the sample index of dispersion, φ,

$$\varphi = \frac{\text{sample variance}}{\text{sample mean}}.$$

As Bartko et al (1968) have shown, given a sample of size n from a Poisson population, $(n-1)\varphi$ is approximately distributed as χ^2 with $(n-1)$ degrees of freedom, provided the population mean is not too small. Further, $E(\varphi) = 1$ for any sample size. For large n, therefore, the statistic d,

$$d = [2(n-1)\varphi]^{1/2} - [2(n-1)]^{1/2},$$

is approximately normally distributed with zero mean and unit variance.

Table 4.5. Results of tests [a] for spatial autocorrelation in lattices derived from Matui's data.

Value of I	Lattice					
	1		2		3	
	rook	queen	rook	queen	rook	queen
Expected	−0·0008		−0·0017		−0·0017	
Observed	0·0601	0·0731	0·0842	0·0826	0·1056	0·0843
σ_R	0·0147	0·0206	0·0209	0·0293	0·0209	0·0294
Standard deviate under R	4·16	3·59	4·12	2·87	5·13	2·93
	4		5		6	
	rook	queen	rook	queen	rook	queen
Expected	−0·0033		−0·0025		−0·0050	
Observed	0·1327	0·0943	0·0982	0·0645	0·1154	0·0863
σ_R	0·0297	0·0416	0·0259	0·0362	0·0365	0·0512
Standard deviate under R	4·58	2·34	3·89	1·85	3·30	1·79
	7		8		9	
	rook	queen	rook	queen	rook	queen
Expected	−0·0033		−0·0067		−0·0101	
Observed	0·1070	0·0731	0·1097	0·0640	0·0926	0·0424
σ_R	0·0299	0·0419	0·0423	0·0594	0·0517	0·0727
Standard deviate under R	3·69	1·83	2·75	1·19	1·99	0·72

[a] The testing procedure uses the approximation given in equation (2.58).

Note: The null hypothesis of no spatial autocorrelation is rejected in favour of positive spatial autocorrelation (one-tailed test) at the 5% level whenever the standard deviate exceeds 1·645.

Key: rook ≡ rook's case; queen ≡ queen's case; R ≡ using randomisation.

Table 4.6. Results of the X^2 goodness-of-fit test between observed quadrat counts and simple Poisson expectations for lattices 1–9.

Item	Lattice number								
	1	2	3	4	5	6	7	8	9
X^2	7·87	7·21	3·68	8·56	12·04	19·50	6·97	27·72	17·28
df	3	4	4	6	5	7	6	8	9
α	4·9	12·5	45·1	20·0	3·4	0·7	32·4	0·1	4·5

$\alpha = \text{prob}(X^2 > \text{observed} | H_0) \times 100$

The values of φ and d for the nine lattices are as follows.

Lattice	φ	d	Lattice	φ	d	Lattice	φ	d
1	1·17	4·03	4	1·34	3·87	7	1·10	1·21
2	1·27	4·31	5	1·28	3·70	8	1·56	4·28
3	1·19	3·16	6	1·20	1·88	9	1·78	4·67

These results imply the rejection of the simple Poisson model in favour of the compound version, except for lattice 7. Although superficially different, this form of analysis is in fact very similar to the 'variance components' analysis carried out by Moellering and Tobler (1972); see section 5.3.

Finally, if we again examine table 4.5 we can see that, in confirmation of the earlier discussion of choice of quadrat size, the degree of spatial autocorrelation between adjacent quadrats generally declines as quadrat size increases. There is also much less spatial autocorrelation for the queen's case than for the rook's case, which reflects the chequered spatial arrangement of houses in the study area, as seen in figure 4.2.

4.2.4 Implications for settlement geography

As noted earlier, Matui (1932, page 251) and Ginsberg (1958) suggested that the settlement pattern in the Tonami Plain is essentially one of nucleated villages. The better fits of the negative binomial model to the data, as opposed to those obtained with the simple Poisson (tables 4.4 and 4.6), and the values of φ tend to confirm this. However, the tests for spatial autocorrelation indicate that the apparent, rather than the true, contagion version of the negative binomial model is the more plausible. This would seem to argue for a pattern of colonisation which is essentially random (Poissonian), but with varying propensities to settle in different parts of the region (because of different land quality, for example). It is apparent that the hypothesis of nucleated village settlement should not be accepted simply because the negative binomial model provides a good fit. For further discussion of the problem of equifinality, see Harvey (1968).

4.3 The study of pattern by using distances

The basic distance measurements that are possible are of two kinds:

nearest neighbour (NN—the distance from a selected individual to the first, second, ..., nearest individual;

point-to-individual (PI—the distance from a selected point in the plane to the nearest, second nearest, ..., individual. We use 'point' here and in the remainder of the chapter in the sense of position, place, or location, and not in the sense of a concrete object or individual.

To give an idea of future developments, consider a process with very closely knit clusters. We would expect most first NN distances to be small, as most of the nearest neighbours would be fellow cluster members.

Conversely, the PI distances could be large, because we may well select a point well clear of any of the clusters. On the other hand, if the individuals are very regularly spaced, as on a lattice, the first NN distances will be (nearly) equal and consistently larger than the first PI distances. In this way, we might expect the NN and PI distances to enable us to distinguish different spatial patterns.

We shall begin this section with an account of some of the spatial models proposed in the literature and then summarise some of the many distance-based tests available; these subsections rely heavily on the work of Diggle (1979a). We conclude with an analysis of data on the spacing of river towns along the Mississippi.

4.3.1 The Poisson and other processes

As mentioned in section 4.2, the Poisson process provides the standard of comparison for more complex processes, although we must remember the mildly worded caution of Cox and Lewis (1966) that "The Poisson process is a mathematical concept and no real phenomenon can be expected to be exactly in accord with it". To give a feel for what a Poisson pattern (or *random* pattern to use the term popular in ecology) should look like, we present two pictures in figure 4.4. If you look closely you may be able to see distinct clusters in the left-hand picture, but not in the right-hand one!

From equation (4.2) we know that, in any area A, the probability of observing fewer than j individuals is given by

$$\text{prob}[X(A) < j] = \exp(-\lambda A) \sum_{x=0}^{j-1} \frac{(\lambda A)^x}{x!} \,. \tag{4.6}$$

If the selected area is a disc of radius r, so that $A = \pi r^2$, the probability that j or fewer individuals lie within a circle of radius r centred on a

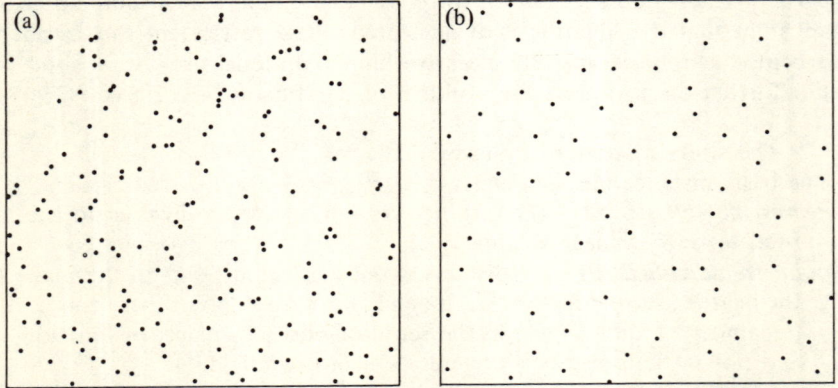

Figure 4.4. (a) Partial realisation of a Poisson process: 200 events independently and identically uniformly distributed in the unit square. (b) Realisation of simple sequential inhibition: 80 events in a square of side 15 units. Source: Diggle (1979a).

randomly selected point is given by expression (4.6). However, if R_j denotes the jth PI distance, we have *for any process*,

$$\text{prob}(R_j > r) = \text{prob}[X(\pi r^2) < j] . \tag{4.7}$$

The duality between counts and distances expressed in equation (4.7) means that once the counts distribution is fully specified, we may derive the distribution for R_j directly. Given expression (4.6), we obtain

$$\text{prob}(R_j > r) = \exp(-\rho u) , \tag{4.8}$$

where $u = r^2$, and $\rho = \pi\lambda$; ρ is the intensity for the unit circle. That is, R_1^2 has an *exponential distribution* with density function

$$f_1(u) = \rho\exp(-\rho u) . \tag{4.9}$$

Likewise, R_j^2 has a *gamma distribution with index j*, whose density function is given by

$$f_j(u) = \rho^j u^{j-1} \frac{\exp(-\rho u)}{(j-1)!} . \tag{4.10}$$

Another way of expressing these findings is to say that $2\pi\lambda R_j^2$ has a chi-squared distribution with $2j$ degrees of freedom. These results for the Poisson process yield the requisite distributions for the tests introduced in the next section.

The results given in expressions (4.8) and (4.9) do not depend upon whether the selected disc has an individual at its centre. Thus, for the Poisson process, the jth NN distance has the same distribution as the jth PI distance. Given the different effects upon NN and PI distances of non-Poisson processes, a random sample of NN and PI distances would be a useful way to test whether a pattern is consistent with the Poisson process. However, it is important to note that the selection of individuals *at random* in the field is effectively impossible, since any rule of the type 'select the

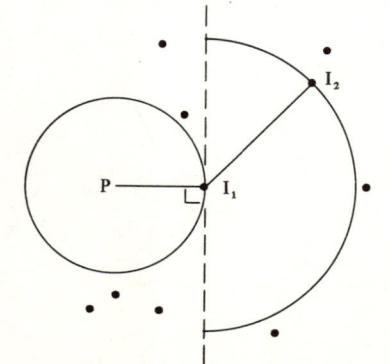

P arbitrary point
I_1 nearest individual to P
I_2 T-square nearest neighbour to I_1

Figure 4.5. T-square sampling.

individual nearest a randomly selected point' is biased in favour of more isolated individuals. To avoid this problem in field trials, we can use the T-square sampling procedure of Besag and Gleaves (1973); see figure 4.5. A point, P, is selected at random and the distance to the nearest individual, I_1, is measured, yielding the first PI distance. A second measurement is then made, to the nearest individual, I_2, to I_1, but only the half-plane formed by the line perpendicular to PI_1 is searched. It follows that the squared distance $(I_1I_2)^2$ has an exponential distribution with parameter 2ρ. In the analysis of mapped data this difficulty does not arise because all NN distances are measured.

By analogy with the discussion in section 4.2.1, the main departures from the Poisson process which we might expect are clustering and heterogeneity; some form of local inhibition is another common departure from the Poisson scheme.

Clustering The *Poisson cluster process*, also known as the *centre-satellite* process, was first developed by Neyman and Scott (1958) and by Warren (1962). It is made up of three components:
C1 Parents, or centre events, are generated by a Poisson process with intensity λ per unit area.
C2 At each centre we generate, independently, a random number of satellites with a common probability distribution.
C3 Each satellite is independently displaced from the centre according to a dispersal distribution.
Postulates C1 and C2 make up the assumptions for the generalised Poisson distributions of section 4.2.1. As an example of the centre-satellite assumptions, we may take the cluster size in C2 to follow a logarithmic series distribution, whereas the dispersal distribution, C3, is bivariate normal. This model is shown diagrammatically in figure 4.6.

The theoretical properties of this process are difficult to obtain, although Bartlett (1975, chapter 1), Warren (1971), Diggle (1975), and Warren and Batcheler (1979) have obtained results for special cases.

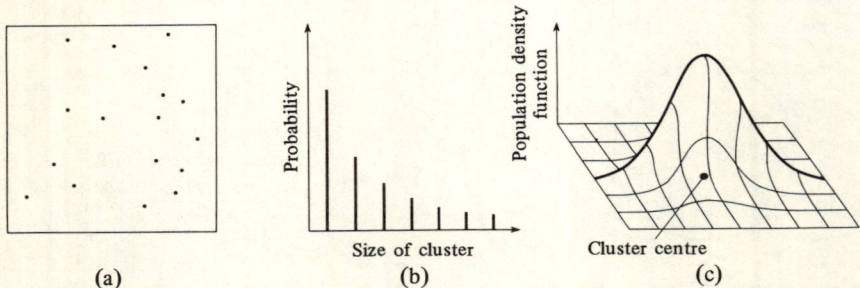

Figure 4.6. Components of the centre-satellite model. (a) Partial realisation of the Poisson process for individual cluster centres. (b) Logarithmic series distribution for the size distribution of each cluster. (c) Bivariate normal distribution for dispersal from cluster centre. (Based on Haggett et al, 1977, page 435.)

However, the process is very amenable to computer simulation since we may successively generate:
(a) the centres (uniformly distributed over the study area for the Poisson process);
(b) the number of satellites (log-series distribution or other appropriate assumption); and
(c) the satellite locations (bivariate normal, double exponential, or other dispersal distribution).

Heterogeneity The other kind of process outlined in section 4.2.1 was the heterogeneous, or apparent contagion, scheme. The full version of this rests upon two postulates:
H1 The presence or absence of individuals at any point in the plane, u say, is determined by the Poisson process with local intensity $\lambda(u)$.
H2 The intensity $\lambda(u)$ is itself a random process.
These processes are known as *doubly stochastic Poisson* processes; a full description is given by Matérn (1971). Although it is by no means obvious, it may be shown that each doubly stochastic scheme corresponds to a specific centre-satellite scheme and vice versa (the proof depends upon characteristic functionals and is beyond the scope of this book; see Bartlett, 1975, pages 6-8). A pictorial demonstration of the result is given by Diggle (1979a, section 2.3). As noted in section 4.2.2 we can only distinguish the two processes if we are prepared to make additional assumptions.

Inhibition The third main departure from the Poisson processes is the possibility that individuals are not allowed to be too close together. This minimum distance between individuals may reflect simple physical constraints (only one tree can grow at a given spot) or may depend on other more complex factors (such as the minimal distance between barnacles on a rock, strangers in a bar, people in bed).

Most existing models for inhibitory processes involve the planar Poisson scheme together with a rule for deleting individuals which are 'too close together'. For example, Matérn (1960) has proposed two such rules based upon a minimal separation distance (MSD):
1. Any individual which is less than the MSD from any other individual is deleted (whether or not the second individual has been deleted already).
2. Each individual is allocated a birth time, and an individual is retained if no other individual within the MSD of it has an earlier birth time. Again Matérn included individuals which may have been deleted earlier.

To these two schemes, Diggle et al (1976) added a third version which is similar to Matérn's second option, in that locations are generated sequentially, but their scheme allocates individuals only to those regions which are *not* within the MSD of any *existing* individual. A fourth option, outlined by Ripley (1977), is to simulate Poisson schemes and accept only those for which no two points are less than the MSD apart.

Some theoretical progress has been made on the two Matérn processes, although the more attractive version proposed by Diggle et al (1976) seems intractable. However, the sequential placement scheme is readily simulated.

These inhibitory processes are special cases of the Markov point processes introduced by Ripley and Kelly (1977). They also arise naturally in the study of competition between individuals; for discussion of this topic, the reader should consult Cormack (1979), especially section 7.

4.3.2 Tests using distances

Occasionally, spatial pattern may be evaluated in terms of a one-dimensional set of data such as a line transect. In a single dimension, result (4.1) is replaced by a Poisson distribution with mean λr, where r denotes the distance sampled. The gaps or *spacings* between successive observations then follow an exponential distribution, and various tests of randomness (or the Poisson process) have been devised; see Pyke (1965). The test we have singled out for further use was suggested by Durbin (1965) in the discussion on Pyke's paper.

First we scale the observations to lie on the interval [0,1] which is partitioned into n nonoverlapping segments, $g_1, ..., g_n$. These segments may be ranked to give the order statistics,

$$g_{(1)} \leq g_{(2)} \leq ... \leq g_{(n)} \, .$$

Durbin proposed the statistic

$$S = 2n - 2 \sum_{j=1}^{n} j g_{(j)} \, . \tag{4.11}$$

Under the null hypothesis of a Poisson process this has

$$E(S) = \tfrac{1}{2}(n-1) \, , \quad \text{and} \quad \text{var}(S) = \tfrac{1}{12}(n-1) \, .$$

Unless n is very small, the distribution of S may be approximated by the normal; large positive values of $[S - E(S)]$ correspond to regular spacing whereas large negative values imply clustering. This test is used to examine population levels in administrative units by Cliff, Haggett, Ord, Bassett and Davies (1975, chapter 3) and Cliff and Robson (1978).

For two-dimensional schemes, we return to the Poisson process described in section 4.3.1. Both for PI and for NN distances it may be shown (cf Pielou, 1977, pages 148–151) that

$$E(R_j) = \frac{j}{2^{2j}} \binom{2j}{j} (\rho \pi)^{-\tfrac{1}{2}} \, , \qquad E(R_j^2) = j\rho^{-1} \, ,$$

$$\text{var}(R_j) = j\rho^{-1} - [E(R_j)]^2 \, , \quad \text{and} \quad \text{var}(R_j^2) = j\rho^{-2} \, .$$

Thus, the PI and NN distances can be used to formulate tests of the Poisson process, and several such tests are presented in table 4.7. It should be noted that all these assume the availability of n independent (sets of) observations. Tests for mapped data will be discussed in section 4.4.

The original Clark–Evans test requires prior knowledge of the intensity λ and it is not particularly powerful. The Pielou–Mountford and Skellam statistics assume that an estimate of λ is available from a quadrat sample. A variant of the Clark–Evans method, particularly useful when directional differences are suspected, is the sectoral (or regional) method proposed by Dacey and Tung (1962). At each sampling point, k distinct sectors of angle $2\pi/k$ radians are formed, and the nearest individual in each sector is sought.

In general, it has been found that tests based upon the squared distances are more sensitive to departures from the Poisson process so all later tests use $u_i = r_i^2$. The Hopkins–Moore test is potentially powerful, but suffers from the defect that it is rarely possible to select *individuals* at random. The Holgate tests perform similarly but appear to be less powerful than the Besag–Gleaves alternatives which use T-square sampling; some power comparisons are given in the papers by Holgate and by Besag and Gleaves.

Table 4.7. Distance tests of randomness.

Statistic	Distribution under H_0	Reference
$2\sqrt{\lambda}\Sigma r_i/n$	approx. $N[1, (4-\pi)/(\pi n)]$	Clark and Evans (1954)
$\pi\lambda\Sigma u_i/n$	approx. $N[1, (\lambda A + n + 1)/(n\lambda A)]$	Pielou (1959) Mountford (1961)
$2\pi\hat{\lambda}\Sigma u_i$	approx. χ^2_{2n}	Skellam (1951)
$\Sigma u_i/\Sigma(u_i + u_i^*)$	beta (n,n) or approx. $N[\frac{1}{2}, 1/(8n+4)]$	Hopkins (1954) Moore (1954)
$\Sigma u_i/\Sigma u_{i2}$	beta (n,n) or approx. $N[\frac{1}{2}, 1/(8n+4)]$	Holgate (1965)
$\Sigma(u_i/u_{i2})/n$	approx. $N(\frac{1}{2}, 1/12n)$	Holgate (1965)
$\Sigma u_i/\Sigma(u_i + 0\cdot 5v_{iT})$	beta (n,n) or approx. $N[\frac{1}{2}, 1/(8n+4)]$	Besag and Gleaves (1973)
$\Sigma\{u_i/(u_i + 0\cdot 5v_{iT})\}/n$	approx. $N(\frac{1}{2}, 1/12n)$	Besag and Gleaves (1973)
$-\frac{1}{2}\Sigma \ln\{u_i/(u_i + 0\cdot 5v_{iT})\}$	χ^2_{2n}	Besag and Gleaves (1973)

Notation:
r_i is the distance from a random point to the nearest individual;
$u_i = r_i^2$;
u_{i2} is the squared distance from a random point to the second nearest individual;
u_i^* is the squared distance from a random individual to its nearest neighbour;
v_i is the squared distance between the nearest individual to a random point and its nearest neighbour;
v_{iT} as for v_i but with the nearest neighbour restricted to T-square sampling.

Source: Cormack (1979).

Recently, Cox and Lewis (1976) have extended the Besag–Gleaves tests by using a curved (circular segment) boundary rather than the half-plane to exclude the area already sampled. Their statistic seems to have similar power to those of Besag and Gleaves, but it is rather less practical for field use.

Diggle (1977b) suggests a two-stage procedure for use with T-square sampling. If the first of the Besag–Gleaves statistics in table 4.7 is used to test for randomness, and H_0 is not rejected, we may conclude that true contagion (or small clusters as described in section 4.2.2) is not present. However, the test lacks power against alternatives where the spatial pattern appears locally random but is heterogeneous at a larger scale (termed *random-heterogeneous* by Diggle).

Thus, when the Besag–Gleaves statistic does not produce a significant result, Diggle proposes the statistic

$$M^* = \frac{4n\{n\ln(\sum z_i/n) - \sum \ln z_i\}}{(13n+1)}, \qquad (4.12)$$

where

$$z_i = u_i + \tfrac{1}{2}v_{iT}, \qquad i = 1, ..., n,$$

in the notation of table 4.7. Under the null hypothesis of spatial homogeneity, the distribution of M^* is approximately chi-squared with $(n-1)$ degrees of freedom.

4.3.3 Reflexive pairs

An interesting question to ask of (first) nearest neighbours is whether or not two individuals I_1 and I_2 are reflexive; that is, is I_1 the first NN of I_2 and vice versa? For a planar Poisson process the probability of a pair being reflexive (cf Pielou, 1977, pages 158–159) is

$$\int_0^\infty 2\pi r\lambda \exp\left[-\lambda r^2\left(\frac{\sqrt{3}}{2}+\frac{\pi}{3}\right)\right]\exp[-\pi\lambda r^2]\,dr = \frac{6\pi}{3\sqrt{3}+8\pi} = 0\cdot 6215. \qquad (4.13)$$

The first exponential term in the integrand represents the probability that the shaded area in figure 4.7 is empty for each r and we then integrate over all r.

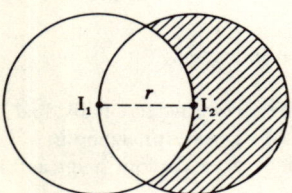

Figure 4.7. Required empty area (shaded) if I_1 is to be nearest neighbour of I_2, given that I_2 is nearest to I_1.

In his study of reflexive pair relationships, Dacey (1969b) extended results (4.13) to reflexive tth order nearest neighbours in k-dimensional space. In conjunction with the earlier analysis by Clark (1956), Dacey suggests that a deficiency of reflexive pairs implies clustering, although this interpretation has been a matter of debate; see Pielou (1977, page 159) and Haggett et al (1977, pages 442-446) for further discussion.

4.3.4 Spacing of towns along the Mississippi

To show how some of the methods discussed in this section may be applied, we shall now consider a set of data first analysed by Dacey (1960), using the reflexive pairs procedure, and which is reproduced in figure 4.8; these data refer to the spacing of major towns along the central Mississippi valley. The numbers of tth order reflexive pairs ($t = 1, 2, 3$) are presented in table 4.8, the maximum possible being seventeen. Since the counts increase in multiples of two, the counts for $t = 2$, and $t = 3$, are as near as they could be to the expected value, and the count for $t = 1$ is only one pair off [note that the proportion for $t = 1$ differs from that given in equation (4.13), since the current analysis is based on one dimension, not two]. This evidence would seem to suggest little departure from the Poisson scheme, although the deficiency for $t = 1$ may be slightly suggestive of grouping.

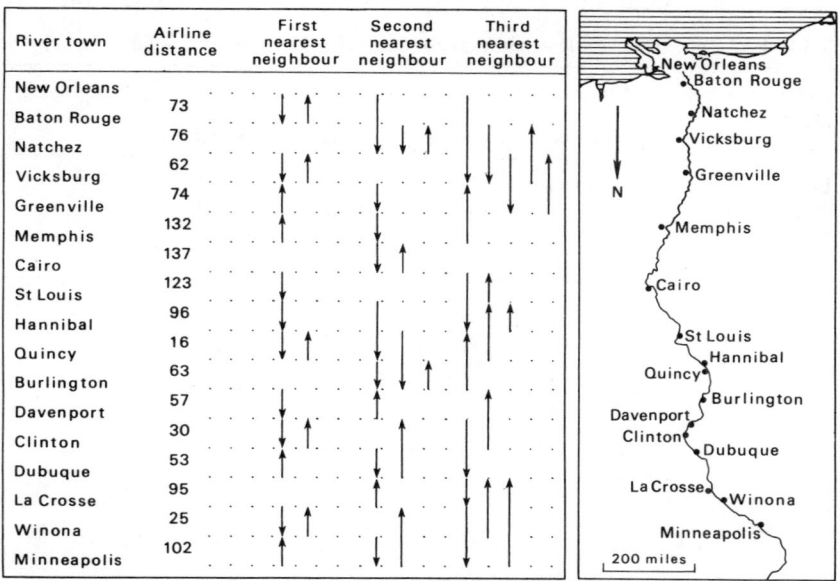

Figure 4.8. Spacing of river towns in excess of 20000 population in the central Mississippi valley. Arrows point to the order neighbours of the various towns. Source: Dacey (1960, page 60).

To reinforce the analysis, we next evaluated S, given by equation (4.11), with the two end towns fixed so that the number of spacings n is equal to 16. This yields $S = 10 \cdot 76$, against $E(S) = 7 \cdot 5$, and $\text{var}(S) = 1 \cdot 25$; the standard-deviate value is $+2 \cdot 92$, which is highly significant and strongly suggestive of an element of regular spacing.

In the univariate case, we may evaluate the expected values of the order statistics, $g_{(j)}$; see Cliff, Haggett, Ord, Basset, and Davies (1975, pages 32–36) for details. The expected proportion attributable to the jth region (1 = smallest, n = largest) is

$$E(g_{(j)}) = \left(\frac{1}{n} - \Delta\right) \sum_{i=1}^{j} (n+1-i)^{-1} + \Delta ,$$

where Δ is the minimum threshold spacing.

In figure 4.9, we have plotted the observed values of $g_{(j)}$ against their expected values for the Mississippi data, taking $\Delta = 12 \cdot 0$ which is the value of the minimum variance unbiased estimator,

$$\tilde{\Delta} = \frac{n^2 g_{(1)} - 1}{n(n-1)} .$$

The combination of points above the 45° line for low j values, and well below the line for the top three values of j, is symptomatic of regular spacing; the occurrence of very high spacings is very much less than we

Table 4.8. Reflexive nearest neighbours for towns in the Mississippi Valley (Dacey, 1960).

Order of nearest neighbour	Number		Proportion	
	observed	expected	observed	expected
First	10	11·3	0·588	0·667
Second	6	6·3	0·353	0·370
Third	4	4·6	0·235	0·272

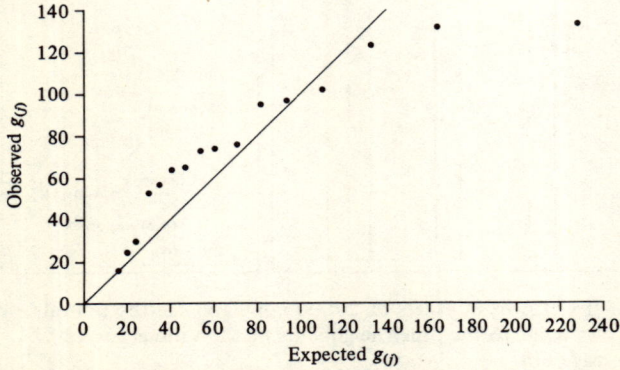

Figure 4.9. Observed and expected order statistics for the Mississippi valley towns data.

would expect, even allowing for a minimum threshold. The discrepancies are even more marked if the plot is drawn for $\Delta = 0$ (pure Poisson process).

Finally, we evaluated the spatial autocorrelation coefficient, I [equation (1.15)], with weights w_{ij} applied as $w_{ij} = 1$ when intervals i and j are adjacent (have a common town as an end point), and $w_{ij} = 0$ otherwise. The resulting statistic has the value $I = 0 \cdot 225$, with a standard deviate of $+1 \cdot 94$, which is significant at the 5% level in a one-tailed test of positive autocorrelation. That is, there is some evidence to indicate that successive spacings are similar. However, it is apparent from the data that there are only three towns in the 488 miles between Greenville and Hannibal. Referring back to the discussion of section 4.2.3, this implies an element of heterogeneity.

The final picture which emerges is one of predominantly regular spacing overlaid with heterogeneity as the valley winds its way through relatively densely and sparsely populated states. The heterogeneity induces an element of clustering which counteracts the overall regular pattern when we come to look at reflexive pairs. Nevertheless, the reflexive pairs analysis is useful in hinting at the possibility of clustering when other analyses strongly suggest regular spacing. Thus reflexive pairs analysis will rarely offer a complete picture on its own, but may lead to further insight about the nature of the process.

4.3.5 Estimation of intensity

If we sample at random, then we may wish to estimate the overall intensity of the process. For example, there is a considerable literature in forestry devoted to the estimation of the number of trees and the amount of timber in an area of forest (cf Pielou, 1977, pages 150–152; Warren and Batcheler, 1979).

Suppose we take n sample quadrats of size a yielding the observations $x_1, ..., x_n$. If the intensity (for a unit area) is λ and the observations have mean $\bar{x} \; (= \Sigma x_i/n)$, the estimator $\hat{\lambda}$,

$$\hat{\lambda} = \frac{\bar{x}}{na},$$

is always unbiased and, for the Poisson process with intensity λ, has variance

$$\text{var}(\hat{\lambda}) = \frac{\lambda}{na}. \tag{4.14}$$

If the total region studied has area A, the unbiased estimator for total numbers, $\hat{\lambda}_A$, is given by

$$\hat{\lambda}_A = A\hat{\lambda}, \text{ with Poisson process variance,}$$

$$\text{var}(\hat{\lambda}_A) = \frac{A^2 \lambda}{na}.$$

In applications, it is important to plan the sampling so that edge effects are avoided. In addition, if different parts of the study area are thought to have very different intensities, stratified random sampling should be used. When the process is markedly clustered, *the variance of the estimator may be much greater than that given in equation (4.14)*.

Quadrat sampling may be time-consuming. Thus the investigator may prefer to use distance-based methods, both for convenience and to gain some information about the spatial pattern. When the process is Poissonian, the jth order neighbour has a density function given by equation (4.10), which yields the maximum likelihood estimator for ρ, based upon n observations $V_{1j}, ..., V_{nj}$, as

$$\hat{\rho} = \frac{j}{\bar{V}_j}, \qquad (4.15)$$

where $\bar{V}_j$ is the variate corresponding to the mean of the n observations on the jth nearest-neighbour (NN) squared distances. Further, if n independent sets of observations are available on the first, second, ..., jth neighbours, the maximum likelihood estimator for ρ is still given by equation (4.15); see Thompson (1955). Unfortunately such estimators are *not* robust; that is, they become badly biased when the underlying process is not Poissonian (Persson, 1971). Indeed, when it is realised that many such statistics are listed in table 4.7 as *tests of randomness*, it is hardly surprising that they are not robust as estimators of intensity.

Typically, the PI and NN distances yield estimators which are biased in opposite directions. If U and V denote first PI and NN squared distances respectively, and samples of size n are available for each, Diggle (1975; 1977a) shows that the expression,

$$\rho^* = [\bar{U}\bar{V}]^{-\frac{1}{2}}, \qquad (4.16)$$

provides an estimator which is fairly robust against a variety of departures from the Poisson. As noted in section 4.3.1, it is not always possible to sample NN distances and so some other scheme such as T-square sampling must be used; the estimator expression then becomes

$$\rho^* = [\bar{U}\bar{V}/2]^{-\frac{1}{2}},$$

to allow for the half-plane search on NN distances. Other modifications to allow for clusters have been introduced by Batcheler in a series of papers; for details and a general review see Warren and Batcheler (1979).

The difficulty with distance-based estimators is that the distribution of distances is determined by the spatial process. One way of circumventing this difficulty is to use Dirichlet cells (known also as Thiessen or Voronoi polygons—see Matérn, 1979; Haggett et al, 1977, pages 436-439; Getis and Boots, 1978). The Dirichlet cell for an individual may be defined as the set of points nearer to that individual than to any other individual in the study region.

It is apparent that the Dirichlet cells are nonoverlapping and that they completely cover the study area. Thus, if a location is selected at random, it must fall in exactly one Dirichlet cell. Suppose we take a random sample of n observations and let δ_i denote the number of selected locations which fall into the ith Dirichlet cell, which has area A_i. Ord (1977) shows that the estimator $\hat{\lambda}$, given by

$$\hat{\lambda} = \frac{1}{n}\sum_i \frac{\delta_i}{A_i} \, , \tag{4.17}$$

is unbiased, where the summation is taken over all i for which $\delta_i \geqslant 1$. For forest sampling, the Dirichlet cells may be replaced by a combination of exclusion-angle and tree-size measurements. Further theoretical and empirical results are given in Ord (1977).

4.4 Analysis of mapped data

When the study area is completely mapped, overall intensity estimation is not of interest. Instead, we use such data to examine the spatial pattern. We shall consider *first-order* (or mean) methods for the study of trends, and *second-order* (or covariance) methods for the study of interactions, as well as distance-based statistics.

4.4.1 First-order (or moment) methods for intensity surfaces

We assume that the intensity at a location u is $\lambda(u)$ and that it is 'slowly varying' so that trends exist in the data. When the study area is partitioned into nonoverlapping areas, and the random variate X_i for area i denotes an overall count or measurement for the area, some form of trend-surface analysis may be employed (cf section 8.8). An appropriate generalisation for heterogeneous Poisson counts is described by Kooijman (1977; 1979); we shall next summarise Kooijman's approach when the coordinates, rather than quadrat counts, are given.

Suppose that the location of each individual in the study area is known, with coordinates $u_1, ..., u_n$, say, for the n individuals on the map. At each location u_i, defined by (u_{1i}, u_{2i}), we assume that the Poisson process operates with intensity $\lambda(u_i)$, which is a function of the coordinates and of one or more unknown parameters. If the map describes all points in some set M, or $u \in M$, the density function for the intensity at u, conditioned on the mapped area, M, is

$$\lambda(u) \left[\iint_{u \in M} \lambda(u) \, du_1 \, du_2 \right]^{-1} , \tag{4.18}$$

and the log-likelihood, given observations at $u_1, ..., u_n$, is

$$\sum_{i=1}^{n} \ln\lambda(u_i) - n \ln\left[\iint_{u \in M} \lambda(u_i) \, du_1 \, du_2 \right] .$$

For example, when the map is rectangular, it may be reduced to the unit square by a suitable transformation of the distances. By taking

$$\lambda(u) = c_0(1 + c_1 u_1 + c_2 u_2), \qquad (4.19)$$

Kooijman finds that the maximum likelihood estimators for c_1 and c_2 are the solutions of the equations

$$\sum_{i=1}^{n} u_{ji}(1 + \hat{c}_1 u_{1i} + \hat{c}_2 u_{2i})^{-1} = n(2 + \hat{c}_1 + \hat{c}_2), \qquad j = 1, 2. \qquad (4.20)$$

Since c_0 drops out of expression (4.18), only the two parameters, c_1 and c_2, are relevant. Assumption (4.19) is one of the few cases where the likelihood equations are tractable, but it suffers from the defect that $\lambda(u)$ could go negative. An alternative form is

$$\lambda(u) = c_0 \exp(c_1 u_1 + c_2 u_2),$$

which stays nonnegative and yields maximum likelihood estimators for c_1 and c_2 as

$$\hat{c}_j = \ln\left(\frac{\bar{u}_j}{1 + \bar{u}_j}\right), \qquad j = 1, 2,$$

where

$$n\bar{u}_j = \sum_{i=1}^{n} u_{ji}.$$

However, the extension to higher order polynomials requires numerical integration.

4.4.2 Distance-based methods

As in section 4.3, we could use NN and PI distances to examine the spatial pattern of points. However, if we examine all n of the NN distances on a map, it is quite clear that these will not form n independent observations; they will be *interdependent* in a complex way. Hence, the proposal by Clark and Evans (1954) that their test (see table 4.7) could be used to search for randomness is not strictly valid, since the distribution of the statistic under the Poisson null hypothesis differs from that for n independent observations; the same comment applies to all other statistics based on the full n observations. Despite this theoretical objection, Diggle (1975) has found that the null distribution of the Clark-Evans statistic for mapped data is close to that for independent observations, at least in the tail areas. The test could therefore be used, even though results on the power of the test are not encouraging.

In order to make better use of the data, we look at empirical distributions for the PI and NN distances. These empirical forms provide unbiased estimators of the distribution function at any point, despite the pattern of dependencies. The NN empirical distribution is formed by looking at the distances to the first nearest neighbour for all individuals, reflexive pairs duly being recorded twice. We refer to this empirical distribution function as $\hat{F}_{NN}(r)$. For the PI distances, we lay down a fine, regular, $k \times k$ grid over the map, containing a total of m points, where $m = k^2$. If the map is not regular, we ignore points falling outside the grid. Then, from each of the m points, we measure the distance to the nearest individual; that is, the first PI distance. This gives the empirical distribution function, $\hat{F}_{PI}(r)$. As m increases, so does the amount of information, although very little will be gained once m reaches a reasonable size. Diggle (1979a, section 4.3) suggests that k of order $n^{\frac{1}{2}}$ seems appropriate. As noted earlier, a PI distance greater than r implies that there is a disc of radius r, centred on the selected point, which does not contain any individuals. In the terminology of Ripley (1977), the disc is said to form a *test-set*, and $\hat{F}_{PI}(r)$ approximates the proportion of points on the map that are more than r from the nearest individual.

If we specify a pattern-producing process, then we may generate simulated realisations of the process and use these to perform Monte-Carlo tests of fit, as in section 3.2. To show how the approach works, let us consider an example from Diggle (1979a). The data were first given in Strauss (1975), and figure 4.10 shows the locations of sixty-two redwood seedlings in a square plot. This square is a subset of the whole data set, selected by Ripley (1977). As might be expected, the seedlings show a marked tendency to cluster. Apparently the locations of the original redwoods which had been cleared from the plot were not available (but can be inferred!). The empirical distributions, $\hat{F}_{PI}(r)$ and $\hat{F}_{NN}(r)$, are plotted in figure 4.11 and show the clear departures from randomness which one would expect for such clustered data.

An alternative method of analysis for mapped data is to consider all $\frac{1}{2}n(n-1)$ distances between pairs of individuals. However, the larger distances do not appear very informative, as is shown in figure 4.12. The *envelope* plots in this diagram represent the smallest and largest values of $\hat{F}(r)$, for each $F_0(r)$, obtained from nineteen simulations of sixty-two

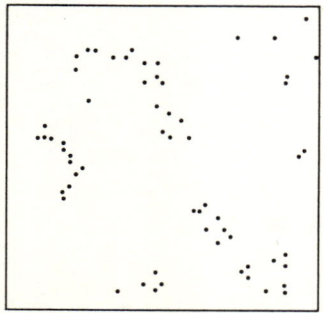

Figure 4.10. Location of sixty-two redwood seedlings in approximately 75 foot square plot [extracted from Strauss (1975) by Ripley (1977)].

individuals located independently on the square according to the uniform distribution; that is, the Poisson process conditioned onto the mapped area. For any one value of $F_0(r)$, we could reject the Poisson scheme at the 5% level in favour of clustering if $\hat{F}(r)$ lay above the upper envelope; clearly, tests for different values of $F_0(r)$ will be dependent. Nevertheless, the main impetus for the rejection of the Poisson assumption comes from the lower 10% of distances which will contain many of the NN distances.

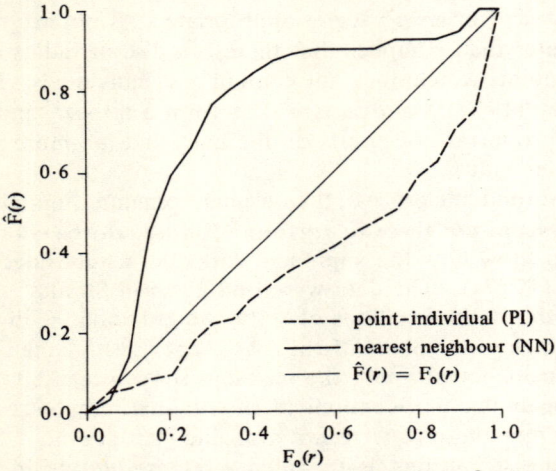

Figure 4.11. Empirical distribution functions of point-individual and nearest-neighbour distances for redwood seedlings data. $F_0(r)$ denotes common marginal distribution function for the Poisson null hypothesis. Source: Diggle (1979a).

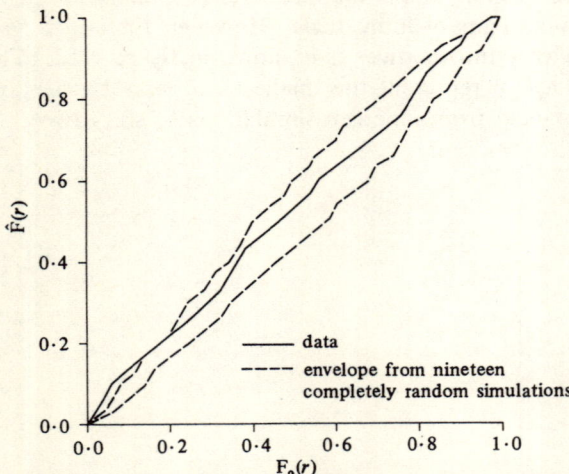

Figure 4.12. Analysis of redwood seedlings data using distances between all pairs of individuals. $F_0(r)$ denotes marginal distribution function under complete spatial randomness. Source: Diggle (1979a).

4.4.3 Second-order analysis

For a process with overall intensity λ, Ripley (1977) introduced the function

$$\lambda K(r) = \left\{ \begin{array}{l} \text{expected number of other individuals} \\ \text{within a distance } r \text{ of any given individual} \end{array} \right\}.$$

Independently, Kooijman (1977; 1979) had developed essentially the same form of analysis. It can be shown that $K(r)$ may be estimated from the data map by locating discs of radius r about each of the n individuals and counting the number of discs containing at least one other individual. Due allowance needs to be made for edge effects so that, for a region of area A with n individuals, we use the estimator,

$$\hat{K}(r) = \frac{A(n-1)}{n} \sum_{\substack{i=1 \\ i \neq j}}^{n} \sum_{j=1}^{n} k(i,j) \,,$$

where $k(i, j)$ is defined as follows:

For two points i, j located at u_i and u_j and distance d apart, consider a circle of radius d centred at u_i. Then $[k(i,j)]^{-1}$ is the proportion of the circumference of that circle which lies within the study region.

For the Poisson process,

$$K(r) = \pi r^2 \,.$$

A plot of $K(r) = \pi r^2$ is shown for the redwood data in figure 4.13. Again, the Poisson process is clearly rejected.

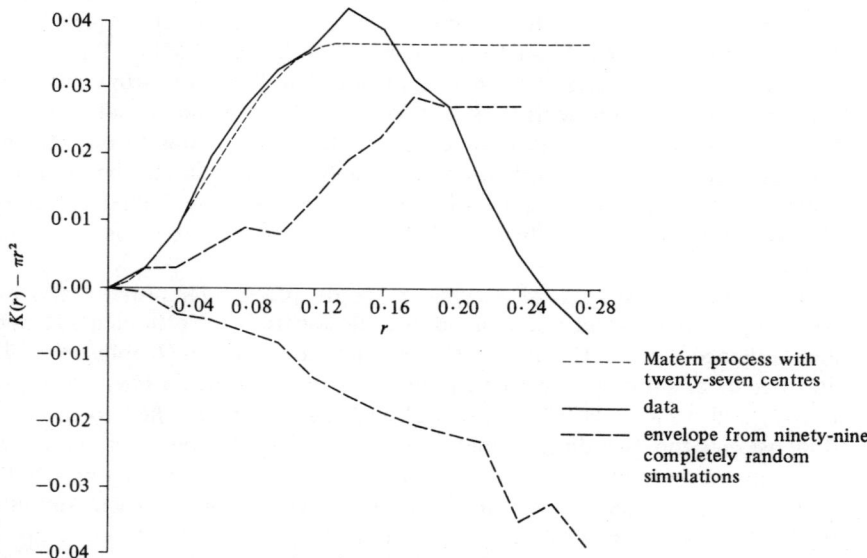

Figure 4.13. Second-order analysis of redwood-seedlings data: the function $K(r) - \pi r^2$. Source: Diggle (1979a).

As a constructive alternative to the Poisson scheme, Ripley (1977) suggested a Matérn process (Matérn, 1960), which is a centre-satellite process with offspring distributed around the parent on a circle of fixed circumference. It can be seen from figure 4.13 that the $K(r)$ function for a Matérn process with twenty-seven centres is very close to that of the data for smaller values of r. Diggle (1978; 1979a; 1979b) shows that the discrepancies for larger r are within the bounds of sampling variation because the simulation envelopes widen drastically for larger r values. To overcome this problem, Besag (1977a), in the discussion on Ripley's paper, suggests plotting $\sqrt{K(r)}$, which has a variance function that is nearly constant for increasing r. Thus the Matérn process seems a reasonable model, and we can only regret the nonavailability of the locations of the original trees as this would have enabled a more exact evaluation to be made. As before, we note that this analysis alone cannot argue for clustering against heterogeneity, although the prior evidence for clustering is very strong.

4.4.4 Spectral analysis

The literature on this topic is vast, and a detailed treatment is outside the scope of this book. Accordingly, we shall do no more here than indicate some of the major lines of inquiry. A review of the main approaches is contained in Haggett et al (1977, pages 390-413).

In a single dimension, the spectral analysis of spatial data follows the path well charted by time-series analysis (cf Jenkins and Watts, 1968). Tobler (1969b) used spectral analysis to study the spacing of settlements along US Route 40 from Baltimore to San Francisco; further work on this data set was carried out by Rayner and Golledge (1973). Tobler argued that the power spectrum was an outcome from a positive autoregressive process (see section 6.2.2) for town sizes, which he attributed to the spatial organisation produced by central place competition. Rayner and Golledge's work implied, however, that this result should be credited mainly to the population smoothing model employed by Tobler to prepare the Route 40 data for analysis, and that settlement spacing was essentially random.

Two-dimensional spectral analysis proceeds in a similar fashion, and has been applied to several topics including the location of settlements (Rayner and Golledge, 1972); the exploration of mineral resources (Robinson, 1975); and the description of the tree canopy in a forest (Ford, 1976). Bartlett (1963; 1964) has shown how spectral analysis may be applied to one-dimensional and two-dimensional point patterns by laying a fine grid over the mapped data and coding each cell as one or zero depending on whether or not it contains an individual. The spectral estimators are then computed in the usual way. Rayner and Golledge have applied this technique in their study of settlements in North Dakota. The separate settlements constituted the individuals in the analysis. The resulting power spectrum revealed a distinct 'banding' in the location of towns. In North Dakota, settlements are aligned mainly east to west across the state with roughly a 6-8 mile gap between the settlement bands.

4.4.5 Analysis in space and time

If we view time as an additional dimension, then the methods we have discussed could be extended to cover spatiotemporal processes. Indeed, repeated computation of the I_{s-t} index of section 1.6 for different time intervals and distances provides a spatiotemporal version of Ripley's K function (section 4.4.3); as such, it is a direct extension of the ideas of Mantel (1967). Such a tool would be useful[5] in the study of the spread of an epidemic, where high K values for small distances and times would indicate strong local contagion and a virulent disease, whereas lower K values for small distances would be suggestive of a disease which did not spread evenly along an advancing front, but jumped over intervening regions containing susceptibles. Such differences in behaviour depend upon the velocity with which the infection spreads; this is discussed in Mollison (1978), and we have reviewed his work in detail in section 3.3.3.

A somewhat different extension of the I_{s-t} index is that of Green (1979) who measured the numbers of pairs of nesting sites of Blue Tits and Great Tits within a given distance. From this study, Green concluded that the Tits appear to exhibit intraspecific, but not interspecific, interference in the choice of nesting sites.

4.5 Conclusions

In this chapter, we have reviewed the different approaches to the analysis of point pattern data. Although the twin mechanisms of clustering and heterogeneity may be indistinguishable given a single realisation of the data, ancillary information may be available concerning the size of possible clusters or the nature of the variations in the mean level of the process. In such cases, contiguous quadrat data may enable us to separate the hypotheses, as indicated in section 4.2.3.

The theory and application of distance sampling is presented in section 4.3 where a variety of distance-based tests are presented. Some of these methods are used to analyse the spacings of towns along the Mississippi river valley.

When mapped data are available, different distance methods become appropriate, and these are summarised in section 4.4, which includes an analysis of a pattern of redwood seedlings. The chapter concludes with a brief review of other methods and possible extensions.

---

[5] The authors are indebted to Dr D Mollison for this suggestion.

5

Spatial correlograms and related methods

5.1 Introduction
In this chapter, various data analytic methods will be introduced which are of value in assessing the spatial scale of a process. In place of the single set of weights used to evaluate the I statistic in earlier chapters (section 1.4), we shall now consider several sets which refer to sites one, two, ... steps apart. The term 'step apart' is capable of several interpretations, but the essential feature is that relationships between sites are considered at several different distances; see section 5.2.1 for further details. In section 5.2, the spatial correlogram will be introduced and we shall use it to analyse data on the spread of measles in Cornwall.

A somewhat different approach is the analysis of contiguous quadrats by a hierarchical analysis of variance procedure. This scheme is outlined in section 5.3 and is applied to Matui's data already described in section 4.2.3. In section 5.4, we extend this analysis to the bivariate case to assess variations in correlation structure with distance. Land-use data from the *Atlas of London* (Jones and Sinclair, 1968) are analysed as an example.

Partial correlograms are developed in section 5.5; reexamination of the land-use data with correlogram methods will enable us to compare the information conveyed with that from the analysis of variance.

Finally, in section 5.6, we shall briefly examine different approaches to the estimation of the weighting matrix.

5.2 Spatial correlograms
5.2.1 Theory
Although the interaction between sites may be strongest between immediate neighbours, often the strength of interaction will vary in a complex way with distance. Thus outbreaks of measles usually spread to contiguous areas, but they may jump from one major urban area to another, initially leaving the intervening rural areas largely uninfected. To detect such variations in the spatial pattern, we define a *spatial correlogram* by analogy with the correlogram used in time-series analysis (Kendall, 1976, page 70).

Consider a system of n sites with random variables $X_1, ..., X_n$ and let the sites i and j be gth-order *neighbours* (or g spatial steps apart). Various definitions of neighbourliness are possible. Thus, two sites i and j may be g steps apart in either of the following cases.
(a) If the shortest path from i to j on the graph connecting adjacent sites has g edges; that is, the path passes through $(g-1)$ intervening sites ($g \leq D$, where D is known as the *diameter* of the graph). Then the set $C(g)$ contains all pairs of subscripts whose sites are separated by $(g-1)$ intervening sites, or

$C(g) = \{(i, j)$ such that i and j are separated by $(g-1)$ intervening sites$\}$.

(b) If the distance, d_{ij}, between sites i and j falls in the gth distance class $a_{g-1} < d_{ij} \leq a_g$, so that the set $C(g)$ contains all pairs of subscripts for which d_{ij} is within these limits, or

$$C(g) = \{(i,j) \text{ such that } a_{g-1} < d_{ij} \leq a_g\}, \quad g = 1, ..., D,$$

where $a_0 = 0$, and $a_D \geq \max_{(i,j)} d_{ij}$.

Clearly the method of graph construction and the choice of distance function depend upon the investigation, so that the definition is very broad. The shortest paths for each pair of sites, as described in (a), may be evaluated using the matrix powering algorithm described in Haggett et al (1977, pages 319-320). If the variates refer to areas rather than to point locations, we may still construct graphs based upon common edges (and, possibly, vertices), or measure distances from convenient reference points such as the area centroids.

Given the classification of each pair of sites into one of the sets $C(g)$, we may define the weighting matrices $W(g)$ with elements

$$w_{ij}(g) = \begin{cases} 1, & \text{if } (i,j) \in C(g), \\ 0, & \text{otherwise.} \end{cases}$$

Then the gth-order sample spatial autocorrelation is given by

$$I(g) = \frac{n}{S_0(g)} \frac{z^T W(g) z}{z^T z}, \qquad (5.1)$$

where

$$z^T = (z_1, ..., z_n), \quad z_i = x_i - \bar{x}, \quad i = 1, ..., n,$$

and

$$S_0(g) = \sum_{(2)} w_{ij}(g).$$

Alternatively, this may be written as

$$I(g) = \frac{n}{S_0(g)} \sum_{(i,j) \in C(g)} w_{ij}(g) z_i z_j \bigg/ \sum_{i=1}^{n} z_i^2.$$

We observe that the symmetric form of W in equation (5.1) means that each term appears twice in the summation. The means and variances for these measures follow directly from equations (1.37)-(1.39). The plot of $I(g)$ against g yields the *spatial correlogram.*

Kooijman (1976) used the correlogram to formulate a 'maximised' I statistic as follows. Consider the weighting matrix W with elements w_{ij} where

$$w_{ij} = b_g \quad \text{when } (i,j) \in C(g),$$

so that the overall I statistic may be written as

$$I = \sum_{g=1}^{s} b_g I(g) ,$$

where all the other constants have been absorbed into the $\{b_g\}$. When there are m_g pairs in $C(g)$, appropriate constraints on the $\{b_g\}$ are

$$\sum_{i=1}^{s} m_g b_g = 1 ,$$

and

$$\sum_{g=1}^{s} m_g b_g^2 = 1 \quad \text{(or } \alpha \text{ in general)} . \tag{5.2}$$

As before, the symmetric formulation implies that each pair is counted twice, so that $\sum_g m_g = n(n-1)$.

It follows (Kooijman, 1976, page 117) that the maximum value of I subject to the constraints is, to a very close approximation,

$$I_{\max} = -\frac{1}{n-1} + \left[\sum_g m_g [I(g)]^2 - \left(\frac{n}{n-1}\right) \right]^{\frac{1}{2}} ,$$

where

$$b_g = \frac{n + I_{\max} + (n^2 - n - 1)I(g)}{n + n(n-1)I_{\max}} .$$

The effect of using α rather than 1 in constraints (5.2) is to multiply the value of $I_{\max}$ by a term in α; subsequent tests based on $I_{\max}$ are not affected. The appendix to Kooijman's paper gives the first four moments of $I_{\max}$ under randomisation, allowing tests to be performed in the same way as in chapter 2.

For the most part, we shall use the correlogram as a data analytic tool, although the Kooijman procedure using the correlogram gives a valuable overall test for spatial dependence.

5.2.2 A regional application: measles in Cornwall

We can illustrate the spatial correlogram approach by reexamining the Cornish measles data considered previously in section 1.7.2. The number of measles cases reported in each week, 1966 (week 40) to 1970 (week 52), was recorded for each of the twenty-seven local authority areas comprising the county of Cornwall (see figure 1.7). The local authority areas are known as General Register Office (GRO) districts for the purposes of disease reporting. The GRO map shown in figure 1.7 may be converted into the weighting matrix, $\mathbf{W}(1)$ by putting the typical element of $\mathbf{W}(1)$, $w_{ij}(1)$, equal to one if the ith and jth GROs are contiguous, and $w_{ij}(1)$ equal to zero otherwise. The map was thus reduced to a binary planar graph, and the nonzero elements of $\mathbf{W}(1)$ represent those GROs which are first nearest neighbours (one spatial lag apart). Second, third, ... up to eighth (spatially lagged) nearest neighbours were then determined by using the matrix powering algorithm mentioned above, and the weighting matrices $W(2)-W(8)$ so defined. The diameter of the graph was eight lags. Having specified the $\{W(g)\}$, we tested for spatial autocorrelation between the GROs, for $g = 1, 2, ..., 8$ spatial lags, using $I(g)$ given in equation (5.1). The data series was 222 weeks long, and the analysis was carried out for each of the 222 weeks. We conventionally called 1966 (week 40), week 1 and 1970 (week 52), week 222. Binary only weights were used, in accordance with the discussion in section 5.2.1. During the time period studied, two major epidemic waves affected the area, in weeks 1-50 (wave 1) and in weeks 186-204 (wave 2). To summarise the results, the average value of $I(g)$ in standardised score form at each spatial lag for waves 1 and 2 is reproduced in figure 5.1. No clear pattern of spatial autocorrelation was found during the period between the waves.

From figure 5.1(a), it is evident that positive autocorrelation predominates at spatial lags 1, 5, 6, and 8, and negative autocorrelation at lags 2-4. The positive spatial autocorrelation at lag 1 and negative autocorrelation at lags 2-4 suggests that measles outbreaks are clustered spatially (cf Haggett, 1972, and sections 1.1 and 1.7.2). The interesting feature of the average correlogram for the first epidemic wave is the marked positive autocorrelation at lags 6 and 8. To help interpret this we determined the numbers of urban-urban, rural-rural, and urban-rural links at each spatial lag 1-8. These counts, with the expected numbers (in brackets) under

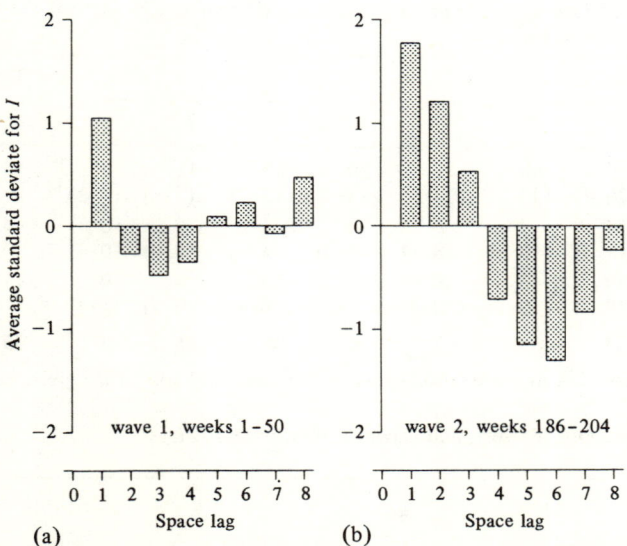

Figure 5.1. Average correlograms for I: (a) weeks 1-50, and (b) weeks 186-204. Source: Cliff, Haggett, Ord, Bassett, and Davies (1975, page 171).

the assumption of independence between link type and spatial lag, are reproduced in table 5.1. This table shows that spatial lags 1 and 2 are mainly urban–rural and rural–rural links, whereas lags 4–8 comprise mainly urban–urban links. In addition, lags 6–8 include predominantly those GROs on the main transport routes. A picture therefore emerges of similar levels of measles cases in (1) noncontiguous urban areas and (2) contiguous rural–urban and rural–rural districts. Recall from section 1.7.2 the spread model for measles epidemics, in which we postulated initial outbreaks of measles in urban areas (a central place effect), followed by movement of the disease from the towns into surrounding rural areas by a spatial diffusion process. The evidence of the analysis here would seem to support the proposed model.

If we now turn to the second epidemic wave, the spatial clustering of measles outbreaks was again confirmed by the positive autocorrelation at lags 1–3, but there was negative autocorrelation at lags 4–8 [figure 5.1(b)]. This suggests that the central place effect was not identifiably present in the second epidemic as it was in the first. As a possible explanation of this behaviour, the first epidemic peaked in the winter of 1966, whereas the second reached its maximum in the summer of 1970. The massive tourist traffic in the Southwest during the summer months may have produced a much more rapid wave-like dissemination of the disease down the Cornish peninsula than occurred in the first epidemic, thus swamping any central place effect. A further contributory factor may have been school holidays, with the consequent lack of congregation of children in town schools.

Table 5.1. Observed and expected numbers of links at different spatial lags for the GRO map of Cornwall, England.

Link type	Spatial lag								Totals
	1	2	3	4	5	6	7	8	
Urban–urban	1 (13·6)	20 (28·4)	33 (32·3)	31 (24·5)	20 (17·9)	18 (13·2)	11 (5·8)	3 (1·2)	137
Rural–rural	13 (4·5)	14 (9·3)	7 (10·6)	6 (8·1)	4 (5·9)	1 (4·3)	0 (1·9)	0 (0·4)	45
Urban–rural	21 (16·9)	39 (35·3)	43 (40·1)	26 (30·4)	22 (22·2)	15 (16·4)	4 (7·2)	0 (1·4)	170
Totals	35	73	83	63	46	34	15	3	352

Expected numbers assuming independence between link type and spatial lag given in brackets.
Source: Cliff, Haggett, Ord, Bassett, and Davies (1975, page 172).

5.3 The spatial scale of a process

In the previous section, we saw the value of the correlogram in evaluating the different spatial scales at which a process may operate[6]. We now examine the scale of a process in more detail, using data available on a regular grid of contiguous quadrats; the methods are extendable to cover irregular patterns of cells.

5.3.1 Greig-Smith's method

The use of contiguous quadrats to search for spatial pattern at different scales was first developed in the ecological literature by Greig-Smith (1964). Although we shall be concerned with two-way arrays, the method is often used for line-transect data (Kershaw, 1964). Suppose that the grid has 2^{2m} cells and is square ($2^m \times 2^m$). We can define ($m+1$) levels by taking cells one, four, sixteen, ..., at a time as in figure 5.2.

If X_i denotes the random variable corresponding to the ith cell, we may postulate a hierarchical model for the m levels as

$$X_i = \mu + \epsilon_1(i) + ... + \epsilon_m(i) , \qquad (5.3)$$

where

(a) the means of the $\{X_i\}$ are taken to be equal to μ;
(b) the random variables $\epsilon_j(i)$ have zero means and variances $\sigma_\epsilon^2(j)$, denoting the variance attributable to the jth level of the hierarchy.

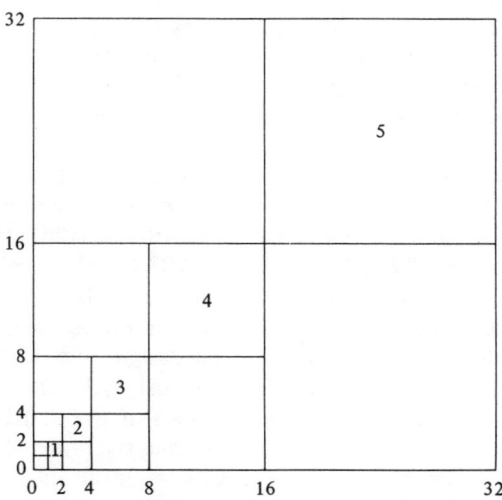

Figure 5.2. Aggregation scheme for cells to define higher order levels as a function of lower order levels.

---

[6] That is, at a local spatial scale the contagious spread of measles among GROs in Cornwall, and at a higher spatial scale the spread between urban areas.

The magnitudes of the variances $\sigma_\epsilon^2(j)$ indicate the importance to be attached to that level of the hierarchy. These variances may be estimated as follows.

If s cells are combined at each stage of the hierarchy, write

$$X_i(j) = \sum_{r \in A_{ij}} X_r(j-1) \, , \tag{5.4}$$

where A_{ij} denotes the set of s cells at level $(j-1)$ which are combined into one cell (the ith) at level j. The jth sum of squares, SS_j, corresponding to the jth row of the analysis of variance table may be written as

$$SS_j = \sum_i [X_i(j) - s^{-1} X_i(j+1)]^2 \, ,$$

whence the mean square at level j, MS_j, is

$$MS_j = \frac{SS_j}{n_j - n_{j+1}} \, ,$$

where $n_j = s n_{j+1}$. Finally, the jth component variance may be estimated as

$$\hat{\sigma}_\epsilon^2(j) = \frac{(s MS_j - MS_{j-1})}{s-1} \, , \qquad j = 1, \ldots, m \, . \tag{5.5}$$

Although the parameters $\sigma_\epsilon^2(j)$ are nonnegative, it is possible for the estimates to be negative unless restricted. Several rules are possible, such as the one used in the numerical example below.

The analysis may be extended to cover unequal groups at each stage of the hierarchy, either for the conventional analysis of variance (Moellering and Tobler, 1972) or for the variance components models (Kendall and Stuart, 1976, chapter 35). When $n = s_1 s_2 \ldots s_m$, all that is required is to replace s by s_j in equation (5.5).

The particular example which we now look at in more detail is taken from Moellering and Tobler (1972). These authors examined the data set shown in figure 4.1. It gives the number of isolated houses per 1000 m² quadrat in Hukuno Town, Tonami Plain, Japan (Matui, 1932). Moellering and Tobler took this quadrat size as their smallest spatial scale (level one) and looked at the variance component at this scale, and for various higher scales obtained by amalgamating quadrats in a balanced design. To do this, they deleted the last eight columns of data from the original (30 × 40) matrix shown in figure 4.2 and repeated rows one and two of the matrix at the bottom. This resulted in a (32 × 32) matrix for analysis, and the amalgamation scheme was as in figure 5.2. Their results are given in table 5.2. As they state, "the importance of the scale-variance components diminishes quickly as the cell size gets larger. The interpretation is clearly that the distribution of population in Western Hukuno is attributable to local site conditions." This conclusion is consistent with the earlier analysis in section 4.2.

To reinforce Moellering and Tobler's analysis, the formulae given in equations (5.5) yield

$\hat{\sigma}_\epsilon^2(1) = 0.79$, $\hat{\sigma}_\epsilon^2(2) = 0.043$, $\hat{\sigma}_\epsilon^2(3) = 0.024$,
$\hat{\sigma}_\epsilon^2(4) = -0.001$, $\hat{\sigma}_\epsilon^2(5) = 0.0023$.

The preponderance of variance at the first level supports Moellering and Tobler's conclusions. The negative estimate for $\sigma_\epsilon^2(4)$ arises from sampling variation. This can be handled by setting $\hat{\sigma}_\epsilon^2(4) = 0$ and pooling the fourth and fifth levels to yield MS(4 and 5) = 1·39, and $\hat{\sigma}_\epsilon^2(4 \text{ and } 5) = 0.001$.

The model established in equation (5.3) is an hierarchical analysis of variance (AV) scheme of the random components (or model II) kind. Thus, it would be possible to base an hierarchical variance analysis on table 5.2. However, as Bartlett (1971) has cautioned, in a discussion on Greig-Smith (1971), once the null hypothesis H_0: $\sigma_\epsilon^2(j) = 0$ has been rejected at one level j, the standard AV procedure does not provide a valid test at any level $k > j$. Valid tests, at successive levels of the hierarchy, have been developed by Mead (1974) using a randomisation argument. An alternative test procedure, which is based upon Scheffe's method of multiple comparisons, is given by Zahl (1974). This method uses overlapping blocks and has the advantage that the significance levels of the various tests are known.

Table 5.2. Scale variances from Matui's data.

Source of variation	Sum of squares	Percentage of TSS	df	Mean square
Level 1	614·25	69·37	768	0·79
Level 2	184·80	20·98	192	0·96
Level 3	64·54	7·29	48	1·34
Level 4	15·27	1·73	12	1·27
Level 5	5·58	0·63	3	1·86
Total	885·45	100·00	1023	—

Notes: TSS ≡ total sum of squares; df ≡ degrees of freedom.
Source: Moellering and Tobler (1972, page 44).

5.3.2 Other approaches

A weakness of the Greig-Smith method is that it provides variance estimates for very few distinct levels, even when the number of data points is large. Further, it relies upon a single mode of hierarchical grouping on a grid of (possibly arbitrary) orientation; for further discussion of some of the difficulties, see Pielou (1977, pages 141–143).

To overcome the positional problems, Usher (1969; 1975) considered a method whereby the average variances for *all* blocks of size 1, s, s^2, ... are examined. This operates by the use of different starting positions on the grid [for example, starting with columns 1, 2, ..., $(s-1)$]. The resulting estimators are no longer independent, but appear to have smaller variances. Hill (1973) has taken this approach one step further by evaluating the variances at all possible block sizes, thereby providing (correlated) estimates at all intermediate block sizes. It should be noted that both methods were developed in the context of line-transect data, but they are readily extended to two-dimensional arrays. In addition to Monte Carlo studies by Usher and Hill, Ludwig (1979) gives a detailed numerical comparison of the different approaches. He finds that Hill's method appears to provide the most accurate evaluation of the scale and intensity of any spatial patterns, although it is not immune from criticism. Cormack (1979, section 4.3) also provides a useful overview of these methods.

The different terms in Hill's method may all be written as sums of squares and cross products of the observed values (taken about the overall mean). Thus the variance estimates at different levels are expressible (approximately) in terms of the autocorrelations, so that the correlogram is a viable alternative to these methods. Indeed, it yields much the same information and provides an alternative view of the data; we shall present an example using the correlogram in section 5.4.

5.3.3 The variogram

When a process is defined continuously over space (so that the n sites are sampling points in a continuum), direct study of the autocovariance structure seems to pay better dividends than the modelling approach to be outlined in chapter 6. To allow for the possibility of a nonstationary process, Matheron (1971) recommends use of the *variogram*; this is written as

$$\gamma(h) = E\{[X(u+h) - X(u)]^2\},$$

where u represents the coordinates of one point and $u+h$ describes another set of coordinates. When the process is isotropic (direction invariant) the argument, h, can be replaced by the distance, d, between the points.

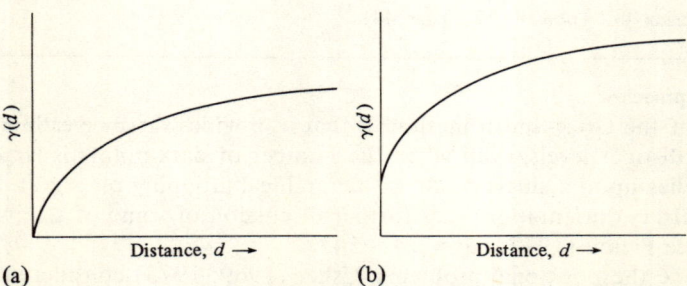

(a)   (b)

Figure 5.3. Typical forms of the variogram: for (a) stationary, and (b) nonstationary processes.

When the mean, μ, and variance, σ^2, of $X(u)$ are constant for all u, it follows that

$$\gamma(h) = 2\sigma^2[1 - \text{corr}(h)] ,$$

where $\text{corr}(h)$ is the spatial autocorrelation for points h apart. However, when the process is nonstationary (in that it is subject to local 'white noise'), $\gamma(h)$ has a discontinuity at the origin; the theoretical details are given in section 6.3.5. Typical shapes of the variogram in the two cases, assuming isotropy, are shown in figure 5.3. Estimation procedures for the variogram are considered in section 6.3.5.

5.4 Scale and correlation between processes
5.4.1 Model formulation

We now suppose that two random variables, X_i and Y_i, are observed for each cell of the grid. The same hierarchical pattern is employed as before (figure 5.2), but we must extend the model to allow for (1) variation which is common both to X and to Y, and (2) variation which affects only one of the two variables. To achieve this, we partition the jth-level random component into three parts:

Z_j, the jth-level *factor* which is common both to X and to Y,
ϵ_j, the *'noise'* component which affects X only, and
δ_j, the *'noise'* component which affects Y only.

The use of the term *factor* is consistent with normal usage in factor analysis. Thus we suppose that X_i and Y_i may be written in terms of the linear expressions,

$$\left. \begin{array}{l} X_i = \mu_x + \phi_1 Z_1(i) + ... + \phi_m Z_m(i) + \epsilon_1(i) + ... + \epsilon_m(i) , \\ Y_i = \mu_y + \theta_1 Z_1(i) + ... + \theta_m Z_m(i) + \delta_1(i) + ... + \delta_m(i) , \end{array} \right\} \quad (5.6)$$

where all terms and assumptions are defined in table 5.3(a).

This framework allows a wide variety of models to be developed. In particular, unless restrictions of the sort imposed for models 1 and 2 and discussed below are specified, the framework implies that dependence is built into the model by 'sharing' errors, as shown in figure 5.4. For example, from equation (5.6) and dealing only with the $\{\epsilon\}$ terms,

$$X_1 = \epsilon_1(1) + \epsilon_2(1) + ... ,$$
$$X_2 = \epsilon_1(2) + \epsilon_2(2) + ... ,$$

but $\epsilon_2(1) \equiv \epsilon_2(2)$ by specification of the model; see also equation (5.3).

We can now examine some special cases of the model.

Model 1 Suppose we impose the additional restrictions given in table 5.3(b), that is, the model reduces to

$$\left. \begin{array}{l} X_i = \mu_x + \phi \xi_i + \epsilon_1(i) , \\ Y_i = \mu_y + \theta \xi_i + \delta_1(i) , \end{array} \right\} \quad (5.7)$$

where $\xi_i \equiv Z_1(i)$, and we have dropped the subscripts on θ and ϕ for convenience. The model is in the form of a structural relationship (Kendall and Stuart, 1979, chapter 29) if we regard θ and ϕ, but not the $\{\xi_i\}$, as unknowns to be estimated. It is a functional relationship if we also wish to estimate the $\{\xi_i\}$. Alternatively, the model may be interpreted as a single (common) factor underlying two variates with θ and ϕ as factor loadings. Let us examine the model in the factor context with the

Table 5.3. Assumptions for scale-components models.

(a) All models

Item	Mean	Variance	Correlation structure	Other
μ_x, μ_y	—	—	—	constant over all areas
ϕ_j, θ_j	—	—	—	parameters at all levels
$Z_j(i)$	0	1	uncorrelated over all areas i and levels j	factors
$\epsilon_j(i)$	0	$\sigma_\epsilon^2(j,i)$	uncorrelated (1) with $Z_j(i)$ for all areas, levels; (2) with each other for all areas, levels; (3) with any $\delta_j(i)$	random noise components
$\delta_j(i)$	0	$\sigma_\delta^2(j,i)$	as for $\epsilon_j(i)$	random noise components

(b) Additional restrictions

Model 1
$\theta_j = \phi_j = 0$, levels 2 and above.
$$\sigma_\epsilon^2(j,i) = \begin{cases} \sigma_\epsilon^2, & \text{over all areas, level 1;} \\ 0, & \text{level 2 and above.} \end{cases}$$
$$\sigma_\delta^2(j,i) = \begin{cases} \sigma_\delta^2, & \text{over all areas, level 1;} \\ 0, & \text{level 2 and above.} \end{cases}$$

Model 2
$\theta_j = \phi_j = 0$, levels 2 and above.
$$\sigma_\epsilon^2(j,i) = \begin{cases} B_i \sigma_\epsilon^2 & \text{in area } i, \text{ level 1}, B_i \text{ known;} \\ 0, & \text{level 2 and above.} \end{cases}$$
$$\sigma_\delta^2(j,i) = \begin{cases} B_i \sigma_\delta^2 & \text{in area } i, \text{ level 1}, B_i \text{ known;} \\ 0, & \text{level 2 and above.} \end{cases}$$
$\text{var}(\xi_i) = B_i$.

Model 3
$\sigma_\epsilon^2(j,i) = \sigma_\epsilon^2(j)$, for all areas and levels;
$\sigma_\delta^2(j,i) = \sigma_\delta^2(j)$, for all areas and levels.

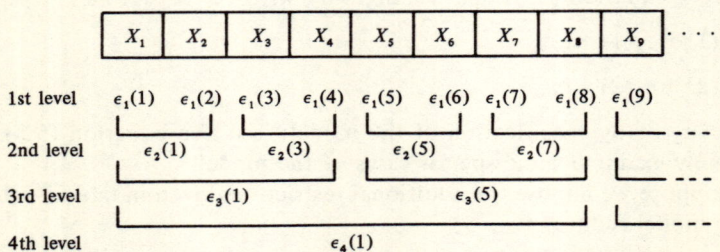

Figure 5.4. Illustration of the way dependence is built into a scale-components model by sharing errors as cells are aggregated.

constraint $\sum \xi_i = 0$. It follows directly that

$$\left. \begin{array}{ll} E(X_i) = \mu_x, & E(Y_i) = \mu_y, \\ \text{var}(X_i) = \phi^2 + \sigma_\epsilon^2, & \text{var}(Y_i) = \theta^2 + \sigma_\xi^2, \\ \text{and} & \\ \text{cov}(X_i, Y_i) = \theta \phi. & \end{array} \right\} \quad (5.8)$$

When X_i and Y_i are bivariate normal, the parameters $\{\theta, \phi, \sigma_\xi^2, \sigma_\epsilon^2\}$ cannot be estimated separately without additional information (Kendall and Stuart, 1979, pages 403-404). However, the correlation between X and Y can be estimated using the conventional sample correlation coefficient in the usual way.

Model 2 The restrictions imposed are those given in table 5.3(b). The resulting estimators are:

$$\hat{\mu}_x = \frac{\sum A_i X_i}{\sum A_i}, \qquad \hat{\mu}_y = \frac{\sum A_i Y_i}{\sum A_i},$$

$$\hat{\text{cov}}(X, Y) = \frac{\sum A_i (X_i - \hat{\mu}_x)(Y_i - \hat{\mu}_y)}{\sum A_i}, \qquad \hat{\text{var}}(X) = \frac{\sum A_i (x_i - \hat{\mu}_x)^2}{\sum A_i},$$

with a similar expression for $\hat{\text{var}}(Y)$. Here, $A_i = B_i^{-1}$, and the expectations for these estimators are given by expressions (5.8). These results hold for any weighting function, B_i, but a natural choice for data recorded as an average over a spatial unit of area a_i is $B_i = a_i^{-1}$. This form of weighting function has been used by Robinson (1956) and by Thomas and Anderson (1965) to weight the conventional correlation coefficient to try to overcome the dependence of values of the coefficient upon the size of the areal units used (see section 5.4.4). Clearly the extension to regression analysis should employ the same weighting function.

Models 1 and 2 are of only limited interest because they require that there should be no spatial autocorrelation between geographical units larger than the first level of the hierarchy. The descriptions given here are intended firstly to provide a link with earlier investigations, and secondly to pave the way for the more realistic model discussed below, which does not carry this restriction.

Model 3 The restrictions are again given in table 5.3(b). The variance component for X at level j is given by $\sigma_\epsilon^2(j) + \phi_j^2$, which may be estimated by

$$\hat{\sigma}_\epsilon^2(j) + \hat{\phi}_j^2 = \frac{\{s\,\text{MS}_j(x) - \text{MS}_{j+1}(x)\}}{s - 1}, \qquad j = 1, ..., m. \quad (5.9)$$

This is the same expression as (5.5), save that an x has been inserted to indicate the variable of interest. The reason that expressions (5.5) and (5.9) are identical is that the factor and noise components collapse into one when only a single random variable is considered. In similar vein, the

jth-level sum of cross products may be written as

$$\text{SCP}_j = \sum_i \{X_i(j) - s^{-1}X_i(j+1)\}\{Y_i(j) - s^{-1}Y_i(j+1)\}, \qquad j = 1, ..., m,$$

whence the mean cross product is given as

$$\text{MCP}_j = \frac{\text{SCP}_j}{(n_j - n_{j+1})},$$

with n_j defined as before. Finally, the covariance component $\theta_j \phi_j$, is estimated by

$$\widehat{\theta_j \phi_j} = \frac{\{s\text{MCP}_j - \text{MCP}_{j+1}\}}{s - 1}, \qquad j = 1, ..., m. \tag{5.10}$$

At any given level, we have three statistics (two variances and one covariance) but four parameters $[\theta_j, \phi_j, \sigma_\epsilon^2(j), \text{ and } \sigma_\delta^2(j)]$. However, this does not matter provided we confine attention to the estimation of the three *components*.

Example 5.1 Suppose there are k (= 4) levels, s (= 2) areas combined, and $\theta_j = \phi_j = 1$ for $j = 1, ..., 4$. Also let $\sigma_\epsilon^2(1) = \sigma_\delta^2(1) = 6$, and $\sigma_\epsilon^2(j) = \sigma_\delta^2(j) = 0$, $j \geqslant 2$. Then we obtain

	$j = 1$	2	3	4
var$\{X(j)\}$ = var$\{Y(j)\}$	10	6·5	4·25	2·625
corr$\{X(j), Y(j)\}$	0·4	0·54	0·65	0·71

The reduction in variance and increase in correlation as areas are aggregated was noted in the spatial case by Yule and Kendall (1965, pages 310-312 and earlier editions); that is, the degree of correlation between the variables is a function of the size of area considered, as has been discussed by Robinson (1956) and Thomas and Anderson (1965). What is supplied by equations (5.5) and (5.9) is a spectral decomposition of the total (single unit) variances and covariance into components for the particular cell sizes considered. If a suitable functional form describing the change in correlation with size can be found, it might be calibrated from the estimated values.

5.4.2 Negative estimates

To avoid estimates which imply negative values for any $\sigma_\delta^2(j)$ or $\sigma_\epsilon^2(j)$, we may modify the estimators given in equations (5.5) and (5.9) as follows:
(a) If both variance estimators are greater than zero but the modulus of the correlation exceeds unity, scale the covariance down by a factor c_1 and each variance up by a factor c_2. These factors c_1 and c_2 are chosen such that the modulus of the correlation is reduced to unity.
(b) If one variance is negative, replace that variance (corresponding to X, say) by

$$\widehat{\text{var}}\{X(j)\} = \frac{[\text{cov}\{X(j), Y(j)\}]^2}{\widehat{\text{var}}\{Y(j)\}}.$$

(c) If both variances are negative, set both variances and covariance to zero.

In all cases, the investigator may prefer to set the appropriate estimators to zero or to aggregate these values with the next level of the hierarchy. These negative estimates arise from the use of equations (5.5) and (5.9) and can be avoided by deriving the maximum likelihood estimators subject to a series of boundary constraints. The numerical details tend to be very messy, which is why ad hoc rules such as the above are sometimes used. In section 5.4.4 we have quoted negative estimates as they arise, without any attempts to 'doctor' the analysis.

5.4.3 An alternative approach

In place of equations (5.4) we might employ the usual regression model for X and Y; that is

$$X_i = Z_i^T\beta + \epsilon_i \; ; \quad Y_i = Z_i^T\gamma + \delta_i \; . \tag{5.11}$$

The trend-surface models (with polynomial or sinusoidal terms) are special cases of model (5.11) (see sections 6.2.1 and 8.8). Clearly the covariance between the residuals will depend upon the value specified for $\text{cov}(\epsilon_i, \delta_i)$, and lead to Zellner's (1962) 'seemingly unrelated' regression equations. When ϵ_i and δ_i are uncorrelated the spatial variation in X and Y can be 'explained' entirely by other variables. If a regression model is analysed using model 3 given above, and the Z are spatially autocorrelated, the results will, quite reasonably, reflect correlation at different distances (or for different sized units). The all-important question of interpretation is whether the correlation is intrinsic or whether it is the result of a common dependence upon other variables. The answer to this question must depend upon geographical theory rather than upon statistical methods.

It is possible to envisage other schemes which interrelate spatial autocorrelation and factor analytic methods. However, this development is beyond the scope of the present discussion.

5.4.4 Empirical application: land use in London

As an example of the methods we have described in this section, let us now consider data on land use drawn from the *Atlas of London* (Jones and Sinclair, 1968, sheets 43-45). The data set comprises a 24 × 24 lattice of cells[7], each of which is a square of side 500 metres. For each cell, the area of floor space (in thousands of square feet[8]) in office, commercial, and industrial use in 1962 is given. The maps are redrawn here in choropleth form in figure 5.5.

---

[7] The northwest corner of the grid was located at grid reference 255860.
[8] Our apologies are offered to those readers who are more purist over the matter of units.

Two empirical features of note, which we would expect from bid-rent theory (Alonso, 1964), are that
(1) high-demand areas, such as the City of London, tend to make intensive use of the land available, so that total floorspace in neighbouring cells and the type of use to which it is put, tend to be positively correlated;
(2) there is an element of competition at a local spatial scale between the different potential uses, resulting in some grid squares being dominated by one particular land use.

The data were analysed using model 3. Successive aggregations took in 4, 4, 4, 3, and 3 cells of the previous level; the results are presented in

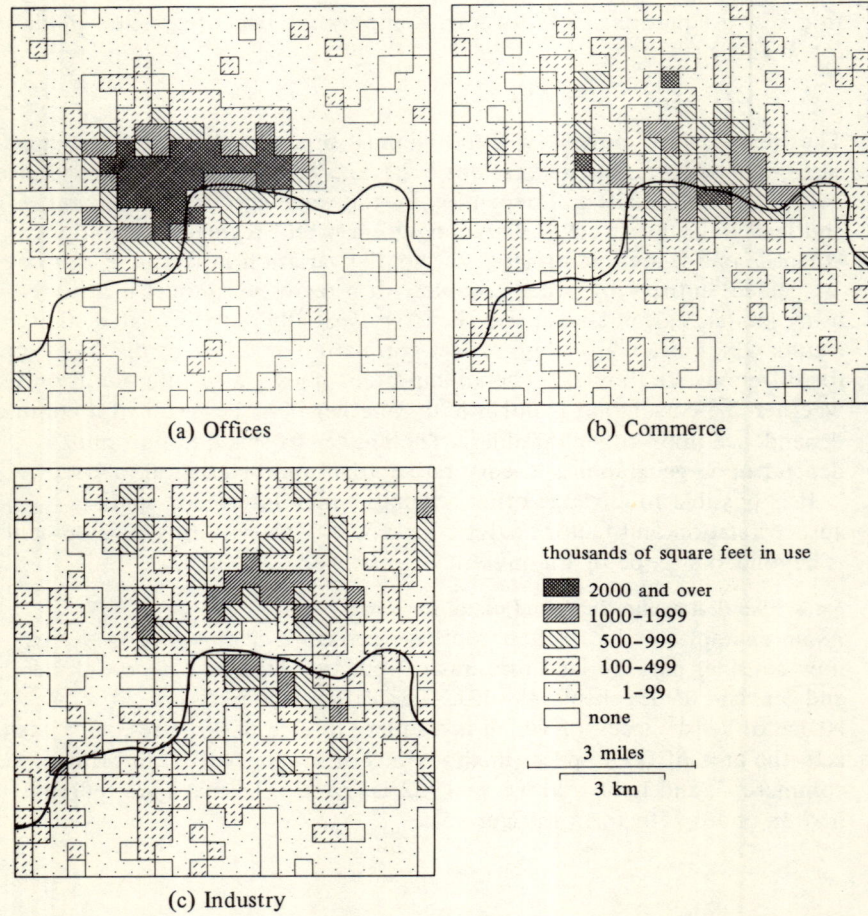

Figure 5.5. Choropleth map of the area of floorspace occupied by three different activities in part of the former London County Council area, 1962. Grid squares are of side 500 metres. Source: based on Jones and Sinclair (1968).

table 5.4. The features of increasing correlation and decreasing variance, noted in the example in section 5.4.1 are present again, although the correlation function flattens out from level 3 onwards in table 5.4(b). The 'correlation coefficients' for individual levels based upon the corresponding *components* are given in figure 5.6. These coefficients provide an even sharper contrast. Despite the fact that the components estimates are unrestricted, so that 'variances' and 'correlations' may wander beyond their feasible regions, the general pattern is clear. For smaller cells, the competitive influence dominates (negative correlation between land uses) whereas the common features of neighbourhood effects (positive correlation

Table 5.4. Scale analysis of London land-use data. Aggregation steps are $4 \times 4 \times 4 \times 3 \times 3$; $n = 576$.

Level	Var(X)	Var(Y)	Cov(X, Y)	X component	Y component	XY component	Corr (X, Y)
(a) $X \equiv commerce,\ Y \equiv offices$							
1	116712	886554	60691	60220	248162	−25371	0·19
2	71547	700432	79719	18738	318847	−10598	0·36
3	46203	414767	92425	29746	195119	71664	0·67
4	17557	197011	41853	2969	105111	6910	0·71
5	9211	78547	26008	9644	58588	31090	0·97
(b) $X \equiv industry,\ Y \equiv offices$							
1	88954	886554	24856	52025	248162	−7390	0·09
2	49935	700432	30399	10801	318847	−8734	0·16
3	32080	414767	38335	15104	195119	28170	0·33
4	16289	197011	19192	7286	105111	10470	0·34
5	7923	78547	7958	7701	58588	6319	0·32

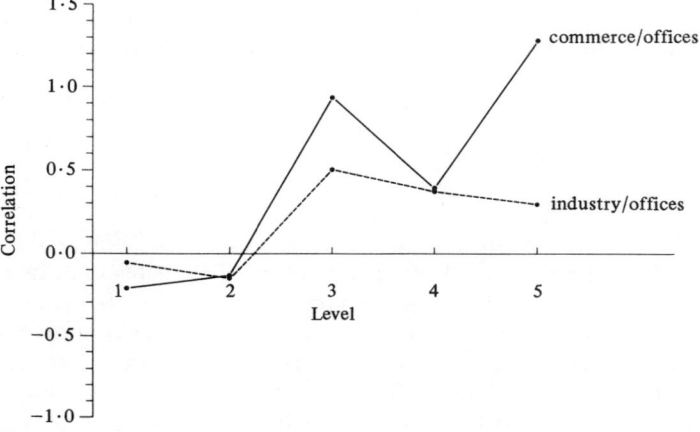

Figure 5.6. 'Correlation coefficients' based upon scale components for the London land-use data.

between land uses) are seen for levels 3 and above. Obviously, as cells are aggregated, mixed land-use patterns are more likely to be found. Clearly, any talk of a single correlation coefficient between variables measured on modifiable units is erroneous. The advantage of the method of analysis we have proposed is that the spectral decomposition enables the variations in the process with different spatial scales to be determined, whereas such variations are masked by the simple correlation coefficient.

5.5 Model identification

In chapter 6, we shall describe different spatial models and related estimation procedures. By analogy with time series (Box and Jenkins, 1970, chapter 6), we need to identify a model prior to estimation. The tools used by Box and Jenkins for model identification are the *correlogram*, already discussed, and the *partial correlogram*. In this section we shall develop the spatial partial correlogram and illustrate its uses, particularly with regard to model identification prior to estimation, and we shall need to draw heavily on the work of Bennett (1975) and Martin and Oeppen (1975).

5.5.1 The partial correlogram

We defined the correlogram in equation (5.1) as

$$I(g) = \frac{n}{S_0(g)} \frac{z'W(g)z}{z'z}$$

for $g = 1, 2, ...$, and it would appear that partial correlograms could be defined in terms of the statistic $I(g)$. Indeed, if we are interested in the *covariance* structure of the process (see section 6.2), then this definition is appropriate. However, if we wish to *identify a model* then we must take account of the number of sites g steps away from a given site. To make the discussion more concrete, consider a series of conditional autoregressive schemes (section 6.2.4) of order $g = 1, 2, ..., D$. That is, for the ith site, we specify the conditional mean and variance of the gth order scheme as

$$\left.\begin{aligned} E(X_i | X_j = x_j, j \neq i) &= \rho_1 \sum_{j \in C(1)} x_j + \rho_2 \sum_{j \in C(2)} x_j + ... + \rho_g \sum_{j \in C(g)} x_j , \\ \text{var}(X_i | X_j = x_j, j \neq i) &= \sigma_g^2 . \end{aligned}\right\}$$

(5.12)

For each such scheme, the $\{\rho_j\}$ are unknown parameters and the sets $C(j)$ are as defined in section 5.2.1. That is, we group all the sites one, two, ..., g steps away from site i, up to the maximum of D steps. If the appropriate form of expressions (5.12) involves nonzero coefficients ρ_j only up to g_0 ($\rho_j = 0$ for $j > g_0$), then we say that the correct model is a g_0th-order autoregressive scheme. In such circumstances, we would expect estimates of the ρ_j to be close to zero for $j > g_0$.

In section 6.3.1, it is shown that the least squares estimators are consistent for the conditional autoregressive scheme, so that we could fit

autoregressive schemes of order g for $g = 1, 2, ...,$ by this means. If we let b_{jg} denote the estimate of ρ_j in the gth-order autoregressive scheme, $j = 1, ..., g$, and $g = 1, ..., D$,
then the partial correlogram is the plot of
b_{gg} against g ($g = 1, ..., D$).

To obtain successive values of b_{gg}, we do not need to fit the next order model (5.12) at each stage, but may proceed directly by Durbin's method, which was developed originally for time series (Durbin, 1960).

Let r_g be the gth modified spatial autocorrelation, defined as

$$r_g = \sum_{i=1}^{n} z_i z_i(g) \bigg/ \left\{ \sum_{i=1}^{n} z_i^2 \sum_{i=1}^{n} [z_i(g)]^2 \right\}^{1/2} , \qquad (5.13)$$

where $z_i = x_i - \bar{x}$, and the $\{z_i(g)\}$ are given by the vector $z(g)$, $z(g) = \mathbf{W}(g)z$. That is,

$$r_g = \frac{S_0(g)I(g)}{n} \left[\frac{\sum z_i^2}{\sum z_i^2(g)} \right]^{1/2} ;$$

see section 1.5.2. Given the $\{r_g\}$, then the $\{b_{gg}\}$ are evaluated successively as follows:

$$b_{11} = r_1 ,$$

$$b_{g+1,g+1} = \left(r_{g+1} - \sum_{j=1}^{g} r_{g+1-j} b_{jg} \right) \bigg/ \left(1 - \sum_{j=1}^{g} r_j b_{jg} \right) ,$$

$$b_{j,g+1} = b_{jg} - b_{g+1-j,g} b_{g+1,g+1} ,$$

for $j = 1, ..., g$, and $g = 1, ..., D$.

When the covariance structure is of primary interest, the same computational procedure is followed, but with $I(g)$ in place of r_g. In the time-series case, $x_i(g) = x_{i-g}$, so that $I(g)$ and r_g are usually indistinguishable unless n is very small; for computational convenience $I(g)$ is commonly used. In spatial studies there will be several terms in each summation in expression (5.12), so that, typically,

$$\sum z_i^2(g) < \sum z_i^2 ;$$

this extra damping in the correlogram formed from the $I(g)$ distorts the resulting partial correlogram when used for model identification.

5.5.2 Computational aspects

For irregular lattices, it is necessary to identify the sets $C(g)$ for each i (as outlined in section 5.2.1) so that r_g may be evaluated from equation (5.13). However, for regular lattices of sites, some simplification is possible provided that the lattice is sufficiently large to allow boundary effects to be ignored. Suppose that the sites form a rectangular grid with p rows, q columns ($p \times q = n$), and observed values x_{ij} at site (i, j), $i = 1, ..., p$, $j = 1, ..., q$. Let the overall sample mean be $\bar{x}$ and set $z_{ij} = x_{ij} - \bar{x}$.

We can now specify the sample covariances as a two-dimensional array

$$c_{uv} = \sum_{j=1}^{q-v} \sum_{i=1}^{p-u} z_{ij} z_{i+u, j+v} , \qquad (5.14)$$

with a similar form for $c_{u,-v}$, and specify the two-dimensional correlogram as

$$I(u, v) = \frac{c_{uv}}{c_{00}} . \qquad (5.15)$$

Turning to equation (5.13), suppose we say that sites are g steps away if $|u| + |v| = g$; that is, we use a 'city block' metric or the rook's case, as defined in section 1.4.2, to specify first-, second-, ... order neighbours. When $g = 1$, the numerator of equation (5.13) is, apart from boundary terms, proportional to

$$\text{Numr}(1) = \tfrac{1}{4} \sum \sum z_{ij}(z_{i-1,j} + z_{i+1,j} + z_{i,j-1} + z_{i,j+1}) = \tfrac{1}{2}(c_{10} + c_{01}) ;$$

the limits of summation all derive from equation (5.14) and are omitted. In like manner, the second term in the denominator becomes

$$\text{Denr}(1) = \tfrac{1}{16}(4c_{00} + 4c_{11} + 4c_{1,-1} + 2c_{20} + 2c_{02}) .$$

Thus

$$r_1 = \frac{\text{Numr}(1)}{[c_{00} \text{Denr}(1)]^{\frac{1}{2}}} . \qquad (5.16)$$

For higher values of g,

$$\text{Numr}(g) = \tfrac{1}{4} \sum \sum z_{ij}(z_{i-g,j} + z_{i+g,j} + z_{i,j-g} + z_{i,j+g}) = \tfrac{1}{2}(c_{g0} + c_{0g}) ,$$

$$\text{Denr}(g) = \tfrac{1}{16}(4c_{00} + 4c_{gg} + 4c_{g,-g} + 2c_{2g,0} + 2c_{0,2g}) ,$$

and

$$r_g = \frac{\text{Numr}(g)}{[c_{00} \text{Denr}(g)]^{\frac{1}{2}}} .$$

For other adjacency patterns similar expressions are readily developed.

5.5.3 Spatiotemporal correlograms

If we refer back to equation (5.14), it is apparent that we could develop a correlogram both in space and time by using one subscript for each dimension, and Bennett (1975) and Martin and Oeppen (1975) have developed such correlograms for modelling spatiotemporal processes; see also Pfeiffer and Deutsch (1980) and Hooper and Hewings (1981). However, when we define the partial correlogram, two approaches are possible:
(a) Bennett uses Durbin's method but requires a prespecified ordering in which the spatial and temporal 'lagged' terms enter the model;
(b) Martin and Oeppen use the standard definition of partial correlations (Kendall and Stuart, 1979, section 27.5) so that the partial coefficient for spatial lag g and temporal lag k is conditioned on all spatial terms less than g and all temporal terms less than k.

Scheme (b) is more comprehensive although scheme (a) offers possibilities for developing a spatiotemporal model in a stepwise fashion. Once all the autocorrelations have been computed, a standard stepwise regression program could be used in this context.

5.5.4 Correlograms and partial correlograms for the London land-use data

It is evident from figure 5.5 that there is a high degree of spatial clustering for all the three activities in central London. It is particularly marked in the case of office use. Such a spatial concentration of activity implies that the data are nonstationary, and this nonstationarity increases as the degree of clustering increases. On *a priori* grounds, we should therefore expect problems to arise in the correlogram analysis which we did not encounter in the variance decomposition approach discussed in section 5.4.4. This indeed proved to be the case.

We computed the autocorrelation function (ACF) and partial autocorrelation function (PACF), for each of the land uses shown in figure 5.5, from equation (5.13) and the linked relationships given in section 5.5.1. To search for directional biases in the data, several forms of W were tried:
(1) $w_{ij}(g) = 1$ if cells i and j were lag g apart on a row of the lattice, and $w_{ij}(g) = 0$ otherwise (east–west autocorrelation);
(2) $w_{ij}(g) = 1$ if cells i and j were lag g apart on a column of the lattice, and $w_{ij}(g) = 0$ otherwise (north–south autocorrelation);
(3) $w_{ij} = 1$ if cells i and j were lag g apart at a vertex, and $w_{ij}(g) = 0$ otherwise (bishop's case, by analogy with chess—northeast–southwest and northwest–southeast autocorrelation);
(4) $w_{ij} = 1$ if cells i and j were lag g apart on a row or column of the lattice, and $w_{ij}(g) = 0$ otherwise (rook's case—north–south and east–west autocorrelation);
(5) $w_{ij} = 1$ if cells were lag g apart on a row, column, or vertex, and $w_{ij}(g) = 0$ otherwise (queen's case, no directional bias).
We took $g = 1, ..., 12$.

The results obtained are shown in figure 5.7. Note that in equation (5.13) we have defined $z_i = x_i - \bar{x}$. This use of a common mean, whatever the value of g, can be sufficient to produce PACFs which range outside ±1. For the nonstationary office data this happened and we have omitted these meaningless results. The results for commerce in figure 5.7 show that the ACF tends to die away, but that the PACF does not. This is suggestive of a moving-average process (section 6.2.5) rather than an autoregressive scheme. The industry data give PACFs which are very small after the first lag, indicating a first-order autoregressive model (see section 6.2.2).

When nonstationary data are analysed using correlogram methods, there are two precautions which should be considered. First we can replace the common mean, $\bar{x}$, in equation (5.13) by the mean of the data at each lag. This ensures that the PACF damps inside the range ±1, but does not remove the nonstationarity. Second, trends in the data can be removed

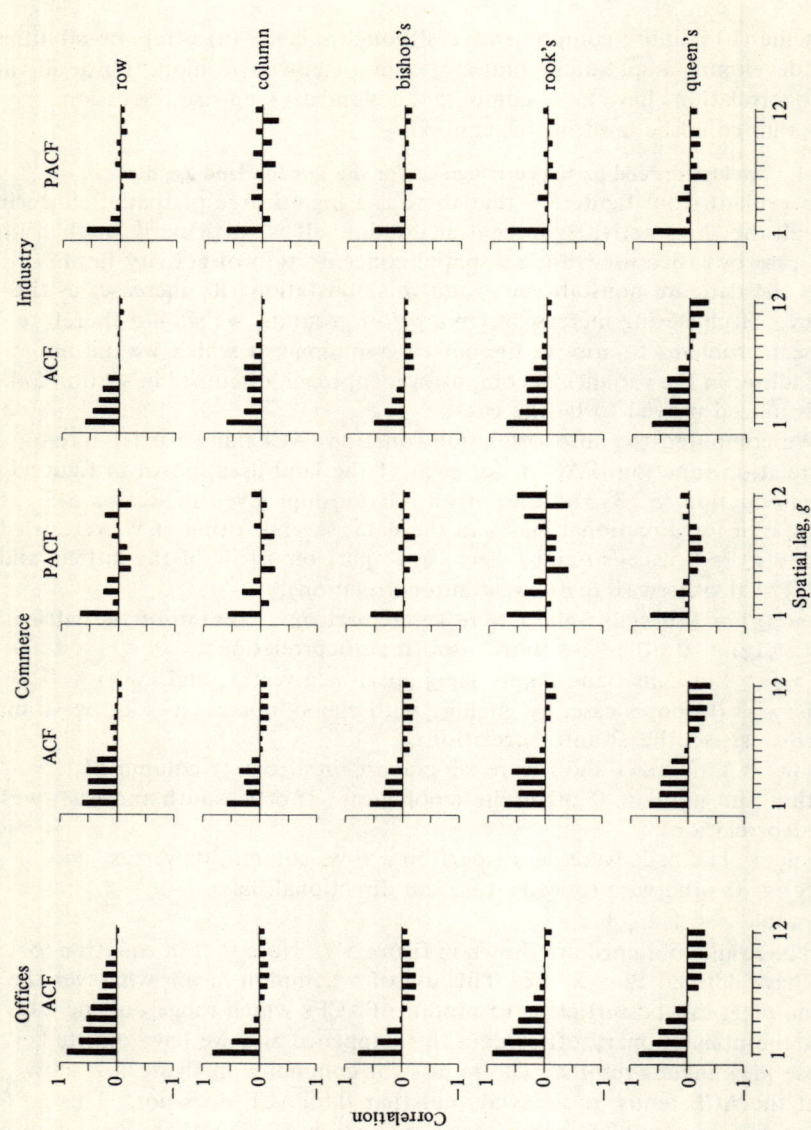

Figure 5.7. Spatial correlograms and partial correlograms for London land-use data.

prior to analysis using equation (5.13). See Chan et al (1977) and section 7.4 for a discussion of trend-removal methods and their effects upon the correlogram.

We have already noted from the covariance analysis of section 5.4.4 that the office use is negatively correlated with industry and commerce at small spatial scales. This is evident in figure 5.5, where offices display a much greater degree of spatial concentration than industry or commerce. It is reflected in the larger correlations and slower damping of the ACFs for offices in the row only and rook's cases (figure 5.7), compared with those for industry and commerce. Note also from figure 5.5(a) that the clustering extends further in an east–west than in a north–south direction. Hence the ACF for offices (row only) damps much more slowly than that for offices (column only).

We may conclude from the analysis that all three activities have similar levels of square footage devoted to them in neighbouring areas at the local (small) spatial scale (positive ACFs for the first few lags). This is very marked for offices. The spatial clustering of uses at this scale implies that grid squares may often be dominated by one particular activity (see figure 5.5), as we noted in the variance analysis of section 5.4.4.

5.6 Estimation of the weights

If we wish to test a specific hypothesis, then the weights are determined in accordance with the alternative hypothesis (see section 6.4.3). When the alternative hypothesis takes the general form 'spatial autocorrelation is present', then the $I_{\max}$ statistic of Kooijman (section 5.2.1) may be used to calculate the weights, given some prior structure. Again, if the covariances or the weights can be specified as a parametric function (of distance, common boundary, etc) then the parameters may be estimated by maximum likelihood or a similar procedure (see section 6.3). However, all such schemes must involve an element of prior restriction, since n observations can hardly hope to estimate $\frac{1}{2}n(n-1)$ terms in the weighting matrix otherwise!

When n is not too large and observations are available for each site at several time periods, other approaches become feasible. For example, in accordance with the approach of Arora and Brown (1977), suppose that the random variable for site i at time t is $X_i(t)$ and that

$$\text{cov}[X_i(t), X_j(t)] = \sigma_{ij} , \qquad \text{cov}[X_i(t), X_j(s)] = 0 , \quad t \neq s ; \qquad (5.17)$$

we assume that n observations are available, $\mathbf{x}^T(t) = [x_1(t), ..., x_n(t)]$, for t time periods, $t = 1, ..., T$. When the $\{X_i(t)\}$ are normally distributed, it follows that the maximum likelihood estimators for σ_{ij} are given by

$$T\hat{\sigma}_{ij} = \sum_{t=1}^{T} z_i(t) z_j(t) ,$$

where

$$z_i(t) = x_i(t) - \bar{x}_i, \qquad T\bar{x}_i = \sum_{t=1}^{T} x_i(t).$$

Typically, temporal autocorrelation will be present also, and more general results are desirable in this area. Arora and Brown (1977) extend their analysis to linear regression models with an error structure specified by expression (5.17). They also consider linear models with random coefficients, which provide an alternative explanation for the presence of spatial autocorrelation in the residuals.

An interesting extension to the use of a parametric weighting function such as,

$$w_{ij} = d_{ij}^{-\beta},$$

is given by Gatrell (1979). Instead of using the usual distances between points, Gatrell carried out an initial analysis of the data by multidimensional scaling and then evaluated the weights in terms of the distances imputed by the scaling algorithm. Gatrell applied his method to an analysis of the major Swedish cities and found evidence of 'spatial' autocorrelation in the generated interaction space. Since autocorrelation was not apparent when the original map was analysed, this study stresses the importance of selecting a meaningful distance metric.

5.7 Conclusions

In this chapter, spatial correlograms and partial correlograms have been defined. Their use in model identification, and in the determination of the spatial scales at which processes are operating, has been discussed and illustrated. Analysis of variance methods provide an alternative approach to the problem of identifying the spatial scale of a process, and suitable models have been described and applied. A basic difficulty with any correlation analysis is the dependence of the coefficients upon the size of the areal units for which the data have been collected. A spectral decomposition method which overcomes this feature has been presented. Finally, ways of estimating the elements in the weighting matrix, **W**, have been briefly considered.

6

Models for spatial processes

6.1 Introduction
In earlier chapters we have looked at a variety of tests designed to search for spatial autocorrelation in data. However, the detection of spatial pattern is often not an end in itself, but instead interest focuses upon the development of spatial models. In turn these models provide alternative hypotheses against which the null hypothesis of no spatial autocorrelation can be tested, which enables us to compare the performance of the tests against selected alternatives.

In section 6.2 we shall develop models for spatial processes both by using the autoregressive–moving-average approach and by means of the covariance function. Then, in section 6.3, we shall consider various estimation procedures for these models. In sections 6.4 and 6.5, the discussion reverts to tests of hypotheses, and we shall examine the performance of different statistics theoretically, empirically, and by Monte Carlo studies. The conclusions are given in section 6.6.

6.1.1 Reaction or interaction?
When we develop a model for a spatial process, we must always ask whether the levels of the process at two (neighbouring) sites reflect interaction (between the sites) or reaction to some other variable. The case is rarely open and shut. For example, two adjacent trees may compete for resources (sunlight and nutrients), thus displaying between-tree interactive effects; but they will also react to the general availability of nutrients within the scope of their root systems, thus displaying a regression-like dependence for, say, tree size on the environment. Similarly, two adjacent supermarkets will compete for trade, and yet their turnover will be a function of general factors such as the distribution of population and accessibility.

Notwithstanding the general rule that 'everything affects everything else', it is often useful to assess whether the dominant effects are caused by reaction to external forces or by interaction between (neighbouring) individuals. When reaction is the major influence, a regression model is appropriate, whereas interactive effects suggest the need for a model with a spatially dependent covariance structure. Since the major theme of this book is spatial autocorrelation, we naturally tend to emphasise interactive models in this chapter, but the importance of regression models should not be lost. Indeed, it may be appropriate to fit a regression model and then examine the residuals for spatial dependence (see chapter 8), or to fit a model which incorporates both regression and spatial autocorrelation (as in chapter 9).

As noted in section 1.5.2, the presence of spatial autocorrelation may be attributable either to trends (or gradients) in the data or to interactions;

and if gradients are suspected then a regression model is appropriate. Thus, significant autocorrelation in the original data does not imply one model rather than the other. The choice of model must involve the scientific judgement of the investigator *and* careful testing of the assumptions.

6.2 Model specification

To specify a model involving spatial interaction, we must incorporate the spatial dependence into the covariance structure either explicitly or implicitly by means of an autoregressive and/or moving-average structure. Suppose that the process is observed at n sites ($i = 1, ..., n$) and that Y_i describes the variate at the ith site. Let

$$E(Y_i) = \mu_i, \qquad \text{var}(Y_i) = v_{ii}, \qquad \text{and} \quad \text{cov}(Y_i, Y_j) = v_{ij}, \quad i \neq j,$$

so that the variate vector Y, $Y = (Y_1, ..., Y_n)^T$, has

$$E(Y) = \mu = (\mu_1, ..., \mu_n)^T, \tag{6.1}$$

and

$$\text{cov}(Y) = V = \{v_{ij}\}. \tag{6.2}$$

An explicit approach would involve specifying a functional form for μ and V in terms of a limited number of parameters, whereas an autoregressive scheme requires a (linear) model relating Y_i to its neighbouring variates in some way. When Y is multivariate normal, there is a link between the two specifications as is shown in theorem 6.1 (see chapter appendix). The model may be specified either in the *strict sense* or in the *wide sense*, depending on the extent of the assumptions we are willing to make. A *strict-sense* specification involves the joint distribution of the Y, whereas a *wide-sense* specification requires knowledge only of the first two moments, μ and V. For the most part we shall consider Y to be multivariate normal, when knowledge of μ and V determines the form of the distribution. However, the distinction is important for nonnormal schemes.

So far, we have not imposed any conditions upon μ and V. Since they correspond to a total of $n + \frac{1}{2}n(n+1)$ parameters, it would not be possible to make any useful inferences about the structure from a set of n observations. Of course, the position is different if we have observations over several time periods and n is not too large, as happens in some econometric studies. Restricting attention to a single set of observations, we must assume that μ and V are determined by some smaller set of parameters (usually the size of this set will be taken not to depend upon n). This structure should enable us to model the dependence of Y_i upon neighbouring sites. A central concept in the specification of such models is that of *spatial stationarity*, by which we mean that the average of the Y_i is the same for all i, and that Y_i and Y_j depend upon each other only by virtue of their relative positions and not through their absolute locations.

Models for spatial processes

The notion of stationarity is of crucial importance in spatial models, as in time series; therefore, a formal definition is worthwhile. We distinguish two cases, depending on whether the set of possible sites is finite or forms a continuum.

Definition 6.1 (For a finite set of sites or for a spatially discrete process) Consider a spatial process Y_i defined at sites i ($i = 1, ..., n$) and let the coordinates of site i be $x_i = (x_{1i}, x_{2i})^T$. Then the process is *spatially stationary* if, for all i,

$$E(Y_i) = \mu, \quad \text{var}(Y_i) = \sigma_Y^2,$$
$$\text{and}$$
$$\text{cov}(Y_i, Y_j) = \sigma_Y^2 c(x_i - x_j), \tag{6.3}$$

where c is a *correlation* function, depending on the relative locations of the two sites.

Definition 6.2 (For a continuum of sites or a spatially continuous process) Consider a spatial process $Y(x)$ defined at all points $x = (x_1, x_2)^T$, in the plane. Then the process is *spatially stationary* if, for all x,

$$E[Y(x)] = \mu, \quad \text{var}[Y(x)] = \sigma_Y^2,$$

and, for any pair of points in the plane,

$$\text{cov}[Y(x_i), Y(x_j)] = \sigma_Y^2 c(x_i - x_j). \tag{6.4}$$

The conditions imposed in the two cases are the same *but* they apply for all points x in the continuous case, whereas they apply only at the n sites in the first definition.

Example 6.1
(a) If we were to conduct a study of population growth in major cities, given the locations of the n major cities to be considered, it might be plausible to use definition 6.1. Clearly definition 6.2 would not be acceptable since rural areas (or lakes!) would have totally different population-growth characteristics.
(b) If we were studying rainfall definition 6.2 could be acceptable provided, for example, that the topography was not too variable. This statement still holds even if our data were restricted to n sites at which rain gauges were available, since definition 6.2 implies that definition 6.1 holds for the n sites in question.
(c) Diffusion in space from a single source is an example of a process that is not spatially stationary. Two sites a distance d apart, but equidistant from the source, are likely to be more similar in their degree of reception of the item being diffused at a given time than are two sites distance d apart, but where one is distance d nearer the source than the other. For a more general discussion of diffusion processes, the reader should see Haggett et al (1977, chapter 7) and Cliff et al (1981, chapter 2).

(d) Haining (1981a; 1981b) discusses the patterns of spatial dependence in rural and urban areas. When the study area is wholly rural, interactions may be expected to occur only between neighbouring regions, so that the covariance function shows a steady decay with increasing distance between regions. However, when we consider different towns in an urban hierarchy, central place effects are important and interactions become size-related and function-related, rather than distance-related. Haining (1981b) presents a method for testing for central place effects using the I statistic.

Example 6.2
Bearing in mind that 'near things are more related than distant things', we might define the correlation function in terms of the distance between two sites. For example, the distance between sites i and j according to the usual Euclidean metric is

$$d_{ij} = \{(x_{i1} - x_{j1})^2 + (x_{i2} - x_{j2})^2\}^{1/2} .$$

For sites on a regular grid the Manhattan or city-block metric might be appropriate, to account for topographic and/or directional differences, where

$$d_{ij} = |x_{i1} - x_{j1}| + |x_{i2} - x_{j2}| .$$

Once the distance metric is given, we might use a distance-decay correlation function, such as

$$c(d_{ij}) = \exp\{-\alpha d_{ij}\} , \qquad (6.5)$$

or

$$c(d_{ij}) = (d_{ij} + \gamma)^{-\alpha} , \qquad (6.6)$$

where γ and α are positive constants.

Again, on a regular grid (with steps of unit length along the rows and columns) we might specify the correlation function as

$$c(0) = 1 , \qquad c(1) = \alpha , \qquad c(d) = 0 , \quad \text{for } d > 1 ,$$

where $|\alpha| < 1$. That is

$$\text{cov}(Y_i, Y_j) = \begin{cases} \alpha , & \text{if } i \text{ and } j \text{ are one step apart,} \\ 0 , & \text{otherwise.} \end{cases}$$

This scheme represents a spatial first-order moving average (see section 6.2.5).

If the process is both stationary and direction-invariant, it is said to be *isotropic*. In example 6.2, a process using the Euclidean metric is isotropic, whereas one using the city-block metric is not.

Bearing in mind that we propose to specify models through their mean and covariance properties, we can now proceed to examine the different possible formulations for spatial models.

6.2.1 Trend-surface analysis

If we elect to use a 'reactive' model rather than an 'interactive' one, we often formulate a regression model based upon equations (6.1) and (6.2) with the mean as a function of regressor variables, $x_1, ..., x_k$ say,

$$\mu_i = \mu(x_{i1}, ..., x_{ik}),$$

and a simple covariance structure such as

$$v_{ii} = \sigma^2, \quad \text{and} \quad v_{ij} = 0, \quad \text{for all } j \neq i, \quad i = 1, ..., n.$$

That is, the model is nonstationary in the mean but (trivially) stationary in covariance structure. Discussions of regression analysis are available elsewhere (cf Draper and Smith, 1966; Johnston, 1972) and it is not our intention to pursue the topic here. However, an interesting special case of the technique arises when the regressor variables refer to the spatial coordinates of the sites (site i has coordinates x_{i1} and x_{i2}) so that we may assume a linear form such as

$$\mu_i = \sum_{s=0}^{p} \sum_{r=0}^{q} \beta_{rs} x_{i1}^r x_{i2}^s, \tag{6.7}$$

such that $p+q \leq k$ denotes the highest order polynomial terms. A regression model with mean given by equation (6.7) is known as a *kth-order trend-surface* model. The parameters are estimated by ordinary least squares in the usual way; an early application of the model appears in Student (1914). Orthogonal polynomials (Kendall and Stuart, 1973, pages 372-376) or cosine functions (Fourier series) may be used in place of the polynomial form; for further details see Watson (1971; 1972) and Whitten (1974; 1975).

Trend-surface analysis is a flexible technique for describing trends, although the authors would argue that the level of k should be such as to remove broad trends rather than to pick up very localised variations (which could originate from interactions). An example of the trend-surface method will be given in section 8.8, followed by an analysis of the residuals for spatial autocorrelation. A trend-surface method of a factor analytic type, for Poisson counts rather than normal data, is described in Kooijman (1979).

6.2.2 Autoregressive models

We now turn to models of an 'interactive' type. Here, primary interest focuses upon the covariance structure, either directly or through an autoregressive framework. To see how we may develop such a scheme, it is useful to consider the first-order autoregressive model for the time series Y_t ($t = ..., -1, 0, 1, ...$), which may be described in three equivalent ways (putting $|\rho| < 1$ to ensure stationarity):

(1) $\quad Y_t = \rho Y_{t-1} + \epsilon_t,$ \hfill (6.9)

with

$$E(Y_t) = 0, \quad E(\epsilon_t) = 0, \quad \text{var}(\epsilon_t) = \sigma^2,$$

and

$$\text{cov}(\epsilon_t, \epsilon_{t-s}) = 0, \qquad \text{cov}(Y_{t-s}, \epsilon_t) = 0, \qquad \text{for } s > 0;$$

(2) $\qquad \text{E}(Y_t | y_{t-1}, y_{t-2}, \ldots) = \rho y_{t-1},$ (6.10)

and

$$\text{var}(Y_t | y_{t-1}, y_{t-2}, \ldots) = \sigma^2;$$ (6.11)

(3) $\qquad \text{E}(Y_t) = 0, \qquad \text{and} \quad \text{cov}(Y_t, Y_{t-s}) = \sigma_Y^2 \rho^{|s|},$ (6.12)

where $\sigma_Y^2 = \sigma^2/(1-\rho^2)$.

In this case the three formulations, which we refer to respectively as (1) *simultaneous* autoregressive, (2) *conditional* autoregressive, and (3) *covariance*, are determined readily one from another and each has an intuitive appeal. However, in the spatial case the *multilateral* dependence of the observations has two important repercussions:
first the specification of linear conditional means and constant conditional variances will usually imply multivariate normality (see Kendall and Stuart, 1979, page 375);
second a simple covariance structure does not possess a convenient autoregressive form, or vice versa.
Bearing in mind these rather sobering results, we now look at spatial models defined by analogy with descriptions (1)-(3) above.

6.2.3 Whittle's model

The seminal paper on spatial models is that of Whittle (1954), who proposed a *simultaneous* autoregressive scheme of the form

$$Y_i = \sum_{j \neq i} g_{ij} Y_j + \epsilon_i, \qquad i = 1, \ldots, n,$$ (6.13)

where the $\{\epsilon_i\}$ are uncorrelated error terms with $\text{E}(\epsilon_i) = 0$, and $\text{var}(\epsilon_i) = \sigma_i^2$, $i = 1, \ldots, n$. We have taken $\text{E}(Y_i) = 0$ for all i as it is the covariance structure which is of interest; however, the restriction is easily relaxed. In matrix notation, model (6.13) may be rewritten as

$$Y = GY + \epsilon,$$ (6.14)

where

$$Y = (Y_1, \ldots, Y_n)^\text{T}, \qquad \epsilon = (\epsilon_1, \ldots, \epsilon_n)^\text{T}, \qquad \text{and} \quad G = \{g_{ij}\};$$

G is an $n \times n$ matrix. We may then write

$$\text{E}(\epsilon) = 0, \qquad \text{and} \quad \text{var}(\epsilon) = \Sigma = \begin{bmatrix} \sigma_1^2 & & 0 \\ & \sigma_2^2 & \\ & & \ddots \\ 0 & & & \sigma_n^2 \end{bmatrix},$$

Models for spatial processes

so that Σ is a diagonal matrix. From equation (6.14),
$$Y = (I-G)^{-1}\epsilon ,$$
so that, since $E(Y) = 0$,
$$\mathrm{var}(Y) = V = E(YY^T) = (I-G)^{-1}E(\epsilon\epsilon^T)(I-G^T)^{-1}$$
$$= (I-G)^{-1}\Sigma(I-G^T)^{-1} . \qquad (6.15)$$
No distributional assumptions are necessary to arrive at expression (6.15), but if we do assume normality for ϵ it follows that Y is multivariate normal (MVN) with covariance matrix V, or
$$Y \sim \mathrm{MVN}(0, V) .$$
If $Y^* = Y + \mu$, then $Y^* \sim \mathrm{MVN}(\mu, V)$, so that nonzero means are introduced readily into the original model (6.14).

Example 6.3
Suppose that the n sites lie on a regular grid as in figure 6.1; two subscripts are used, as in the diagram, to describe the relative locations of the sites. An isotropic first-order simultaneous autoregressive scheme would be
$$Y_{ij} = \rho(Y_{i-1,j} + Y_{i+1,j} + Y_{i,j-1} + Y_{i,j+1}) + \epsilon_{ij} ,$$
with modifications at the edges of the study area where necessary and with $|\rho| < \frac{1}{4}$ to ensure stationarity. The matrix G could be written in the form ρW, where the elements of W would be one or zero depending upon whether the two sites concerned were one step apart on the grid or not.

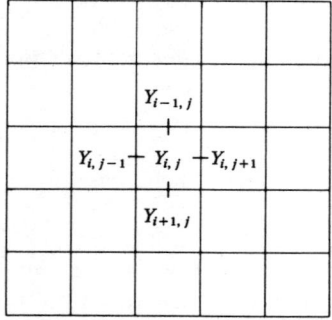

Figure 6.1. Dependencies (shown by joins) among counties in an isotropic first-order simultaneous autoregressive scheme.

6.2.4 The conditional approach

In the manner of Bartlett (1971) and Besag (1974) we may specify the conditional autoregressive model as follows. We let $y_i^* = \{y_j, j \neq i\}$, that is y_i^* denotes y after deletion of y_i. As before we take the $\{Y_i\}$ to have zero means. The model is

$$E(Y_i|y_i^*) = \sum_{j \neq i} g_{ij} y_j \,, \tag{6.16}$$

$$\text{var}(Y_i|y_i^*) = \sigma_i^2 \,, \tag{6.17}$$

for $i = 1, ..., n$. It follows from theorem 6.2 (stated in the chapter appendix) that, when the conditional distributions are each normal, the joint distribution is

$$Y \sim \text{MVN}(0, V)$$

where

$$V^{-1} = D(I - G) \,, \qquad D^{-1} = \Sigma = \begin{pmatrix} \sigma_1^2 & & 0 \\ & \ddots & \\ 0 & & \sigma_n^2 \end{pmatrix}, \qquad \text{and} \quad G = \{g_{ij}\}.$$

Clearly DG must be symmetric, so that equations (6.16) and (6.17) must be specified subject to $\sigma_j^2 g_{ij} = \sigma_i^2 g_{ji}$ for all $i \neq j$.

Example 6.4

As in example 6.3, consider the regular grid shown in figure 6.1. If we let the conditional mean for Y_{ij} depend only and equally upon its four immediate neighbours, and y_{ij}^* denotes y after deletion of y_{ij}, then we have

$$E(Y_{ij}|y_{ij}^*) = \rho(y_{i-1,j} + y_{i+1,j} + y_{i,j-1} + y_{i,j+1}) \,,$$

where $|\rho| < \frac{1}{4}$ to ensure stationarity.

The models in examples 6.3 and 6.4 seem analogous to the *simultaneous* and *conditional* schemes given for time series in equations (6.9) and (6.11). However, they are, in fact, *different* as we shall now show.

For simplicity, let $\sigma_i^2 = \sigma^2$ for all i. Then the *simultaneous* scheme has the covariance matrix,

$$V = \sigma^2 (I - G)^{-1} (I - G^T)^{-1} \,, \tag{6.18}$$

where G need not be symmetric, but the *conditional* scheme has

$$V = \sigma^2 (I - G)^{-1} \,, \tag{6.19}$$

where G must be symmetric to ensure the symmetry of V. These two contrasting forms will be used repeatedly later in this chapter, and in section 6.3 we shall compare the merits and drawbacks of each scheme. At this stage it is worth noting that, in the normal case, any *simultaneous* scheme with defining matrix G_s may be expressed as a *conditional* scheme with defining matrix

$$G_c = G_s + G_s^T - G_s G_s^T \,.$$

Models for spatial processes

Example 6.5 (Besag, 1974)
The simultaneous scheme for a regular grid, developed in example 6.3, may be expressed as a conditional scheme with expectations:

$$(1+4\rho^2)\mathrm{E}(Y_{ij}|y_{ij}^*) = 2\rho(y_{i-1,j}+y_{i+1,j}+y_{i,j-1}+y_{i,j+1})$$
$$- 2\rho^2(y_{i-1,j-1}+y_{i+1,j+1}+y_{i-1,j+1}+y_{i+1,j-1})$$
$$- \rho^2(y_{i-2,j}+y_{i+2,j}+y_{i,j-2}+y_{i,j+2}) \,.$$

These autoregressive linkages are shown in figure 6.2. The reason for the difference between the two models is to be found in the *dependence* between ϵ_i and Y_j in Whittle's model (6.13); these terms are *independent* in the time-series case and in the conditional scheme.

In section 6.2.2, we mentioned that the conditional specification imposes strict conditions upon possible forms of the joint distribution of the $\{Y_i\}$ and, in particular, that equations (6.16) and (6.17) will imply joint normality in general. The most general possible form of the joint distribution determined by the conditional statements is given in a remarkable theorem due to Hammersley and Clifford; this is presented as theorem 6.3 in the chapter appendix. For a recent discussion of these alternatives and their relevance to spatial modelling, see Haining (1979).

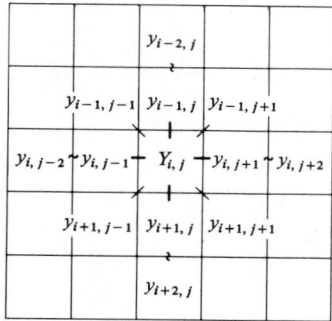

Figure 6.2. Autoregressive (AR) and moving-average (MA) schemes. Under the AR scheme, joins indicated by the heavy lines have weight 2ρ, those indicated by the light lines have weight $-2\rho^2$, and those indicated by the tildes have weight $-\rho^2$. For the MA scheme, the nonzero covariances in example 6.6 are 2θ, $2\theta^2$, and θ^2 for pairs of cells joined by heavy lines, light lines, and tildes, respectively.

6.2.5 Moving-average models
By analogy with time series (Box and Jenkins, 1970, page 67) it is possible to specify moving-average models as follows. We take $\mathrm{E}(Y_i) = 0$ without loss of generality and let

$$Y_i = \epsilon_i + \sum_{j \neq i} g_{ij}\epsilon_j \,, \tag{6.20}$$

where

$$E(\epsilon_i) = 0, \qquad E(\epsilon_i^2) = \sigma_i^2, \qquad \text{and} \qquad E(\epsilon_i \epsilon_j) = 0 \quad \text{if } i \neq j.$$

Equation (6.20) can be rewritten in matrix form as

$$Y = (I+G)\epsilon,$$

whence

$$\text{var}(Y) = (I+G)\Sigma(I+G^T), \tag{6.21}$$

taking $\Sigma = \begin{bmatrix} \sigma_1^2 & 0 \\ 0 & \ddots & \sigma_n^2 \end{bmatrix}$ as before. The process is spatially stationary provided only that $\Sigma g_{ij}^2 < \infty$, $j \neq i$, and is invertible (representable as an autoregressive scheme) if and only if the largest eigenvalue of G is less than one in absolute value.

These properties are given by Haining (1978a), who has provided a comprehensive development of these schemes. When $\sigma_i^2 = \sigma^2$ (for all i), the elements of expression (6.21) may be written as

$$\text{var}(Y) = \sigma^2(I+G)(I+G^T) \tag{6.22}$$

or

$$\text{cov}(Y_i, Y_j) = \sigma^2 \left(\sum_k g_{ik} g_{jk} + g_{ij} + g_{ji} \right), \qquad i \neq j.$$

An alternative to the Haining model may be developed as follows: let

$$Y_i = \epsilon_{ii} + \sum_{j \neq i} g_{ij} \epsilon_{ij}, \qquad \text{where} \quad \epsilon_{ij} = \epsilon_{ji},$$

and

$$E(\epsilon_{ij}) = 0, \qquad \text{var}(\epsilon_{ij}) = \sigma^2 \quad \text{for all } i \text{ and } j, \text{ and}$$
$$E(\epsilon_{ij} \epsilon_{kl}) = 0, \quad \text{unless } (i,j) = (k,l).$$

Then the covariances have the form

$$\text{var}(Y_i) = \sigma^2 \left(1 + \sum_{j \neq i} g_{ij} g_{ji} \right) = d_{ii}, \quad \text{say,}$$

and

$$\text{cov}(Y_i, Y_j) = \sigma^2 g_{ij}.$$

That is, in matrix form,

$$V = \sigma^2 (D+G),$$

where D is the diagonal matrix with elements d_{ii}. As far as we are aware, this version has not been used in spatial studies, but the simpler covariance structure makes it worth further consideration. The covariance matrices for these two schemes have a similar form to the inverse matrices, V^{-1}, for the autoregressive schemes.

Example 6.6
Consider the Haining first-order moving-average scheme on a regular lattice,
$$Y_{ij} = \epsilon_{ij} + \theta(\epsilon_{i-1,j} + \epsilon_{i+1,j} + \epsilon_{i,j-1} + \epsilon_{i,j+1}) .$$
Then

$\text{var}(Y_{ij}) = 1 + 4\theta^2$,

$\text{cov}(Y_{ij}, Y_{i,j\pm 1}) = \text{cov}(Y_{ij}, Y_{i\pm 1,j}) = 2\theta$, $\quad \text{cov}(Y_{ij}, Y_{i\pm 2,j}) = \theta^2$,

$\text{cov}(Y_{ij}, Y_{i\pm 1, j\pm 1}) = 2\theta^2$, $\quad \text{cov}(Y_{ij}, Y_{kl}) = 0$ otherwise.

The pattern of nonzero autocorrelations is as shown in figure 6.2. It is evident that the autocorrelation between adjacent sites is less than or equal to $\frac{1}{2}$, with the maximum occurring at $\theta = 1$. For the alternative scheme to the Haining model,

$\text{var}(Y_{ij}) = 1 + 4\theta^2$,

$\text{cov}(Y_{ij}, Y_{i,j\pm 1}) = \text{cov}(Y_{ij}, Y_{i\pm 1,j}) = \theta$,

so that the maximum autocorrelation is $\frac{1}{4}$, suggesting that higher order schemes will often be necessary.

6.2.6 Regionalised variables

The various processes specified so far all consider a finite set of sites. In many geographical applications, we shall indeed be considering a finite set of sites or areas (spatial aggregates), so the emphasis is justified. However, in other areas of study, notably geology, the spatial process is continuous and the sites are likely to be sampling locations rather than locations of particular interest. When the process is continuous we must revert to the stronger version of spatial stationarity given in definition 6.2. If sampling is done on a regular grid, the models of sections 6.2.2–6.2.5 may be useful but, in general, direct specification of the correlation function, c, in equation (6.4) seems more natural. Often this will involve a precise parametric form such as functions (6.5) or (6.6), although the version suggested by Kooijman (1976) has considerable merit. He suggests setting

$$c(d_{ij}) = \gamma_k , \quad \text{if } D_{k-1} < d_{ij} \leq D_k ,$$

where the intervals $(D_i, D_{i+1}]$ form a suitable partition of

$$(0, D_{\max} = \max_{i,j} d_{ij}] .$$

The assumption of spatial stationarity is especially irksome in the continuous case and it has been relaxed by Matheron (1971) and his coworkers who have developed the theory of regionalised variables; see also Huijbrechts (1975) and Delfiner and Delhomme (1975). The stationarity assumption is replaced by the weaker condition that, for any two sites x_i and x_j distance d apart,

$$E\{[Y(x_i) - Y(x_j)]^2\} = \gamma(d) ;$$

$\gamma(d)$ is known as the *variogram*; see section 5.3.2. The important feature of the variogram is that $\gamma(d)$ does not approach zero as $d \to 0$, because of the inherent local variation of the process (figure 5.3). Another way of expressing the assumption would be to let

$$Y(x) = Z(x) + \delta(x) , \qquad (6.23)$$

where

$$\text{cov}[Z(x_i), Z(x_j)] = \sigma_Z^2 c(d) , \qquad (6.24)$$

with $c(0) = 1$, and let $\delta(x)$ represent a *white-noise process*, so that

$$\text{var}[\delta(x)] = \sigma^2 , \qquad \text{cov}[\delta(x), \delta(x^*)] = 0 \quad \text{for all } x^* \neq x . \qquad (6.25)$$

A continuous white-noise process is not physically realisable, but given that any sample area occupies a finite area, however small, the formulation is mathematically convenient yet still effectively realisable. It follows from expressions (6.24) and (6.25) that

$$\gamma(0) = 0 ,$$

and

$$\gamma(d) = 2\sigma^2 + 2\sigma_Z^2[1 - c(d)] , \qquad d > 0 .$$

The variogram may be modelled by a simple parametric form or by a scheme of the Kooijman-type suggested above. A cubic in d has sometimes been used (Eagleson, 1967; Matheron, 1971) although the restrictions on the ranges of the coefficients should be noted.

6.2.7 Generation of random variables

As we have seen in earlier chapters, it is not possible to develop small-sample results theoretically, so that any study of the sampling properties of estimators for small numbers of sites will rely upon Monte Carlo methods. Given a normal random number generator, available as a standard facility in most computer software, we can generate n independent, identically distributed random deviates, $\boldsymbol{\epsilon} = (\epsilon_1, ..., \epsilon_n)^T$ say. Then, the values for the random variables of the models described in this section may be calculated as follows:

Simultaneous autoregressive

$$y = \sigma(I - G)^{-1}\boldsymbol{\epsilon} .$$

Conditional autoregressive

Let V have the matrix of eigenvalues Λ and eigenvectors H, so that $VH = H\Lambda$. Then set $T = H\Lambda^{\frac{1}{2}}H^T$. Finally, evaluate $y = T\boldsymbol{\epsilon}$.

Moving-average

$$y = \sigma(I + G)\boldsymbol{\epsilon} .$$

Alternative moving-average and general covariance schemes
Evaluate by the same approach as the conditional autoregressive scheme.

Regionalised variables
Use expression (6.24) to generate the $Z(x)$ variates as for the general covariance scheme. Generate a second set of n values for $\delta(x)$ and form $Y(x)$ from equation (6.23).

In all cases, nonzero means, μ, are obtained by setting $y^* = y + \mu$.

6.3 Estimation procedures
For each of the models outlined in section 6.2, there are a variety of possible estimation procedures for the unknown parameters. Therefore, to keep the discussion within reasonable bounds, we shall concentrate upon the simpler models and consider estimation primarily by ordinary least squares or by maximum likelihood. Before discussing the different models in detail, we shall summarise the main findings, so that the reader less concerned with the mathematical results may gain an overview of the current state of the arts. In this summary, it is assumed that the errors are normally distributed.

Trend-surface analysis This method is a special case of regression analysis, so the least squares estimators are unbiased (and fully efficient when the errors are normally distributed).

Simultaneous autoregression The least squares estimators are inconsistent (Whittle, 1954) and there does not appear to be any viable alternative to maximum likelihood, although Ord (1975) has provided a modified least squares method which yields consistent, if inefficient, estimators. The maximum likelihood estimators are difficult to obtain, save in special cases.

Conditional autoregression The least squares estimators are consistent and correspond to the pseudolikelihood estimators (Besag, 1977b) for the normal cases. The maximum likelihood procedure involves computational problems similar to those of the simultaneous autoregressive model.

Moving averages The sample covariances provide simple but inefficient estimators, and maximum likelihood is rather more intractable than before (Haining, 1978a).

Regionalised variables Unless n is small the maximum likelihood estimators are very tedious to compute. However, the sample variogram could be smoothed or, if the variogram has a parametric form, the parameters could be estimated by least squares.

In the rest of the section we shall examine procedures for these models in more detail and then give some examples.

Throughout this chapter, we shall concentrate upon spatially stationary models. This restriction will be lifted in chapters 8 and 9 when we consider regression schemes with spatial components. Of particular interest among the class of nonstationary spatial models are those of the spatial interaction (or 'gravity') type. Stetzer (1977) has examined alternative estimators for these models when the coefficients vary over space, and in addition has developed tests for spatial autocorrelation among the residuals of the interaction model to check for misspecification.

6.3.1 First-order conditional autoregression

Suppose that the observations $y = (y_1, ..., y_n)^T$ are recorded at n sites for variates $Y = (Y_1, ..., Y_n)^T$ which we take to be jointly normally distributed with known (zero) mean and covariance matrix V. Then the likelihood function (cf Kendall and Stuart, 1979, page 38) is

$$L = (2\pi)^{-n/2} |V|^{-1/2} \exp(-\tfrac{1}{2} y^T V^{-1} y) \, .$$

Henceforth we shall consider the log-likelihood, written as

$$\mathcal{L} = \ln L = \text{const} - \tfrac{1}{2} \ln|V| - \tfrac{1}{2} y^T V^{-1} y \, . \tag{6.26}$$

For the autoregressive model, with error terms ϵ which are independent with identical variances σ^2, V is given by equations (6.18) and (6.19) for the simultaneous and conditional schemes respectively. The moving-average scheme then has V given by equation (6.22). We shall now present the detailed working for a special case of the conditional scheme, based upon Moran (1973) and Ord (1975), and then summarise the known results for other cases.

Consider the single-parameter conditional autoregressive model with $V = \sigma^2 (I - \rho W)^{-1}$; that is $G = \rho W$, where W is a matrix of known weights, specified in accordance with a rule such as those given in section 1.4.2. The log-likelihood in equation (6.26) becomes

$$\mathcal{L} = \text{const} - \tfrac{1}{2} n \ln(\sigma^2) + \tfrac{1}{2} \ln|I - \rho W| - \frac{y^T (I - \rho W) y}{2\sigma^2} \, , \tag{6.27}$$

since

$$|\sigma^2 (I - \rho W)| = \sigma^{2n} |I - \rho W| \, .$$

Differentiating $\mathcal{L}$ with respect to σ^2, we obtain

$$\frac{\partial \mathcal{L}}{\partial (\sigma^2)} = -\frac{n}{2\sigma^2} + \frac{y^T (I - \rho W) y}{2\sigma^4} \, .$$

Setting this equal to zero gives the maximum likelihood estimator

$$n \hat{\sigma}^2 = y^T (I - \hat{\rho} W) y \, , \tag{6.28}$$

where the 'hats' denote estimators for the unknown parameters. That is,

$$n(\text{error variance}) = (\text{residual sum of squares}) \, . \tag{6.29}$$

Result (6.29) holds for all autoregressive and moving-average models when the errors have a common variance σ^2. Also, if we let $y^T W^j y = c_j$, then

Models for spatial processes

(6.29) may be written as

$$\hat{\sigma}^2 = c_0 - \hat{\rho}c_1 .$$

When the Y_j have a common unknown mean, μ, the maximum likelihood estimator is

$$\hat{\mu} = (\mathbf{I}^T\hat{\mathbf{V}}^{-1}\mathbf{I})^{-1}\mathbf{I}^T\hat{\mathbf{V}}^{-1}Y ; \qquad (6.30)$$

that is, a generalised least squares estimator. Provided that the lattice is not too small and/or too irregular, the sample mean, $\bar{Y}$, will often be very close to $\hat{\mu}$. The computational procedure for unknown μ is outlined in chapter 9, as a special case of the mixed regression–autoregressive scheme.

To evaluate $\hat{\rho}$, we need to be able to evaluate the determinant in the log-likelihood expression (6.27). Given n sites, the determinant is a polynomial of order n in ρ which, to make things even worse, is not available in analytic form except for a few special cases and/or for n very small. Substituting expression (6.28) back into (6.27) we see that the problem reduces to that of maximising the function

$$\hat{\sigma}^{-2n}|\mathbf{I} - \rho\mathbf{W}| \qquad (6.31)$$

with respect to ρ. Since only a single parameter is involved, a direct search procedure is reasonably efficient computationally, with the value $\tilde{\rho} = I = y^T\mathbf{W}y/y^Ty$ as a starting value.

The following methods for evaluating this troublesome determinant have been proposed:

Method 1 (Mead, 1967) Evaluate the determinant as a polynomial in ρ. This is satisfactory for the small numbers of sites considered in Mead's experimental studies, but not for larger values of n.

Method 2 (Ord, 1975) Let $\lambda_1 \geq \lambda_2 \geq ... \geq \lambda_n$ be the eigenvalues of $\mathbf{W}$. Then

$$|\mathbf{I} - \rho\mathbf{W}| = \prod_{i=1}^{n}(1 - \rho\lambda_i) . \qquad (6.32)$$

Since $\mathbf{W}$ is symmetric, all the eigenvalues are real; it is apparent from equation (6.32) that the eigenvalues of $\mathbf{W}$ need to be evaluated only once and the search for $\hat{\rho}$ may then proceed to evaluate function (6.31) using (6.32). In general, the eigenvalues of $\mathbf{W}$ must be found numerically, but when the sites form a $p \times q$ ($= n$) rectangular grid, and

$$w_{ij} = \begin{cases} 1, & \text{if } i, j \text{ are immediate neighbours,} \\ 0, & \text{otherwise,} \end{cases}$$

then the eigenvalues take the form

$$\lambda_{rs} = 2\cos\left(\frac{r\pi}{p+1}\right) + 2\cos\left(\frac{s\pi}{q+1}\right), \qquad r = 1, ..., p ; \quad s = 1, ..., q .$$

$$(6.33)$$

Other special cases are discussed in the appendix to Ord (1975). As n increases, this approach becomes unwieldy, although approximate results may be obtained by forcing **W** into a block diagonal structure such as

$$\mathbf{W} = \begin{bmatrix} \mathbf{W}_1 & & & 0 \\ & \mathbf{W}_2 & & \\ & & \ddots & \\ 0 & & & \mathbf{W}_k \end{bmatrix}.$$

That is, we group the sites into subsets with no connections between those subsets (disconnected subgraphs). Since only the largest eigenvalues have any material effect upon the value of function (6.31), this procedure should yield an estimate close to the true maximum likelihood value. However, the approximation remains to be tested systematically.

Method 3 (Whittle, 1954) When the sites lie on a rectangular grid so that the first-order scheme is specified as in example 6.4, then the maximum likelihood estimator for ρ may be determined approximately (provided the grid is not too small) by

minimising $\frac{1}{2}\ln k + \ln(\hat{\sigma}^2)$ (6.34)

where

$$\ln k = \sum_{j=1}^{\infty} \frac{1}{j} \binom{2j}{j}^2 \rho^{2j} .$$ (6.35)

It should be noted that Whittle's original derivation applied to the simultaneous scheme, but the same argument holds for the conditional case except that the factor $\frac{1}{2}$ appears as the coefficient of $\ln k$ in objective (6.34).

The sum (6.35) converges fairly rapidly for $|\rho|$ not too near its maximum value of $0 \cdot 25$. In general, a simple yet accurate approximation (Abramowitz and Stegun, 1965, page 592) is to set

$$\ln k = a_0 - a_1 \gamma - a_2 \gamma^2 + (b_1 \gamma + b_2 \gamma^2) \ln \gamma ,$$

where

$\gamma = 1 - 4\rho^2 , \qquad 0 < \gamma \leq 1 ,$

and

$a_0 = 0 \cdot 363\,3802 , \qquad a_1 = 0 \cdot 294\,7645 , \qquad a_2 = 0 \cdot 068\,6156 ,$

$b_1 = 0 \cdot 156\,1455 , \qquad b_2 = 0 \cdot 026\,2603 .$

When the site locations are large in number and irregularly spaced, none of these methods is easy to apply. Fortunately, as a result of theorem 6.4 (see chapter appendix), the method of ordinary least squares may be used. Provided the lattice is not too small, the least squares estimators reduce to simple functions of the autocorrelations, as the following example shows.

Example 6.7
Consider the first-order scheme on a rectangular lattice with

$$E(Y_{ij}|\cdot) = \rho y_{ij}^*,$$

where

$$y_{ij}^* = y_{i-1,j} + y_{i+1,j} + y_{i,j-1} + y_{i,j+1}.$$

The least squares estimator is

$$\hat{\rho}_{LS} = \frac{\sum y_{ij} y_{ij}^*}{\sum (y_{ij}^*)^2}.$$

If we let $s_{km} = \sum y_{ij} y_{i+k, j+m}$, then, for large lattices, $\hat{\rho}_{LS}$ is approximated by

$$\hat{\rho}_{LS} = \frac{s_{10} + s_{01}}{2s_{00} + s_{20} + s_{02} + 2s_{11} + 2s_{1,-1}}. \tag{6.36}$$

The weakness of the least squares estimator is that $\hat{\rho}$ is not constrained to satisfy $|I - \rho W| > 0$. This is a necessary condition for a valid representation by the first-order scheme, since the covariance matrix must be positive definite. However, since the autocorrelations often decay steadily with distance, the intuitively appealing, but downwards biased, estimator (cf section 1.5.2),

$$\tilde{\rho} = \frac{I}{\max|I|} = \text{corr}(y_{ij}, y_{ij}^*) = \frac{s_{10} + s_{01}}{[8s_{00}(2s_{00} + s_{20} + s_{02} + 2s_{11} + 2s_{1,-1})]^{\frac{1}{2}}}, \tag{6.37}$$

will stay within the feasible region even when $\hat{\rho}_{LS}$ does not.

Besag (1974) and Besag and Moran (1975) have developed a method of coding which, in the normal case, reduces to least squares based on a subset of the possible sites. An advantage of this technique is that it always provides unbiased estimators and exact tests of significance; however, the estimators seem to be somewhat less efficient than those given by ordinary least squares based on all sites. For nonnormal schemes, the pseudolikelihood scheme of Besag (1975) seems the most efficient viable scheme currently available. In general, if $p(y_i|y_i^*, \theta)$ denotes the probability (density) for the nonnormal process, then the pseudo-log-likelihood for the parameter θ is given by

$$\mathcal{L}_p = \sum_{i=1}^{n} \ln p(y_i|y_i^*, \theta).$$

The maximum likelihood estimators may then be obtained by appropriate numerical procedures.

Example 6.8
Haining (1978a) reports a study of crop yields in the high plains area of South Nebraska and North Kansas. The study area comprises forty-two counties, all between 2100 and 2600 square kilometres in size. The spatial pattern of county midpoints closely approximates a 7 × 6 rectangular lattice and will be considered as such. The autocorrelations for wheat yields for the years 1964 and 1969 are shown in table 6.1(a). The autocorrelations decay steadily with distance, although there is some evidence of anisotropy. Nevertheless, a first-order autoregressive scheme seems not unreasonable as an initial model.

From table 6.1(b), we see that $\hat{\rho}_{LS}$ is unacceptable because it exceeds the maximum feasible value of 0·25. The ad hoc estimator, $\tilde{\rho}$, appears to be downwards biased relative to the maximum likelihood estimator which is, however, close to its upper limit. Haining shows that the I values for the residuals are close to zero, suggesting that the model accounts satisfactorily for the first-order autocorrelation. This use of I does not provide an exact test (see section 9.5), but serves as a guideline.

Table 6.1. (a) Autocorrelations at various spatial lags in east-west (s) and north-south (t) directions for wheat yields (bushels/acre) on farms in Southern Nebraska and Northern Kansas, 1964-9. (b) Summary statistics for conditional and simultaneous autoregressive schemes fitted to the data. (Source: Haining, 1978a, pages 220-221.)

(a) *Autocorrelation for*		1964			1969		
		s			*s*		
		0	1	2	0	1	2
	2	0·124	−0·002	−0·110	0·019	−0·146	−0·370
	1	0·533	0·397	−0·014	0·501	0·262	0·025
t	0	1·0	0·450	0·190	1·0	0·536	0·194
	−1	0·533	0·249	0·260	0·501	0·498	0·233
	−2	0·124	0·290	0·265	0·019	0·132	0·191

(b) *Summary statistics for*	1964		1969	
Conditional scheme:	Estimate	SE	Estimate	SE
ML($\hat{\rho}$)	0·23	0·030	0·24	0·020
$\tilde{\rho}$	0·18	—	0·19	—
$\hat{\rho}_{LS}$	0·27	—	0·28	—
SE of ML residuals	2·99		2·44	
Simultaneous scheme:				
ML($\hat{\rho}$)	0·20	0·025	0·20	0·025
SE of ML residuals	2·84		2·33	

6.3.2 First-order simultaneous autoregression

The results for the simultaneous scheme follow by the same line of argument as given in section 6.3.1. In place of equation (6.27), the log-likelihood function becomes

$$\mathcal{L} = \text{const} - \tfrac{1}{2}n \ln(\sigma^2) + \ln|\mathbf{I} - \rho\mathbf{W}| - \tfrac{1}{2}y^T(\mathbf{I} - \rho\mathbf{W}^T)(\mathbf{I} - \rho\mathbf{W})y \ , \qquad (6.38)$$

whence

$$n\hat{\sigma}^2 = y^T(\mathbf{I} - \rho\mathbf{W}^T)(\mathbf{I} - \rho\mathbf{W})y \ ,$$

or

$$\hat{\sigma}^2 = c_0 - 2\rho c_1 + \rho^2 c_2 \ . \qquad (6.39)$$

The reduced form of the likelihood, corresponding to function (6.31), becomes

$$\hat{\sigma}^{-2n}|\mathbf{I} - \rho\mathbf{W}|^2 \ , \qquad (6.40)$$

so that the Mead and Ord approaches may be used as before. Finally, Whittle's approximation now involves the minimisation of

$$\ln k + \ln(\hat{\sigma}^2) \ ,$$

where $\ln k$ is given by expression (6.35).

Unfortunately, as noted by Whittle (1954), the *least squares estimators are inconsistent for the simultaneous model*; see the corollary to theorem 6.4 in the chapter appendix.

Example 6.9
Returning to example 6.8, we can now try a first-order simultaneous scheme. From the results given in table 6.1, we see that the standard error of the maximum likelihood residuals is lower than for the conditional scheme and $\hat{\rho}$ is further away from the boundary value of $0 \cdot 25$. As was noted in example 6.5, the simultaneous scheme may be interpreted as a special case of a third-order conditional model, and it would seem that this hidden flexibility has helped to achieve a better fit in the present case.

Haining (1977) reports the results of an extensive Monte Carlo study of the small-sample properties of the sample autocorrelation function, which could be used to identify an appropriate form of model. The sample autocorrelation function is found to have a useful role to play, although it is suggested that the analysis be supplemented by more formal procedures, such as a likelihood-ratio test (cf section 6.4.2). In a later paper (Haining, 1978b) the same author examines the small-sample properties both of least squares and of maximum likelihood estimators. The combined effects of the small-sample size and boundary effects can lead to a downward bias in the maximum likelihood estimates, although the least squares approach tends to yield overestimates. For values of ρ near the boundary of the parameter space, Haining shows that the least squares procedure may yield infeasible estimates most of the time.

For very small lattices, Ross-Parker (1975) found that the empirical distribution of the maximum likelihood estimator was bimodal. However, this bimodality seems to disappear quickly as the sample size increases.

6.3.3 Higher order autoregressive schemes

Relatively little research has been undertaken for models involving more than one parameter. For the two-parameter conditional scheme

$$E(Y_{ij}|y_{ij}^*) = \alpha(y_{i-1,j} + y_{i+1,j}) + \beta(y_{i,j-1} + y_{i,j+1}) \,,$$

we may use Whittle's approach as outlined in functions (6.34) and (6.35), except that the function is now minimised with respect both to α and to β, and $\ln k$ is evaluated with $\rho^2 = \alpha\beta$. The same approach holds for the comparable two-parameter simultaneous scheme, making the appropriate modifications described in section 6.3.2. More general models for regular grids may be tackled by this method, although the evaluation of $\ln k$ rapidly becomes intractable.

For several-parameter *conditional* schemes, the ordinary least squares approach again comes to our aid (theorem 6.4 in the chapter appendix). That is, if the covariance matrix is of the form

$$\mathbf{V} = \sigma^2 \left(\mathbf{I} - \sum_{j=1}^{k} \rho_j \mathbf{W}_j \right)^{-1} \,,$$

then consistent estimators are found by setting $Z_j = \mathbf{W}_j Y$ and fitting by least squares. The estimators are given by the equations

$$(Y - \rho_1 Z_1 - \ldots - \rho_k Z_k)^T Z_j = 0 \,, \qquad j = 1, \ldots, k \,. \tag{6.41}$$

For the simultaneous scheme, ordinary least squares yields *inconsistent* estimators. When

$$\mathbf{V} = \sigma^2 [(\mathbf{I} - \sum \rho_j \mathbf{W}_j)(\mathbf{I} - \rho_j \mathbf{W}_j^T)]^{-1} \,,$$

consistent, but rather inefficient, estimators are given by the modified least squares equations (following the approach of Ord, 1975)

$$(Y - \rho_1 Z_1 - \ldots - \rho_k Z_k)^T \mathbf{W}_j Z_j = 0 \,, \qquad j = 1, \ldots, k \,; \tag{6.42}$$

where $Z_j = \mathbf{W}_j Y$ as before.

Maximum likelihood estimation for the case $k = 2$ has been studied by Hepple (1976) and by Brandsma and Ketellapper (1979). When $\mathbf{W}_1$ and $\mathbf{W}_2$ are not orthogonal (that is, when $\mathbf{W}_1 \mathbf{W}_2 \neq \mathbf{0}$) a substantial additional computational effort is needed; the reader is referred to Hepple's paper for a full specification of the estimation procedure and details of the numerical methods employed.

6.3.4 Moving-average models

The first-order moving-average scheme with equal error variances has the covariance matrix, from equation (6.21), with $\mathbf{G} = \theta \mathbf{W}$,

$$\text{var}(Y) = \sigma^2 (\mathbf{I} + \theta \mathbf{W})(\mathbf{I} + \theta \mathbf{W}^T) \,. \tag{6.43}$$

Models for spatial processes

From the work of Cliff and Ord (1975b) and Haining (1978a), when **W** is symmetric the log-likelihood expression (6.26) may be represented as

$$\mathcal{L} = \text{const} - \ln|\mathbf{I}+\theta\mathbf{W}| - \frac{n\ln(\sigma^2)}{2} - \frac{1}{2\sigma^2}y^T[(\mathbf{I}+\theta\mathbf{W})(\mathbf{I}+\theta\mathbf{W}^T)]^{-1}y$$

$$= \text{const} - \sum_{i=1}^{n}\ln(1+\theta\lambda_i) - \frac{n\ln(\sigma^2)}{2} - \frac{1}{2\sigma^2}\sum_{i=1}^{n}[z_i(1+\theta\lambda_i)^{-1}]^2 , \quad (6.44)$$

where $z = \mathbf{H}y$, and $\mathbf{H}$ is the matrix of eigenvectors corresponding to **W**; that is $\mathbf{WH} = \mathbf{H}\Lambda$, and Λ is the diagonal matrix of eigenvalues $(\lambda_1, ..., \lambda_n)$ of **W**. The estimator for σ^2 is given by formulation (6.29), and θ can be evaluated from the reduced form of equation (6.43) by a direct search procedure; see Haining (1978a) for further computational details and a comparison with the least squares approach. For the alternative model,

$$\text{var}(Y) = \sigma^2(\mathbf{D}+\theta\mathbf{W}) ,$$

which will reduce to $\sigma^2(\mathbf{I}+\theta\mathbf{W})$ when all the d_{ii} are equal. The maximum likelihood estimators may be developed from an expression similar to (6.44), and simple estimators based upon the covariances are available in each case by solving the equations given in example 6.6 in terms of θ and σ^2.

Table 6.2. (a) Autocorrelations at various spatial lags in east-west (s) and north-south (t) directions for population density (persons/square mile) in Southern Nebraska and Northern Kansas, 1960-70. (b) Maximum likelihood estimates for the moving-average scheme fitted to population data of example 6.10 and wheat yield data of example 6.8. (Source: Haining, 1978a, pages 220-221.)

(a) Autocorrelations for		1960			1970		
		s			s		
		0	1	2	0	1	2
	2	0.098	0.148	0.033	0.049	0.108	0.029
	1	0.213	0.017	0.096	0.137	−0.043	0.077
t	0	1.0	0.211	0.036	1.0	0.152	0.017
	−1	0.213	0.323	−0.023	0.137	0.337	−0.045
	−2	0.098	0.142	−0.002	0.049	0.078	−0.048

(b) *Maximum likelihood estimates*

Population density	Means	LS estimate	ML estimate	SE	SE of residuals
1960	9.37	0.11	0.09	0.05	7.07
1970	8.72	0.05	0.06	0.05	7.32
Wheat yields					
1964	19.74	0.43	0.19	0.03	3.41
1969	29.03	a	0.16	0.04	2.84

[a] Equation (6.45) does not possess real roots.

An (inefficient) moments estimator is given by the first-order autocorrelation. Thus for the first-order isotropic scheme we may solve (see example 6.6)

$$\frac{2\hat{\theta}}{1+4\hat{\theta}^2} = \frac{s_{10}+s_{01}}{2s_{00}} . \qquad (6.45)$$

When the right-hand side of equation (6.45) exceeds $0 \cdot 5$, the equation cannot be solved.

Example 6.10
Haining (1978a) gives the autocorrelation function for population density in 1960 and 1970 over the study area mentioned in example 6.8. This is reproduced in table 6.2. Outside $s, t = \pm 1$, the values are very small and a first-order moving-average scheme such as (6.20) with covariance matrix (6.22) seems appropriate. The results of fitting this scheme both to the population density and to the wheat data are given in table 6.2. The moments estimators seem reasonable for the population data. However, they appear to overestimate grossly the parameter for wheat yields in 1964, and no estimate is available for 1969. Comparison of the residual standard errors in table 6.2 with those given in table 6.1 suggests that the autoregressive scheme fits somewhat better than the moving-average scheme for these data.

6.3.5 Regionalised variables
If we follow Kooijman's (1976) suggestion (as in section 6.2.6) and set $c(d_{ij}) = \gamma_k$ for $D_{k-1} < d_{ij} \leqslant D_k$, then the variogram may be estimated as a step-function with

$$\hat{\gamma}_k = \frac{1}{n_k} \sum_{J_k} (y_i - y_j)^2 .$$

Here $J_k = \{(i, j) \text{ such that } D_{k-1} < d_{ij} \leqslant D_k\}$, and n_k denotes the number of pairs of subscripts in J_k. A more general class of estimators is available as

$$\hat{\gamma}(d) = \frac{\sum f(d-d_{ij})(y_i - y_j)^2}{\sum f(d-d_{ij})} , \qquad (6.46)$$

where $f(d-d_{ij})$ is a weighting function. The optimal form of f is unknown, but clearly it should have a maximum at $d-d_{ij} = 0$ and decline with increasing $|d-d_{ij}|$. A simple choice such as

$$f(u) = \exp(-\theta |u|)$$

seems worthy of exploration, with θ a constant set *a priori*. A very large value for θ will produce a lumpy $\hat{\gamma}$, whereas too small a value will produce a flat, very biased form for $\hat{\gamma}$. Further work is needed to assess the value of such a procedure.

If a functional form is specified for $\gamma(d)$ in expression (6.24), maximum likelihood can be used. However, the need for inversion of the covariance matrix at each stage of the iterative search would make the method very time-consuming unless n was small. Another *ad hoc* possibility would be to minimise

$$\sum_i \sum_j [(y_i - y_j)^2 - \gamma(d)]^2 \qquad (6.47)$$

with respect to the unknown parameters $(\theta_0, \theta_1, ..., \text{say})$. A restricted polynomial such as

$$\gamma(d) = \begin{cases} \theta_0 + \theta_1 d + \theta_2 d^2 + \theta_3 d^3, & d \leq d_0, \\ \gamma(d_0), & d > d_0 \end{cases} \qquad (6.48)$$

has been suggested by Eagleson (1967) and Matheron (1971), whereas an exponential function such as

$$\gamma(d) = \theta_0 - \theta_1 \exp(-\theta_2 d) \qquad (6.49)$$

could be fitted by a nonlinear least squares procedure.

6.4 Choice of tests for spatial autocorrelation

The purpose of this section and the next is to undertake a study of the various tests for spatial autocorrelation to help us decide which test should be used in given circumstances. Much of the material is drawn from Cliff and Ord (1975c), which should be consulted for further details.

The general question that may be asked is 'which test or combination of tests is best?' Naturally, this implies some criterion by which 'best' can be assessed. The standard approach, due to Neyman and Pearson, is to consider the *power* of a test, which is defined as 'the probability of rejecting the null hypothesis, H_0, when it is false'. That is,

power = 1 − probability(type II error) .

The power is examined at given levels of the *size* of the test, where size is defined as 'the probability of rejecting the null hypothesis, H_0, when it is true'. That is,

size = probability(type I error) .

Unfortunately, for the problem in hand, small-sample results cannot be obtained by analytical methods. Therefore three alternative approaches have been used.

(1) *Asymptotic relative efficiency (ARE)*
Instead of computing the full power curve, we can assess the local efficiency of the test as the parameter value(s), under the alternative hypothesis H_1, approach those specified under H_0, and the sample size n goes to infinity. This measure was first used by Pitman (1948) and is reviewed briefly in section 6.4.1. For a fuller discussion, see Kendall and Stuart (1979, chapter 25).

(2) *Field trials*
In section 6.5.1 results are reported for two sets of real data, and the performances of the various statistics are compared.
(3) *Monte Carlo studies*
In section 6.5.2 the power functions of the test statistics are evaluated for several lattices, allowing small-sample comparisons to be made.

When testing time-series data for dependence, the first-order Markov model is a well-established alternative hypothesis, although other orders may be important, as illustrated by the Wallis (1972) procedure for testing for fourth-order (quarterly) serial correlation. In the spatial case, the choice of alternative may be less obvious, particularly when the county system is irregular. If the spirit of the investigation is data-analytic, then the investigator should check for autocorrelation at different spatial scales using the correlogram described in section 5.2. However, if it is desired to test spatial dependence at a particular spatial scale, then we must specify an alternative hypothesis and use one of the statistics introduced in chapter 1.

6.4.1 Asymptotic relative efficiency

Suppose that we wish to test the null hypothesis, H_0: $\psi = 0$, against the alternative hypothesis, H_1: $\psi \neq 0$, where ψ is some parameter. Let h denote any test statistic under consideration and suppose that, without loss of generality, the expected value of h under H_0 is zero; that is

$$E(h|H_0) = 0 \ . \tag{6.50}$$

Further, we shall assume that the variance of h exists, defined as

$$\text{var}(h|H_0) = \omega^2 \ , \quad \text{say,} \tag{6.51}$$

and that h is asymptotically normally distributed. Given certain conditions which are satisfied in all the cases we shall consider (see Kendall and Stuart, 1979, pages 284-286), we may examine alternative hypotheses for which h is asymptotically normally distributed and

$$E(h|H_1) = \psi b + O(\psi^{1+\delta}) \ , \tag{6.52}$$

where $\delta > 0$, O denotes terms of this order or higher in ψ, and b is some function independent of ψ. Then as $n \to \infty$, if we consider alternatives of the form $\psi \propto n^{-\frac{1}{2}}$, the asymptotic *efficacy* of a test based on h is defined as

$$F(h) = \frac{b^2}{\omega^2} \ . \tag{6.53}$$

If we consider two competing test statistics, h_1 and h_2 say, then the *asymptotic relative efficiency* (ARE) of the test based on h_1 to one based on h_2 is defined as

$$\text{ARE}(h_1, h_2) = \frac{F(h_1)}{F(h_2)} \ . \tag{6.54}$$

ARE is a useful measure because it is related to power, as defined in section 6.4, in the following way.

Let $P(h, \psi)$ be the power of a test based on h when the parameter has the value ψ under H_1. Then

$$\mathrm{ARE}(h_1, h_2) = \lim_{n \to \infty} \left[\frac{P(h_1, \psi)}{P(h_2, \psi)} \right]^{1/2}, \qquad (6.55)$$

where $\psi \propto n^{-1/2}$.

Thus the ARE is a measure of the comparative asymptotic local power of a test. This may seem to be a measure of strictly limited value but, in practice, the ARE of two tests often gives a fair guide to their relative power for alternatives not too far from the null hypothesis.

In order to make use of these ideas for tests of spatial autocorrelation, it is assumed that the data are drawn from normal populations (that is, assumption N of sections 1.3.2 and 2.3). We specify a first-order scheme as the alternative hypothesis H_1; that is, for the simultaneous scheme,

$$Y = \rho DY + \epsilon, \qquad (6.56)$$

and, for the conditional scheme,

$$E(Y_i | y_i^*) = \rho \sum_{j \neq i} d_{ij} y_j; \qquad (6.57)$$

the normal distribution being assumed in each case. The means are taken to be zero (known), which will not affect the asymptotic results. The weighting matrix under H_1 is specified as $\mathbf{D} = \{d_{ij}\}$.

Example 6.11

Observations are taken on a regular square grid and the researcher postulates a north-south pattern of spatial autocorrelation under hypothesis H_1 (figure 6.3). Then the appropriate specification for the weights is

$$d_{ij} = \begin{cases} \tfrac{1}{2}, & \text{if county } j \text{ is due north or due south of county } i \text{ and contiguous to it,} \\ 0 & \text{otherwise.} \end{cases}$$

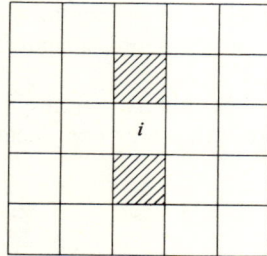

▨ autocorrelated with i under hypothesis H_1

Figure 6.3. Lattice used in example 6.11.

6.4.2 The likelihood ratio test

Let us assume that $Y_1, \ldots, Y_n$ are normally distributed with means zero and covariance matrix $\mathbf{V}$. Then, under the null hypothesis, H_0, $\mathbf{V} = \sigma^2 \mathbf{I}$, and the log-likelihood function is

$$\mathcal{L}_0(\sigma) = \text{const} - n \ln \sigma - \frac{y^T y}{2\sigma^2}.$$

For the alternative hypothesis, H_1, we may use either the simultaneous scheme or the conditional scheme, as outlined previously. Thus, under H_1, we have the log-likelihood

$$\mathcal{L}_1(\sigma, \rho) = \text{const} - n \ln \sigma - \tfrac{1}{2} \ln |\mathbf{C}| - \frac{y^T \mathbf{C}^{-1} y}{2\sigma^2},$$

where $\mathbf{V} = \sigma^2 \mathbf{C}$, and $\mathbf{C}^{-1}$ is given by

$$\mathbf{C}^{-1} = (\mathbf{I} - \rho \mathbf{D})^T (\mathbf{I} - \rho \mathbf{D}), \quad \text{simultaneous}, \tag{6.58}$$

$$\mathbf{C}^{-1} = (\mathbf{I} - \rho \mathbf{D}), \quad \text{conditional}. \tag{6.59}$$

If we consider the alternative $H_1: \rho = \rho_1 \neq 0$, the likelihood ratio test rejects hypothesis H_0 in favour of H_1 when

$$\mathcal{L}_1(\hat{\sigma}_1, \rho_1) - \mathcal{L}_0(\hat{\sigma}_0) > k, \tag{6.60}$$

where k is a suitable constant and $\hat{\sigma}_0$, $\hat{\sigma}_1$ represent the maximum likelihood estimators under H_0 and H_1 respectively. In each case, the criterion reduces to the decision rule

'reject H_0 when $\hat{\sigma}_1^2 / \hat{\sigma}_0^2$ is too small'.

For each scheme

$$n \hat{\sigma}_1^2 = y^T \mathbf{C}^{-1} y,$$

so that the test statistic takes the form

$$\frac{y^T \mathbf{C}^{-1} y}{y^T y}. \tag{6.61}$$

For the conditional scheme, inserting equation (6.59) into this expression yields $1 - \rho_1 (y^T \mathbf{D} y / y^T y)$, so that use of the I statistic corresponds to the likelihood ratio procedure, for fixed ρ_1. For the simultaneous scheme, this is true only if we ignore terms of order ρ_1^2 in $\mathbf{C}$.

If we consider the more general alternative hypothesis with ρ and σ^2 both unknown, then the likelihood ratio procedure is the more complicated form

'reject H_0 if λ, which is equal to $(\hat{\sigma}_1 / \hat{\sigma}_0)^n |\hat{\mathbf{C}}|^{-1}$, is too small', (6.62)

where $\hat{\mathbf{C}}$ is given either by expression (6.58) or by expression (6.59), as appropriate, and ρ_1 is replaced by $\hat{\rho}$. For the simultaneous scheme, the likelihood ratio statistic has been examined by Haining (1977) and we shall consider its properties in more detail in section 6.5.2. However, for the conditional scheme, we can see that λ is a function of I; indeed I and $\hat{\sigma}_1^2$ are jointly sufficient for ρ and σ^2 under H_1. It may be shown that λ is a monotone (decreasing) function of I, so that *I is the likelihood ratio statistic under the conditional scheme.* In agreement with Anderson (1948), we note that a uniformly most powerful test is available only for one-sided alternatives; in this case it is the LR test.

Given the likelihood ratio statistic as a benchmark, we can now proceed to examine the asymptotic relative efficiencies of the test statistics developed earlier in the book.

6.4.3 Comparison of the I and c statistics

Consider the statistic

$$h = \frac{y^T T y}{y^T y}, \tag{6.63}$$

where $\mathbf{T}$ is a matrix of constant coefficients $\{t_{ij}\}$. If $S_0 = \Sigma_{(2)} w_{ij} = n$, then I corresponds to h with $\mathbf{T} = \mathbf{W}$, while for c we set $\mathbf{T} = \Omega - \mathbf{W}$, where Ω is a diagonal matrix with nonzero elements, $\Omega_{ii} = \frac{1}{2} \sum_j (w_{ij} + w_{ji})$. Using the methods of section 2.3, we find that, under hypothesis H_0, and assumption N,

$$E(h|H_0) = \frac{1}{n} \sum_{i=1}^{n} t_{ii} = \frac{1}{n} \operatorname{tr}(\mathbf{T}),$$

and

$$\operatorname{var}(h|H_0) = \frac{1}{n^2(n+2)} \{n \operatorname{tr}[\mathbf{T}(\mathbf{T}+\mathbf{T}^T)] - 2[\operatorname{tr}(\mathbf{T})]^2\}.$$

Further, Cliff and Ord (1975c) have shown that under the alternative hypothesis H_1,

H_1: simultaneous model with $\mathbf{V}^{-1} = (\mathbf{I} - \rho \mathbf{D})^T (\mathbf{I} - \rho \mathbf{D})$, (6.64)

the expected value of h is given by

$$E(h|H_1) = \frac{1}{n} \operatorname{tr}(\mathbf{T}) + \frac{\rho}{n} \operatorname{tr}(\mathbf{D}^T \mathbf{T} + \mathbf{TD}) + O(n^{-1}),$$

recalling that ρ is $O(n^{-\frac{1}{2}})$; $O(n^{-1})$ denotes 'terms of this order or less in n'. Then the efficacy, defined by equation (6.53), is

$$F(h) = [\operatorname{tr}(\mathbf{D}^T \mathbf{T} + \mathbf{TD})]^2 \bigg/ \left\{ \operatorname{tr}(\mathbf{T}^2 + \mathbf{TT}^T) - \frac{2}{n}[\operatorname{tr}(\mathbf{T})]^2 \right\}, \tag{6.65}$$

ignoring terms of lower orders in n.

As noted previously, the efficacy permits us to compute AREs between pairs of statistics. First we consider the effect of two different weighting matrices, $\mathbf{W}_1$ and $\mathbf{W}_2$, upon the ARE of the I statistic. Suppose that I_1

has the weighting matrix $T_1 = W_1 = D$, and that I_2 has the weighting matrix $T_2 = W_2 \neq D$. From equation (6.65), the efficacy for I_1 is

$$F(I_1) = \frac{[\text{tr}(DD^T + D^2)]^2}{\text{tr}(DD^T + D^2)}, \qquad (6.66)$$

whereas for I_2, the efficacy is

$$F(I_2) = \frac{[\text{tr}(D^T W_2 + W_2 D)]^2}{\text{tr}(W_2 W_2^T + W_2^2)}.$$

Thus, the ARE is given by

$$\text{ARE}(I_1, I_2) = \frac{F(I_1)}{F(I_2)} = \frac{\text{tr}(DD^T + D^2)\,\text{tr}(W_2^T W_2 + W_2^2)}{[\text{tr}(D^T W_2 + W_2 D)]^2}. \qquad (6.67)$$

This expression is equivalent to

$$\frac{\sum_{(2)}[w_{ij}(1)]^2 \sum_{(2)}[w_{ij}(2)]^2}{\left[\sum_{(2)} w_{ij}(1) w_{ij}(2)\right]^2}, \qquad (6.68)$$

where $w_{ij}(k)$ is the (i, j)th element of W_k ($k = 1, 2$). By the Cauchy-Schwartz inequality, expression (6.68) cannot be less than one; that is,

$$\text{ARE}(I_1, I_2) \geq 1. \qquad (6.69)$$

We conclude, therefore, that the best I statistic uses the weighting matrix W set equal to D to test the null hypothesis H_0 against the alternative hypothesis specified by equation (6.64). Thus *the investigator should choose, a priori, W to represent the autocorrelation pattern he hypothesises under H_1* (that is, so that $W = D$). Result (6.67) measures the penalty paid if some other W matrix is specified. Exactly the same argument holds for the c statistic.

Example 6.12
If we return to example 6.11, given in section 6.4.1, an appropriate choice of weights was $w_{ij}(1) = d_{ij} = \frac{1}{2}$, if county j was a neighbour due north or south of county i, and $w_{ij}(1) = d_{ij} = 0$, otherwise. However, suppose that the researcher decides to use the weights

$$w_{ij}(2) = \begin{cases} \frac{1}{2}(1-\alpha), & \text{if county } j \text{ is a neighbour due north or south of } i, \\ \frac{1}{2}\alpha, & \text{if county } j \text{ is a neighbour due east or west of } i, \\ 0, & \text{otherwise.} \end{cases}$$

Then the ARE of a test based on I using the second set of weights, compared to a test based on the first set is

$$\text{ARE}(I_2, I_1) = \frac{1}{\text{ARE}(I_1, I_2)} = \frac{(1-\alpha)^2}{(1-\alpha)^2 + \alpha^2},$$

since

$$\sum_{(2)}[w_{ij}(1)]^2 = \tfrac{1}{2}n, \qquad \sum_{(2)}[w_{ij}(2)]^2 = \tfrac{1}{2}n[(1-\alpha)^2 + \alpha^2],$$

and

$$\sum_{(2)} w_{ij}(1) w_{ij}(2) = \tfrac{1}{2} n(1-\alpha).$$

Clearly $\mathrm{ARE}(I_2, I_1)$ has a maximum value of one when $\alpha = 0$, and falls from one to zero as α increases from zero to one. If the rook's case ($\alpha = \tfrac{1}{2}$) is used, then $\mathrm{ARE} = \tfrac{1}{2}$. In this case the pattern of weights appears 'half-correct' and the ARE result is intuitively reasonable.

By the same approach it can be shown that, for any lattice, the efficacy for the c statistic with $\mathbf{W}$ set equal to $\mathbf{D}$ is given by

$$F(c) = \frac{[\mathrm{tr}(\mathbf{D}\mathbf{D}^T + \mathbf{D}^2)]^2}{\mathrm{tr}(\mathbf{D}\mathbf{D}^T + \mathbf{D}^2) + 2\mathrm{tr}(\Omega^2) - 2n} \cdot \tag{6.70}$$

Since Ω is a diagonal matrix with elements $\Omega_{ii} = \tfrac{1}{2}\sum_j (w_{ij} + w_{ji})$, it follows that

$$\mathrm{tr}(\Omega^2) = \tfrac{1}{4}\Sigma(w_{i.} + w_{.i})^2 = \tfrac{1}{4} S_2.$$

Thus from equations (6.66) and (6.70) we have

$$\mathrm{ARE}(c, I) = \frac{1}{1+\xi}, \tag{6.71}$$

where $\xi = (S_2 - 4n)/2S_1$.

It is easily shown that $S_2 \geqslant 4n$, with equality only when $w_{i.} = w_{.i} = 1$, for all $i = 1, \ldots, n$. Therefore, for regular lattices (including time series), ξ is near, or equal to, zero for all n, and always converges to zero as n increases. However, this need not happen for a lattice with irregular weights. (In this context, a regular lattice is one which has the same pattern of weights for every county, except possibly at the boundary.)

Example 6.13
(a) Consider a lattice broken up into blocks of size three as in figure 6.4. In each block of three define the weights as $w_{12} = 1$, $w_{21} = w_{23} = \tfrac{1}{2}$, $w_{32} = 1$, and $w_{ij} = 0$ for all other pairs. If there are m ($= \tfrac{1}{3}n$) blocks, it follows that $S_0 = 3m$, $S_1 = 4 \cdot 5m$, and $S_2 = 13 \cdot 5m$. Therefore $\xi = \tfrac{1}{6}$, and $\mathrm{ARE}(c, I) = \tfrac{6}{7}$ for any m.
(b) For the Eire county system shown in figure 8.1 it may be demonstrated that similar calculations yield $\mathrm{ARE}(c, I) = 0 \cdot 897$.

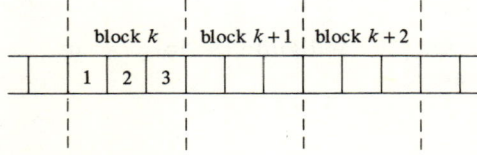

Figure 6.4. Lattice used in example 6.13.

Although such comparisons can be justified only for $n \to \infty$, the following conclusion appears reasonable on the basis of the ARE results: *that the I test is generally better than the c test, although the margin of advantage may be slight.*

For the likelihood ratio statistic λ, $-2\ln\lambda$ is asymptotically distributed as χ^2 with one degree of freedom, so the efficacy of λ may be evaluated using the approach of Kendall and Stuart (1979, pages 284-286). It may be shown that

$$\text{ARE}(I, \lambda) = 1 ,$$

so that the I statistic is asymptotically fully efficient. An alternative derivation of this result, using Lagrange multiplier tests, is given by Burridge (1980). As we have seen already, λ is a monotone function of I for the conditional scheme, so that the two statistics yield equivalent results. However, for the simultaneous scheme, we can expect that the small-sample power of I may be less than that of λ, since λ and I then become increasingly different as $|\rho|$ increases. The apparent difference between small-sample and large-sample results derives from the definition of ARE, which allows $\rho \to 0$ as $n \to \infty$, so that terms in ρ^2 are ignored.

6.4.4 The join-count statistics

Instead of using the I or c statistics we could, of course, use one of the join counts. Three major questions arise.
(1) What is the best choice of weights?
(2) What is the best choice of p, where $p = $ prob(county coded black)?
(3) Should the BB or BW test be used?

Let us assume first that p is known, and so the free sampling model is appropriate. The moments under the null hypothesis H_0 are given in equations (1.25)-(1.28), so to calculate the efficacy we need the expected values under the alternative hypothesis, H_1. Let us suppose that Y_i and Y_j are normally distributed with zero means, unit variances, and correlation specified by equation (6.15) with $\mathbf{G} = \rho\mathbf{D}$. Let county i be coded B if $Y_i \geqslant a$, and W otherwise ($i = 1, ..., n$). If

$$p = P(Y_i \geqslant a) = \int_a^\infty (2\pi)^{-\frac{1}{2}} \exp(-\tfrac{1}{2} y_i^2) dy_i , \qquad i = 1, ..., n ,$$

it follows that

$$P(Y_i \geqslant a, Y_j \geqslant a) = p^2 + \rho d_{ij} [f(a)]^2 + O(\rho^2) , \tag{6.72}$$

where $f(a) = (2\pi)^{-\frac{1}{2}} \exp(-\tfrac{1}{2}a^2)$; see Cliff and Ord (1975c). Let us consider $BB = \tfrac{1}{2}\sum_{(2)} w_{ij} u_i u_j$, where

$$u_i = \begin{cases} 1, & \text{if } y_i \geqslant a , \\ 0, & \text{otherwise.} \end{cases}$$

By using equation (6.72), and ignoring terms in ρ^2 and higher powers, the expected value under H_1 is given by

$$E(BB|H_1) = \tfrac{1}{2}S_0 p^2 + \tfrac{1}{2}\rho[f(a)]^2 \sum_{(2)} w_{ij}(d_{ij} + d_{ji})$$
$$= \tfrac{1}{2}S_0 p^2 + \tfrac{1}{2}\rho[f(a)]^2 \operatorname{tr}(\mathbf{WD} + \mathbf{D}^T\mathbf{W}), \qquad (6.73)$$

since $\rho_{ij} = \rho(d_{ij} + d_{ji}) + O(\rho^2)$. Likewise, for BW

$$E(BW|H_1) = S_0 pq - \rho[f(a)]^2 \operatorname{tr}(\mathbf{WD} + \mathbf{D}^T\mathbf{W}). \qquad (6.74)$$

From equations (6.73), (6.74), and the variances (1.26) and (1.28), the efficacies are

$$F(BB) = \frac{\{[f(a)]^2 \operatorname{tr}(\mathbf{WD} + \mathbf{D}^T\mathbf{W})\}^2}{S_1 p^2 q^2 + S_2 p^3 q}, \qquad (6.75)$$

and

$$F(BW) = \frac{\{[f(a)]^2 \operatorname{tr}(\mathbf{WD} + \mathbf{D}^T\mathbf{W})\}^2}{S_1 p^2 q^2 + S_2 pq(\tfrac{1}{4} - pq)}. \qquad (6.76)$$

By following the argument of section 6.4.3, we can establish that the best choice of weights in the test statistic is to set $\mathbf{W} = \mathbf{D}$. Then on maximising $F(BW)$ with respect to p, we find that the best value for p is $0 \cdot 5$, for any lattice. For BB the results are more involved, as the best value of p varies with τ ($= S_2/4S_1$). At the minimum value of τ, $\tau = 1$ as for a circular time series, $p \approx 0 \cdot 25$ is best, and this slowly falls to $p \approx 0 \cdot 20$ at $\tau = 4$ (queen's case mapped onto a torus). As $1 \leq \tau \leq 4$ for almost all lattices of practical interest, $p = 0 \cdot 20$ or $0 \cdot 25$ is recommended as a simple choice.

When $\mathbf{W} = \mathbf{D}$, result (6.75) reduces to

$$F(BB) = \frac{[f(a)]^4 S_1}{p^2 q(q + 4\tau p)}, \qquad (6.77)$$

from which the best value for p can be found, given τ. Likewise when $\mathbf{W} = \mathbf{D}$, and $p = 0 \cdot 5$, since $f(0) = (2\pi)^{-\frac{1}{2}}$, we have

$$F(BW) = \frac{4S_1}{\pi^2}. \qquad (6.78)$$

Thus,

$$\operatorname{ARE}(BB, BW) = \frac{F(BB)}{F(BW)} \qquad (6.79)$$

can be computed from results (6.77) and (6.78). Since the maximum value of this ARE, for any p and all $\tau \geq 1$, is only $0 \cdot 307$, the BW test is clearly much superior to the BB test.

When nonfree sampling is considered, a quite different picture emerges. If we use the approach given above, we now find that $p = 0 \cdot 5$ is the best value both for the BW and for the BB tests, and that $F(BW) = 4S_1/\pi^2$ as before, while

$$F(BB) = \frac{4S_1}{\pi^2(1+2\xi)},\qquad(6.80)$$

where $\xi = (S_2 - 4n)/2S_1$ as before. Thus, for nonfree sampling,

$$\text{ARE}(BB, BW) = \frac{1}{1+2\xi},\qquad(6.81)$$

which is equal to one for regular lattices, but is less than one for irregular lattices. Thus *we recommend that BW be used in preference to BB*, although the margin of advantage may be slight.

Finally, from equations (6.66) and (6.78), we note that

$$\text{ARE}(BW, I) = \frac{4}{\pi^2},\qquad(6.82)$$

which represents the loss of power when the *BW* test is used for interval scaled data. This compares with an ARE figure of $2/\pi$ for the sign test against Student's *t* when testing for a difference between population means, and suggests that the loss of efficiency is more serious in the present case.

Geary (1954) observed that using three classes (0, 1, and 2, say) rather than two, improved the efficiency of the test procedure. The logical conclusion of such an increase in the number of classes is to use the ranks, $[x = (1, ..., n)^T]$. If I_{rank} denotes the *I* statistic based on ranks, then

$$\text{ARE}(I_{\text{rank}}, I) = 9\pi^{-2} \approx 0.91.$$

Thus the use of a test based on ranks appears to provide a nonparametric procedure of high asymptotic efficiency, although a study of nonnormal populations would be required to confirm this result.

The first two moments, under H_0, for the rank statistic I_{rank}, are

$$E(I_{\text{rank}}) = -\frac{1}{n-1}$$

and

$$E(I_{\text{rank}}^2) = \frac{n(n-1)(5n+6)S_1 - (5n+7)(nS_2 - 3S_0^2)}{5S_0^2(n-1)^2(n+1)}.$$

The distribution of I_{rank} may be approximated by the normal in the same way as before.

A final question which may be asked is whether or not a combination of tests can improve efficiency. Since the *I* statistic has an ARE of one, there can be no gain (asymptotically) in combining this with the *c*, or any other, statistic. Further, Cliff and Ord (1975c) show that the *BW* statistic on its own out-performs any linear combination of the *BW* and *BB* statistics.

6.5 Empirical comparisons
6.5.1 Comparisons using regression residuals

In section 8.7, the data of Taaffe et al (1963) are analysed using both I and the join-count statistics. Certain comparisons between the tests are made there and we shall now explore these results further, bearing in mind, however, that caution is needed in their evaluation since the data used are regression residuals and not original observations.

The Spearman rank correlation, r, between the standard deviates for I and for each of the join-count statistics given in table 8.8 was evaluated, which gave

$$r(I, BB_F) = 0 \cdot 22 , \qquad r(I, BW_F) = -0 \cdot 39 ,$$
$$r(I, BB_{NF}) = 0 \cdot 22 , \qquad r(I, BW_{NF}) = -0 \cdot 70 ,$$

where the subscripts F and NF denote free and nonfree sampling respectively; the negative signs for BW are expected because low BW scores should coincide with high I scores for positive spatial autocorrelation. The better performance of the BW statistic is apparent, particularly under the NF assumption. However, this analysis does not take account of the relative magnitudes of the coefficients, and so the results have been analysed using 2×2 tables with dichotomies corresponding to $z \geqslant 1 \cdot 645$ (significant), or $z < 1 \cdot 645$ (not significant). The results are presented in table 6.3, and from this analysis we conclude that the performance of BW is much better than that of BB, and that the choice of sampling model does not appear to make much difference. It is arguable that the free sampling model should be used with $p = \frac{1}{2}$, given the assumption of normally distributed errors with zero means, but, as noted in section 8.2.4, neither model is strictly valid.

The analysis we have just described is of limited generality because only two distinct data sets have been used, but the results suggest that BW is much better than BB, confirming the asymptotic results (especially for free sampling) given earlier. We believe that further analyses of real data would support these tentative conclusions.

Table 6.3. Comparison of standard deviates for I and join-count test statistics.

	BB_F		BW_F	
	significant	not significant	significant	not significant
I significant	1	4	5	0
not significant	1	5	1	5
	BB_{NF}		BW_{NF}	
	significant	not significant	significant	not significant
I significant	0	5	5	0
not significant	1	5	1	5

6.5.2 Monte Carlo studies

Cliff and Ord (1975c) describe an empirical study which used Monte Carlo methods to construct the power curves of the BB, BW, I, and c statistics for several small lattices. The simultaneous autoregressive model, given by equation (6.13) with $g_{ij} = \rho w_{ij}$ (w_{ij} known), was used as the alternative hypothesis. The variates Y were assumed to be jointly normally distributed. More recently, Haining (1977; 1978a) has followed up this work with an empirical comparison of the likelihood ratio (λ) and I statistics, using the first-order autoregressive and the moving-average schemes as alternative hypotheses under the simultaneous model. The general conclusions are similar from both studies; the λ statistic is more powerful than the I statistic, although the difference is not marked until ρ is relatively large. The moving-average results are discussed in greater detail later in this section, but for more information about the procedures used, the reader should consult the original papers. For the moment the results have been summarised in graphical form (figures 6.5–6.8). As might be expected, there is considerable sampling variation for small lattices, but, as the lattice size increases, the estimated power curves become smoother.

Figure 6.5 gives the power curves for the I, c, BB, and BW statistics on various lattices with $n = 25$. It shows the effect of shape upon these curves, as measured by A/n; A, the reader will recall, being the total number of joins in the system. (For nonbinary weights the ratio n/S_1 may be interpreted in the same way; n/S_1 is equal to 1 for the circle lattice, but equal to 4 for the queen's torus. If the weights are not scaled so that $S_0 = n$, the ratio S_0^2/nS_1 should be used.) Figure 6.6 gives the power curves for the various test lattices with the lattice structure fixed (square lattice, rook's case), but n is allowed to vary.

From figure 6.5 it is evident that the power of each of the statistics decreases as A/n increases. For a given test statistic and value of ρ, power is highest in the 25-cell circle (equivalent to a circular time series) and decreases monotonically, as A/n increases, to the queen's case on a torus. It can be shown that, in a totally connected lattice with $w_{ij} = 1$ for all $i \neq j$, the power of all the spatial autocorrelation test statistics is equal to α ($= 0 \cdot 05$ in this case) for all ρ. In other words, power is inversely related to the degree of connectedness of the lattice.

Clearly there is a conflict between a choice of w_{ij} which (a) maximises power and (b) increases the 'coverage' of possible patterns of spatial autocorrelation. For example, on count (a) the rook's case might be preferred, but on count (b) the queen's case. *The researcher must think carefully about the choice of weights to reflect the alternative hypothesis of primary interest.*

Other conclusions which may be drawn from this study are as follows: (1) From figure 6.6 we note, as would be expected, that power increases as n increases.

Models for spatial processes

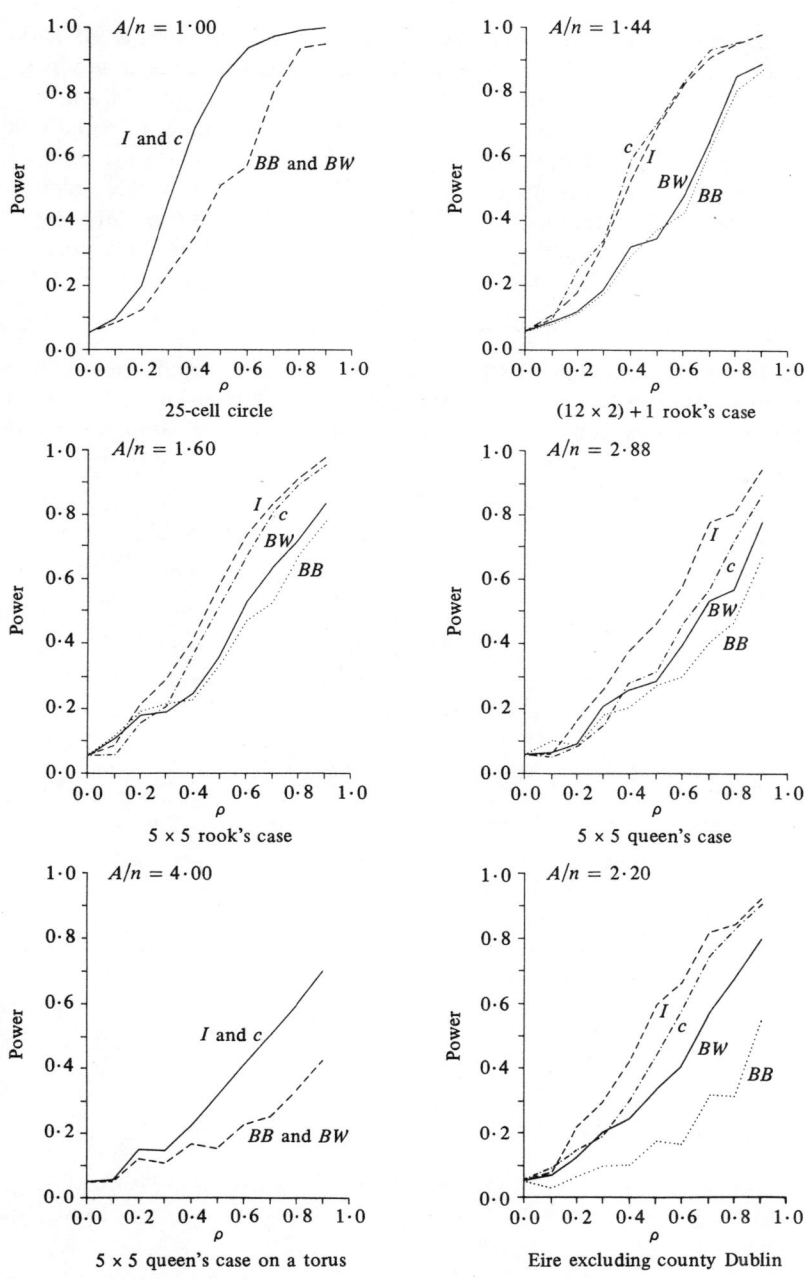

Figure 6.5. Power curves for BB, BW, I, and c statistics for several lattices with $n = 25$ and different A/n ratios. H_1 is a simultaneous first-order autoregressive model obtained from equations (6.13) with $g_{ij} = \rho w_{ij}$ (w_{ij} known).

(2) Using figures 6.5 and 6.6 together, we find that I and c, and BW and BB, have identical power curves when all counties have the same number of joins. See, for example, the 25-cell circle and the 5 × 5 queen's case mapped onto a torus. This confirms the ARE results given in equations (6.71) and (6.81), and the discussion following those equations. In the case of the (12 × 2) + 1 rook's case, where 22 of the 25 counties each have links to three other counties, I and c, and BB and BW differ only slightly. (3) When counties are not equally linked we generally find, in confirmation of results (6.71), (6.81), and (6.82), that

$$\text{power}(I) > \text{power}(c) > \text{power}(BW) > \text{power}(BB) \,. \tag{6.83}$$

From figures 6.5 and 6.6 it is evident that, when ρ is small, there is little to choose between the various coefficients, and relationship (6.83) is sometimes not satisfied by the sample results because of sampling variation.

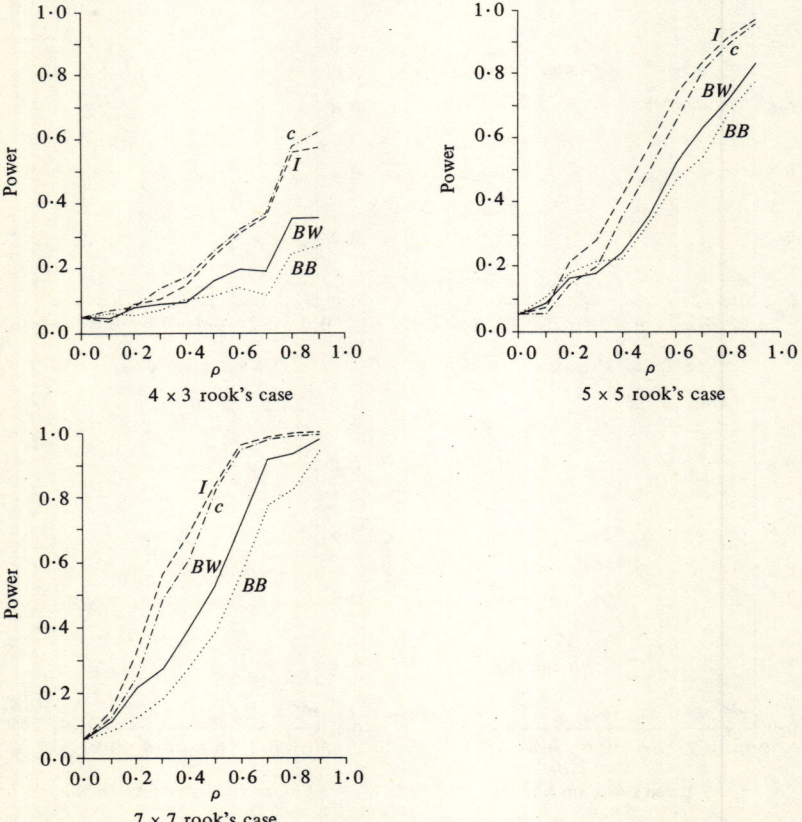

Figure 6.6. Power curves for BB, BW, I, and c statistics for three lattices with fixed **W** and varying n. H_1 is a simultaneous first-order autoregressive model obtained from equations (6.13) with $g_{ij} = \rho w_{ij}$ (w_{ij} known).

(4) In figure 6.7 we have constructed the power curves of I in the 5×5 rook's case for various values of α and β, the probabilities of type I and type II errors. It is evident that the risk of a type II error is high when ρ is low, whatever the value of α. The graph demonstrates well the severe risk of a type II error when extreme values of α are used, and argues for the use of less stringent significance levels in inferential work. Finally, if we look at the position of the 'break-even' line, $\alpha = \beta$, it is clear that a considerable amount of spatial autocorrelation must be present in a lattice of this size before we can hope to detect it without undue risk of either a type I or a type II error. For lattices with the same value of n, but higher values of A/n (or S_0^2/nS_1), diagrams like figure 6.7 give an even bleaker impression.

We now turn to Haining's (1978a) study for the moving-average scheme. Figure 6.8 gives plots of the power curves for the likelihood ratio (λ) and I statistics for various lattices. The general conclusions reached previously continue to hold, but it is evident that λ is somewhat more powerful. This is not surprising, since I is not sufficient for θ. It should be noted that the two statistics are not equivalent under the conditional moving-average scheme. The margin of supremacy of λ over I is not great but it is, perhaps, sufficient to warrant use of the likelihood ratio procedure when the moving-average alternative is proposed. The trade-off between simplicity and efficiency is not one that can be resolved mathematically. However, the pattern of spatial dependence is often sufficiently marked to ensure that either test would have high power.

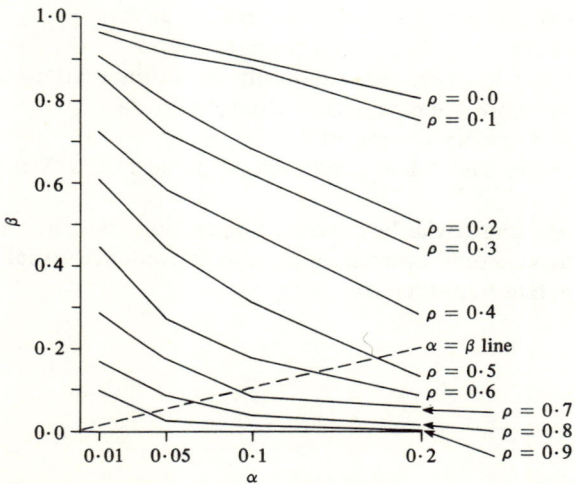

Figure 6.7. Comparison of probabilities of type I and type II errors for I in the 5×5 rook's case lattice. H_1 is a simultaneous first-order autoregressive model obtained from equations (6.13) with $g_{ij} = \rho w_{ij}$ (w_{ij} known).

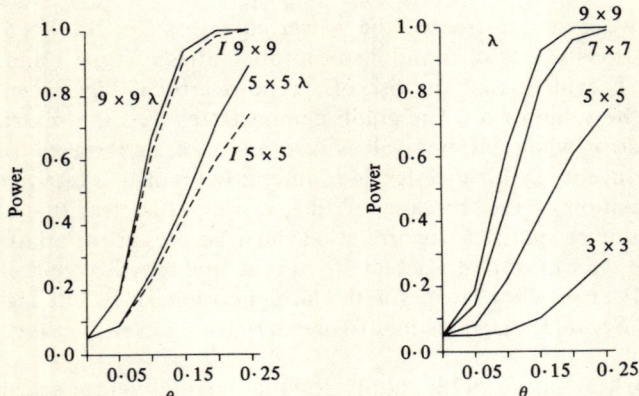

Figure 6.8. Power curves for likelihood ratio (λ) and I statistics for various lattices. H_1 is a first-order moving-average scheme with parameter θ as defined in example 6.6. (Source: Haining, 1978a, pages 213-214.)

6.6 Conclusions

In the first part of this chapter we have developed models for spatial dependence and appropriate estimation procedures. Although the maximum likelihood procedure is not computationally feasible when the number of parameters increases, the different approaches given ensure that a reasonable method is available in most cases.

In the second part of the chapter, we have turned our attention to the selection of a suitable test statistic, given the form of the alternative hypothesis. The conclusions that can be drawn here are as follows:

(1) Given interval scaled data, use I (or λ) in preference to c or the join-count statistics. This result has been validated only for normal data, but lattice shape seems to be more important than the type of data.

(2) For ranked data, use the rank version of I.

(3) For binary data, use the BW statistic and select p as near to $0 \cdot 5$ as is possible.

Last of all, we must make the trivial but crucial observation that the weights in the test statistic should correspond to the dependence structure postulated in the alternative hypothesis.

Appendix on the major theorems used in chapter 6

The purpose of the appendix is to summarise some of the important results used in this chapter, and to give outline proofs or appropriate references.

We assume that $Y = (Y_1, ..., Y_n)^T$, where Y is a vector random variable with mean μ and covariance matrix V, and that Y follows a multivariate normal distribution; that is $Y \sim \text{MVN}(\mu, V)$. We denote the observed values of Y by y, and let $Y_i^* = \{Y_j : j \neq i\}$, $y_i^* = \{y_j : j \neq i\}$. Further, we let C_i be the covariance matrix of Y_i^* and let the covariances between Y_i and Y_i^* be denoted by b_i. Thus, if we reorder the elements of Y so that Y_i appears first as

$$Y(i) = \begin{pmatrix} Y_i \\ Y_i^* \end{pmatrix},$$

the covariance matrix of $Y(i)$, given by shuffling the rows and columns of V, is

$$V(i) = \begin{bmatrix} v_{ii} & b_i^T \\ b_i & C_i \end{bmatrix}. \tag{A6.1}$$

The inverse of this matrix is

$$[V(i)]^{-1} = \begin{bmatrix} d_i & -d_i g_i^T \\ -d_i g_i & C_i^{-1} + d_i g_i g_i^T \end{bmatrix}, \tag{A6.2}$$

where

$$g_i = C_i^{-1} b_i \tag{A6.3}$$

and

$$\frac{1}{d_i} = v_{ii} - g_i^T C_i g_i = v_{ii} - b_i^T C_i^{-1} b_i. \tag{A6.4}$$

Result (A6.2) may be checked by multiplication with (A6.1) to produce the identity matrix.

We now consider our first main results.

Theorem 6.1: Let Y be multivariate normal with mean vector μ and covariance matrix V. The conditional distribution for Y_i, given $Y_i^* = y_i^*$, is normal with mean

$$E(Y_i | y_i^*) = \mu_i + \sum_{j \neq i} g_{ij}(y_j - \mu_j), \tag{A6.5}$$

and variance

$$\text{var}(Y_i | y_i^*) = \frac{1}{d_i}, \tag{A6.6}$$

where $g_i^T = (g_{i1}, ..., g_{i, i-1}, g_{i, i+1}, ..., g_{in})$, and d_i are given by equations (A6.3) and (A6.4) respectively.

Proof: The important steps in the proof relate to the factorisation and inversion of the covariance matrix. We set $\mu = 0$ without loss of generality. The exponent of the conditional distribution for Y_i, given y_i^*, has the form

$$-\tfrac{1}{2}[y^T V^{-1} y - (y_i^*)^T C_i^{-1} y_i^*] . \tag{A6.7}$$

By using matrices (A6.1) and (A6.2), this reduces to

$$-\tfrac{1}{2} d_i (y_i - g_i^T y_i^*)^2 , \tag{A6.8}$$

and the result follows.

Theorem 6.2: (converse of theorem 6.1). If the conditional distribution for each Y_i is normal with mean given by expression (A6.5) and variance given by expression (A6.6) for each i, then the joint distribution for Y is MVN(μ, V), where V is given by matrix (A6.1).

Proof: (see Besag, 1974). The exponent of the conditional distribution has the form

$$-\tfrac{1}{2} d_i (y_i - g_i y_i^*)^2 , \tag{A6.9}$$

where we again set $\mu = 0$ without loss of generality. Given that each conditional distribution is normal, it can be shown that the joint distribution must be normal and the result then follows by comparing the coefficients of each exponent (A6.9) with the appropriate version of matrix (A6.2). An important corollary emerges from the theorem as follows.

Corollary: The inverse of V may be written as

$$V^{-1} = D - H = D(I - G) ,$$

where D is a diagonal matrix with nonzero elements d_i in the ith position on the main diagonal, and H is a symmetric matrix with components $\{h_{ij}\}$, where

$$h_{ii} = 0 , \qquad h_{ij} = d_i g_{ij} = d_j g_{ji} , \qquad i \neq j , \quad i = 1, ..., n .$$

Proof: From equation (A6.2), the first row of the inverse matrix is $(d_i, -d_i g_i^T)$. However, this was obtained by simply reordering the rows and columns of V^{-1}. Hence, all the rows must have this structure.

To present theorem 6.3, the Hammersley–Clifford theorem, we must first introduce the notion of a *clique*. Any set of sites which consists either of a single site, or of an arrangement in which every site is a

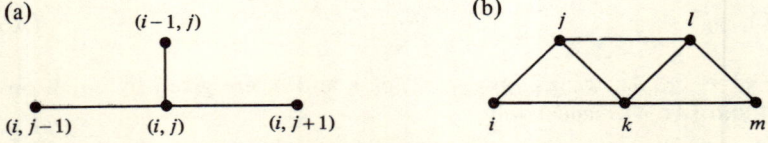

Figure A6.1. Diagram to illustrate definition of cliques in theorem 6.3.

neighbour of every other site, is called a clique. Thus, in figure A6.1, neighbours are joined by solid lines and it is evident that in case (a) cliques are of the form $\{(i, j-1), (i, j)\}, \{(i-1, j), (i, j)\}$, and so on, whereas in case (b), larger cliques such as $\{i, j, k\}$ are possible. Further, let $P(y)$ denote the joint probability distribution of Y, with value $P(0) > 0$ at some convenient origin $\mathbf{0}$. Then, in accordance with Besag (1974), we may state the result as follows.

Theorem 6.3: *(Hammersley-Clifford theorem)*
For any set of sites, $1 \leq i < j < ... < s \leq n$, we may write
$$Q(y) = \ln\{P(y)/P(0)\}$$
as
$$Q(y) = \sum_{(1)} y_i G_i(y_i) + \sum_{(2)} y_i y_j G_{ij}(y_i, y_j) + \sum_{(3)} y_i y_j y_k G_{ijk}(y_i, y_j, y_k) + ...$$
$$+ y_1 ... y_n G_{1...n}(y_1, ..., y_n) , \qquad (A6.10)$$
where the functions $G_{ij...s}$ may be nonnull if and only if the sites $i, j, ..., s$ form a clique. Otherwise, the G functions may be chosen arbitrarily.

Proof: See Besag (1974, pages 197–198). By way of example, we see that the regular lattice in figure A6.1(a) must give rise to forms of $Q(y)$ involving the first two terms in equation (A6.10).

The importance of this result is that it provides a benchmark against which any potential conditional autoregressive scheme may be checked. The scheme is valid only if theorem 6.3 is satisfied. For the normal case, the restrictions may not seem too severe, but Besag gives other cases where the restrictions are very stringent.

Example A6.1 (Besag, 1974, page 201)
Consider a set of n sites ($i = 1, ..., n$) at each of which Y_i may take only the value 0 or 1. If the only cliques available are those involving single sites and pairs of sites, then
$$Q(y) = \sum \alpha_i y_i + \sum \sum \beta_{ij} y_i y_j ,$$
whence
$$P(Y_i = y_i | y_i^*) = \exp\left\{y_i \left(\alpha_i + \sum_j \beta_{ij} y_j\right)\right\} \bigg/ \left[1 + \exp\left(\alpha_i + \sum_j \beta_{ij} y_j\right)\right] .$$
Besag (1974, pages 215–218) gives an example of fitting this model to presence/absence data relating to *Plantago lanceolata* collected on a regular grid. Other nonnormal schemes are described in the same paper (pages 202–203).

When the covariance matrix is of the form
$$\mathbf{V} = \sigma^2 \left(\mathbf{I} - \sum_{j=1}^{k} \rho_j \mathbf{W}_j\right)^{-1} = \sigma^2 \mathbf{A}^{-1} , \qquad (A6.11)$$
the ordinary least squares estimators are consistent as we now show.

Theorem 6.4: When the covariance matrix is given by form (A6.11) and the fourth moments of Y exist, the set of linear estimating equations [cf equation (6.41)]

$$(Y - \hat{\rho}_1 Z_1 - ... - \hat{\rho}_k Z_k)^T Z_j = 0, \qquad j = 1, ..., k, \tag{A6.12}$$

where $Z_j = W_j Y$, provides consistent estimators for the unknown parameters $\rho_1, ..., \rho_k$. These equations are unbiased estimating equations as defined by Durbin (1960).

Proof: For unknown $\rho_1, ..., \rho_k$, the left-hand sides of equations (A6.12) may be written as

$$Y^T \left(I - \sum_{j=1}^{k} \rho_j W_j \right) W_j Y,$$

or as

$$\text{tr}\left[\left(I - \sum_{j=1}^{k} \rho_j W_j \right) W_j Y Y^T \right].$$

By reason of expression (A6.11), this is just

$$\text{tr}(A W_j Y Y^T).$$

If we take expectations,

$$E(YY^T) = \sigma^2 A^{-1},$$

so that

$$E[\text{tr}(A W_j Y Y^T)] = \sigma^2 \text{tr}(A W_j A^{-1}) = \sigma^2 \text{tr}(W_j) = 0, \quad \text{as required.}$$

Thus, equations (A6.12) are unbiased estimating equations.

When the fourth moments of Y exist and the sample size $n \to \infty$, any quadratic form in Y will converge to a definite limit in probability (written as plim). Let

$$\text{plim}\, Y^T W_j Y = \gamma_j; \qquad j = 1, ..., k,$$

and

$$\text{plim}\, Y^T W_i W_j Y = \beta_{ij}; \qquad i, j = 1, ..., k.$$

Then the estimating equations (A6.12) have probability limits which may be written as

$$\begin{bmatrix} \beta_{11} & \cdots & \beta_{1k} \\ \vdots & \ddots & \vdots \\ \beta_{1k} & \cdots & \beta_{kk} \end{bmatrix} \begin{bmatrix} \hat{\rho}_1 \\ \vdots \\ \hat{\rho}_k \end{bmatrix} = \begin{bmatrix} \gamma_1 \\ \vdots \\ \gamma_k \end{bmatrix}. \tag{A6.13}$$

Setting $V = TT^T$ and writing $U_j = W_j T$, we see that

$$\beta_{ij} = \text{tr}(U_i^T U_j).$$

Provided that there are no linear relations among the W_j matrices, this structure will suffice to ensure a unique solution to equations (A6.13). Hence the estimators are consistent.

Corollary: Taking the covariance matrix for the simultaneous scheme as in equation (6.40), a reworking of this argument suffices to show that the ordinary least squares estimators are not consistent for the simultaneous model.

7

Autocorrelation and inferential statistics

7.1 Introduction
Research in subjects such as geography, ecology, and geology has traditionally involved substantial amounts of field work, data collection, and hypothesis testing as prerequisites to inductive theory construction about spatial processes. Much of the hypothesis testing has been carried out using conventional methods of statistical inference. Most of these methods are developed on the basis of assumptions which include a statement like, "let X_i, $i = 1, 2, ..., n$ be n *independent*, identically distributed variates". However, as we have seen, a feature of spatially located data is that the variates *cannot* be assumed to be independent. In the usual situation of positive spatial autocorrelation, an observation carries less information than an independent observation, since it is partly predictable from its neighbours. Thus increasing the number of sampling points in a study area may, in contrast to the position with independent observations, provide little or no extra information. A similar problem arises in time-series analysis, where quarterly observations are not four times as informative as annual observations over the same period.

In this chapter, we shall explore the effect of using spatially autocorrelated observations with methods which assume independence, for the more commonly employed statistical models. In particular, we shall consider Student's t distribution (section 7.2), the X^2 distribution (section 7.5), and the general linear model (section 7.3). Ways of overcoming the problem either by (1) filtering the data to remove the spatial autocorrelation or by (2) modifying the statistical models to allow for the autocorrelation are also discussed.

7.2 Comparison of means
7.2.1 The two-sample case
To illustrate the effect of failure to meet the independence assumption frequently made in statistical analysis, we consider Student's t test for differences between means. Suppose that we have drawn a sample of size n_1 from population X_1, which is $N(\mu_1, \sigma_1^2)$, and a sample of size n_2 from population X_2, which is $N(\mu_2, \sigma_2^2)$. If we take the simplest model and assume that $\sigma_1^2 = \sigma_2^2 = \sigma^2$, then under the null hypothesis, H_0, of no difference between the population means, μ_1 and μ_2, the quantity

$$t = \frac{\bar{x}_1 - \bar{x}_2}{\hat{\sigma}_{\bar{x}_1 - \bar{x}_2}} \qquad (7.1)$$

will follow Student's t distribution with $(n_1 + n_2 - 2)$ degrees of freedom, *provided that the observations are independent*. Here $\bar{x}_1$ and $\bar{x}_2$ are the means of the samples drawn from populations 1 and 2, and $\hat{\sigma}_{\bar{x}_1 - \bar{x}_2}$, the estimated standard error of the differences between sample means, is given

by
$$\hat{\sigma}_{\bar{x}_1-\bar{x}_2} = \left(\frac{n_1 s_1^2 + n_2 s_2^2}{n_1+n_2-2}\right)^{\frac{1}{2}} \left(\frac{n_1+n_2}{n_1 n_2}\right)^{\frac{1}{2}}, \qquad (7.2)$$

where s_1^2 and s_2^2 are the variances of samples 1 and 2.

In 1975, the authors constructed a Monte Carlo experiment to examine the sampling distribution of quantity (7.1) when X_1 and X_2 are autocorrelated by the amounts ρ_1 and ρ_2, respectively, among first nearest neighbours on regular lattices of various sizes (Cliff and Ord, 1975a). We used the simultaneous autoregressive scheme (sections 6.2.2 and 6.2.3)

$$X_i = \rho \sum_j w_{ij} X_j + \epsilon_i, \qquad i = 1, ..., n, \qquad (7.3)$$

to define the spatial interdependence among the $\{x_i\}$. Here, the $\{\epsilon_i\}$ are independent and identically distributed variates with common variance, σ_ϵ^2, and the summation is over the first nearest neighbours of i. The way in which spatially autocorrelated variates are constructed using this model has been described in section 6.2.7. Figure 7.1(a) shows the results obtained for a 7×7 lattice. On the diagram, the fine vertical lines represent the value of $t_{(96)}$ at the significance level $\alpha = 0.01$ (one-tailed test), as given in standard tables of the t distribution; those tables assume that $\rho_1 = \rho_2 = 0$. The heavy vertical lines give the actual values of $t_{(96)}$, $\alpha = 0.01$, determined from the Monte Carlo experiment for various values of ρ_1 and ρ_2. Note the close correspondence of the fine and heavy lines when $\rho_1 = \rho_2 = 0$, as should be the case. It is clear from the diagram that when ρ_1, ρ_2 are both greater than zero, the true probability of a type I error (that is, of rejecting H_0 when it is in fact true) exceeds the nominal level if the conventional tabulated values of t are used. The probability of a type I error increases as the level of autocorrelation increases. When $\rho_1, \rho_2 < 0$, the true probability of a type I error is much smaller than the tabulated nominal value. If one of the variables is positively autocorrelated and the other is negatively autocorrelated, the effects are not self-cancelling; the true probability of a type I error again exceeds the nominal level.

These difficulties with the t test arise because positively autocorrelated observations produce a downwards bias (underestimation) in the estimated standard error which appears in the denominator of t [equation (7.2)]. Negatively autocorrelated observations bias the estimate of the standard error upwards (overestimation).

What approaches are available which might enable the problem to be overcome? An appropriate strategy might be:
(1) before carrying out any analysis, use one of the measures given in chapter 1 to determine whether there is any spatial autocorrelation in the data. If there is not, that is if the data are spatially independent, proceed with the analysis using the conventional models.

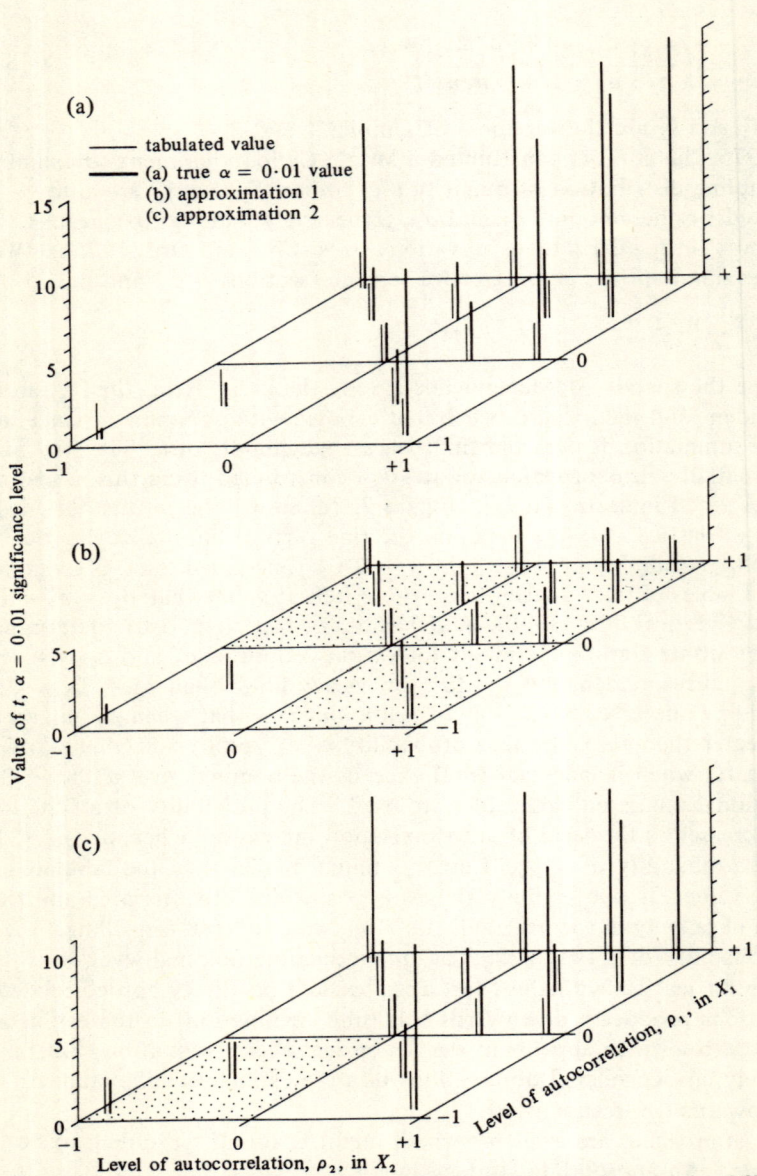

Figure 7.1. Impact of spatial autocorrelation upon Student's t test for differences between means. (a) Comparison of tabulated and true percentage points ($\alpha = 0.01$) for various levels of autocorrelation in X_1 and X_2. (b) Effect of approximation 1, defined in the text, upon percentage points for various levels of autocorrelation in X_1 and X_2. (c) Effect of approximation 2.

(2) If we conclude that the independence assumption cannot be sustained, two main alternatives are open to us. We can either modify the appropriate technique to allow for the correlated observations, or we can try to remove the correlation from the data—make them spatially independent—so that the conventional models can be applied. The formally correct procedure is always to use modified methods. However, the strategy outlined is much simpler if the independence assumption is met, and the loss of efficiency resulting from application of the strategy will generally be slight; data transformation procedures will be considered in section 7.4. Here we consider a way of altering the test statistic itself to overcome the difficulty.

7.2.2 A modified t test

When the variates X_1 and X_2 are first-order spatially autocorrelated by the amounts ρ_1 and ρ_2, respectively, the estimated standard error, $\hat{\sigma}_{\bar{x}_1 - \bar{x}_2}$, in equation (7.1) is given by

$$\hat{\sigma}_{\bar{x}_1 - \bar{x}_2} = s \left(\frac{1}{n_1(1-\tilde{\rho}_1)^2} + \frac{1}{n_2(1-\tilde{\rho}_2)^2} \right)^{1/2} , \qquad (7.4)$$

where

$$s^2 = \frac{(x_1 - \bar{x}_1 \mathbf{1})^T \mathbf{V}_1^{-1}(x_1 - \bar{x}_1 \mathbf{1}) + (x_2 - \bar{x}_2 \mathbf{1})^T \mathbf{V}_2^{-1}(x_2 - \bar{x}_2 \mathbf{1})}{n_1 + n_2 - 2} . \qquad (7.5)$$

In equations (7.3) and (7.4), the notation of section 7.2.1 has been followed. In addition, $\tilde{\rho}_1$ and $\tilde{\rho}_2$ are the estimates of ρ_1 and ρ_2, respectively, and $\mathbf{V}_i^{-1}$ is given by

$$\mathbf{V}_i^{-1} = (\mathbf{I} - \tilde{\rho}_i \mathbf{W}_i^T)(\mathbf{I} - \tilde{\rho}_i \mathbf{W}_i) , \qquad i = 1, 2 . \qquad (7.6)$$

This is the inverse of the covariance matrix of a first-order spatial autoregressive scheme (see section 6.2.3). The elements of $\mathbf{W}_i$ are taken to be scaled so that $\sum_j w_{ij} = 1$. Thus equation (7.6) specifies the spatial covariance structure among the variate values, and so equation (7.4) is feeding this information explicitly into the t-test formula. It is evident that when $\tilde{\rho}_1$ and $\tilde{\rho}_2$ in equation (7.4) are both zero, this equation reduces to equation (7.2), as it should, since both sets of variate values are then spatially uncorrelated.

Strictly speaking, the estimates, $\tilde{\rho}_1$ and $\tilde{\rho}_2$, in the above equations should be obtained by maximum likelihood (ML) estimation. However, the statistic I, given in equation (1.15), can be used to estimate ρ even though it is not consistent, being downwards biased in general[9]. Nevertheless its great virtue, compared with the maximum likelihood estimate, is that it can be evaluated very simply. A further problem in using I is that, as shown in section 1.5.2, the coefficient does not range over $[-1, +1]$ as does the ML estimator. However, we saw in section 1.5.2

---

[9] If the conditional model of section 6.2.4 is used with $\mathbf{V}^{-1} = (\mathbf{I} - \rho_i \mathbf{W}_i)$, the least squares estimators for ρ_i would be consistent. However, the resulting estimates may be close to one, giving unstable estimates of $\mathbf{V}_i^{-1}$.

that max$|I|$ can be calculated for any given lattice as

$$\max|I| = \left[\frac{\text{var}\left(\sum_j w_{ij}z_j\right)}{\text{var}(z_i)}\right]^{1/2}, \tag{7.7}$$

and so we can estimate ρ_i by

$$\tilde{\rho}_i = \frac{I}{\max|I|}, \tag{7.8}$$

which provides a more reasonable assessment on the usual $[-1, +1]$ interval (see section 6.3.1).

The bias present in equation (7.8) used as an estimator is evident from table 7.1, which suggests $|\tilde{\rho}| \leq |\rho|$. To obtain table 7.1, an X variate autocorrelated by a known amount, ρ, was generated for each of the lattices shown. $\tilde{\rho}$ was then calculated. The experiment was repeated 600 times for each value of ρ and each lattice to yield a mean value of $\tilde{\rho}$ (say $\bar{\tilde{\rho}}$) as $\sum \tilde{\rho}/600$. Table 7.1 also shows that despite the bias, equation (7.8) is a reasonable approximation provided that $n \geq 25$.

We may simplify approximation (7.4) even further by taking s in that equation as given in the usual form of the t test, namely

$$s = \left(\frac{n_1 s_1^2 + n_2 s_2^2}{n_1 + n_2 - 2}\right)^{1/2}; \tag{7.9}$$

see equation (7.2). Using the same sort of Monte Carlo experiment as before, approximation (7.4) was evaluated with s given both by expression (7.5) and by expression (7.9). $\tilde{\rho}_i$ was calculated from equation (7.8). The results we obtained for the 7×7 lattices and significance level $\alpha = 0 \cdot 01$ (one-tailed test) are reproduced in figures 7.1(b) and (c). This figure suggests two general conclusions:
(1) If either or both of $\tilde{\rho}_1$ and $\tilde{\rho}_2$ are positive, take the statistic t given in equation (7.1) with $\hat{\sigma}_{\bar{x}_1 - \bar{x}_2}$ given by equations (7.4), (7.8), and (7.9) to

Table 7.1. Comparison of $\bar{\tilde{\rho}}$ and ρ for various lattices. (Source: Cliff and Ord, 1975a, page 730.)

Degree of autocorrelation in the pattern analysed, ρ	Average[a] estimate of the degree of autocorrelation, $\bar{\tilde{\rho}}$			
	3×3 lattice	5×5 lattice	7×7 lattice	10×10 lattice
$-0 \cdot 9$	$-0 \cdot 72$	$-0 \cdot 81$	$-0 \cdot 83$	$-0 \cdot 86$
$-0 \cdot 5$	$-0 \cdot 52$	$-0 \cdot 49$	$-0 \cdot 47$	$-0 \cdot 48$
$0 \cdot 0$	$-0 \cdot 26$	$-0 \cdot 06$	$-0 \cdot 03$	$0 \cdot 00$
$0 \cdot 5$	$0 \cdot 00$	$0 \cdot 34$	$0 \cdot 42$	$0 \cdot 47$
$0 \cdot 9$	$0 \cdot 05$	$0 \cdot 61$	$0 \cdot 76$	$0 \cdot 83$

[a] Based on 600 experiments.

follow Student's t distribution with $(n_1 + n_2 - 2)$ degrees of freedom [approximation 1, say—stippled area in figure 7.1(b)].
(2) If both $\tilde{\rho}_1$ and $\tilde{\rho}_2$ are negative, take t with $\hat{\sigma}_{\bar{x}_1 - \bar{x}_2}$ given by equations (7.4), (7.5), and (7.8) to follow Student's t distribution with $(n_1 + n_2 - 2)$ degrees of freedom [approximation 2, say—stippled area in figure 7.1(c)]. The use of these rules will provide a test of significance which is unlikely to result in serious inferential error. We have been able to show (Cliff and Ord, 1975a) that the test thus modified can, in fact, be safely applied to any lattice for which n_1 and n_2 both exceed 25.

Although it has been possible to modify the t test to allow for correlated observations, the procedure is not straightforward. It is also possible to modify other models in a similar manner, but the task is no simpler and the theory required is often poorly developed, as we shall see in the remainder of this chapter.

7.2.3 Extension to k samples
Exactly the same problems exist when we wish to test for significant difference between several population means (that is, via one way analysis of variance). The theory is reviewed in Griffith (1978).

7.3 Correlation and regression
7.3.1 Correlation
Bivand (1980) has employed the Monte Carlo approach, outlined in section 7.2.1 for the t test, to examine the sampling distribution of Fisher's z transformation of the Pearson product moment correlation coefficient, r. Conventionally, tests of hypotheses about calculated values of r are carried out using this transformation, taking the estimated standard error of the sampling distribution of z, $\hat{\sigma}_z$ say, as $(n-3)^{-1/2}$. Bivand generated pairs of random variables with known levels of autocorrelation in them, and compared the standard error of the empirically generated sampling distribution of z with $\hat{\sigma}_z$. The results he obtained for 5×5 and 7×7 lattices are shown in figure 7.2. In his analysis, Bivand mapped the lattices onto a torus and took the rook's case definition to specify **W**. If we allow for sampling variation, the results suggest that the standard error of z is not badly affected by (1) low and moderate levels of autocorrelation in *both* variables or by (2) high levels of autocorrelation in *one* of the variables. However, if autocorrelation is severe in both variables, the true standard error of z appears to be substantially larger than that given by the usual formula. Bivand also found that the underestimation of the true standard error by $\hat{\sigma}_z$ was much worse for irregular, rather than for regular, lattices.

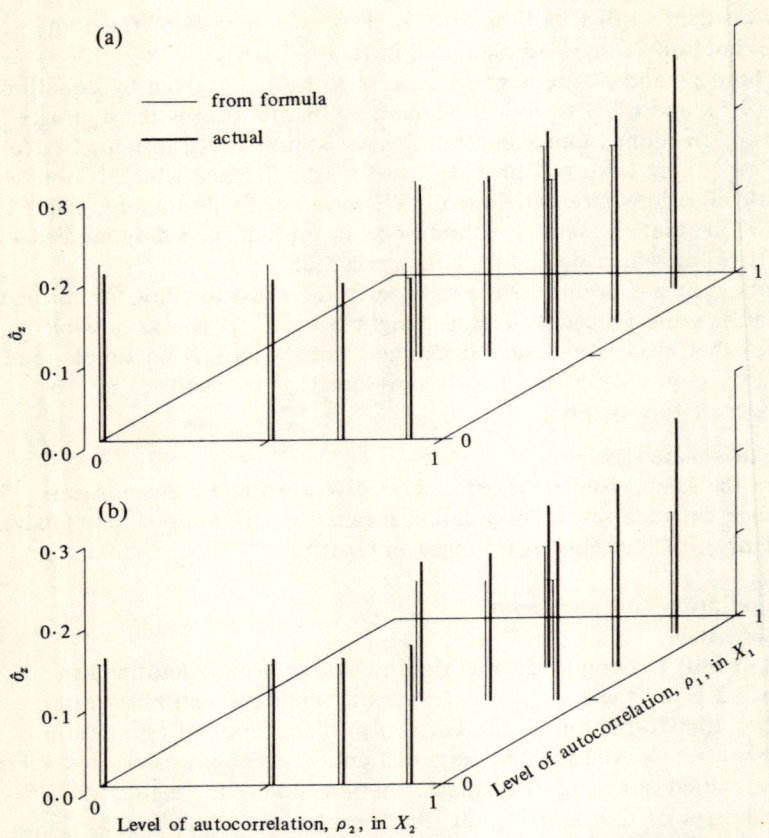

Figure 7.2. Comparison of actual values of $\hat{\sigma}_z$ for Fisher's z transformation with value given by formula $\hat{\sigma}_z = (n-3)^{-1/2}$ for various levels of spatial autocorrelation in X_1 and X_2. (a) A 5×5 lattice, (b) a 7×7 lattice.

7.3.2 Regression analysis

A basic assumption of the familiar linear regression model,

$$\underset{n \times 1}{Y} = \underset{n \times k}{X} \underset{k \times 1}{\beta} + \underset{n \times 1}{\epsilon} , \qquad (7.10)$$

is that the error terms, ϵ, are uncorrelated. For data with a geographical ordering, this requirement implies that the errors must not be spatially autocorrelated. If the errors are positively autocorrelated, we may represent this assumption in terms of the alternative moving-average scheme (section 6.4) as this also covers invertible autoregressive schemes. Thus we assume, in the notation of section 6.4, that

$$\text{var}(\epsilon) = \sigma^2 (I + \sum_{j=1}^{s} \theta_g W_g) . \qquad (7.11)$$

Then, when $k = 1$ (excluding the constant) in equation (7.10) a straightforward extension of Johnston's (1972, page 246) analysis reveals that the least squares estimator of β has variance

$$\text{var}(\hat{\beta}) = \frac{\sigma^2}{x^T x} \left[1 + \sum_{g=1}^{s} \theta_g I(g) \right], \qquad (7.12)$$

where $I(g) = x^T W_g x / x^T x$, taking $S_0(g) = n$; see section 5.2.1 for fuller details of the notation. However, the variance based upon the usual least squares procedure is

$$\text{var}(\hat{\beta}) = \frac{\sigma^2}{x^T x} . \qquad (7.13)$$

We can see that the true $\text{var}(\hat{\beta})$ based on equation (7.12) will be understated by equation (7.13) if the θ_g and $I(g)$ are positive; that is when the x and the ϵ are each positively autocorrelated. Again, in accordance with Johnston (1972, page 248) we find that when

$$e = Y - x\hat{\beta},$$

the least squares estimator for σ^2,

$$\hat{\sigma} = \frac{e^T e}{n-1},$$

has expected value

$$E(\hat{\sigma}^2) = \sigma^2 - \frac{\sigma^2}{n-1} \sum_{g=1}^{s} \theta_g I(g), \qquad (7.14)$$

so that $\hat{\sigma}^2$ is downwards biased when the θ_g and $I(g)$ are both positive.

Example 7.1
If we take $n = 20$, and $\theta_g = I(g) = (\frac{3}{4})^g$, s large, corresponding to a conditional autoregressive scheme with $\rho = \frac{3}{4}$ both for x and for ϵ, then $\hat{\sigma}^2$ is downwards biased by about 7%.

Since the estimate, $\hat{\sigma}^2$, appears in the denominator of the t and F tests used to examine the significance of the regression [Johnston, 1972, equations (2-30) and (2-36)], overstatement of the significance of the regression is likely, (values of t and F will be inflated). In addition, an inflated value for F means that the coefficient of multiple correlation, R^2, will also be inflated [Johnston, 1972, equation (5-60)]. If the error terms are negatively correlated, the reverse occurs (understatement of the significance of any regression; t and F values depressed). From equation (7.11) we can observe that the level of spatial autocorrelation in the $\{X\}$ variables in the regression is important as well. Indeed, equation (7.12) confirms Johnston's view that if the $\{X\}$ variables are spatially independent, "then even if [the $\{\epsilon\}$ are] autocorrelated the bias is not likely to be serious" (Johnston, page 248). Solutions to these problems posed by spatial autocorrelation in regression will be considered in chapter 9.

7.4 Spatial variate differencing

Let us denote the random variable at the ith site by X_i, and suppose that these variates are spatially autocorrelated according to the simultaneous autoregressive scheme; that is

$$X_i = \rho \sum_j w_{ij} X_j + \epsilon_i, \qquad i = 1, ..., n \ . \tag{7.15}$$

The multilateral dependence among observations implied by equation (7.15) means that spatial difference operators can be constructed in several different ways. One of the most commonly used (Lebart, 1969; Martin, 1974) is to define

$$\Delta X_i = X_i - \sum_j w_{ij} X_j \ . \tag{7.15}$$

This difference filter will be effective in eliminating spatial autocorrelation when the autocorrelation parameter, ρ, in equation (7.15) is approximately one. Figure 7.3 shows a simple example of this difference operator applied to values located on a set of contiguous quadrats.

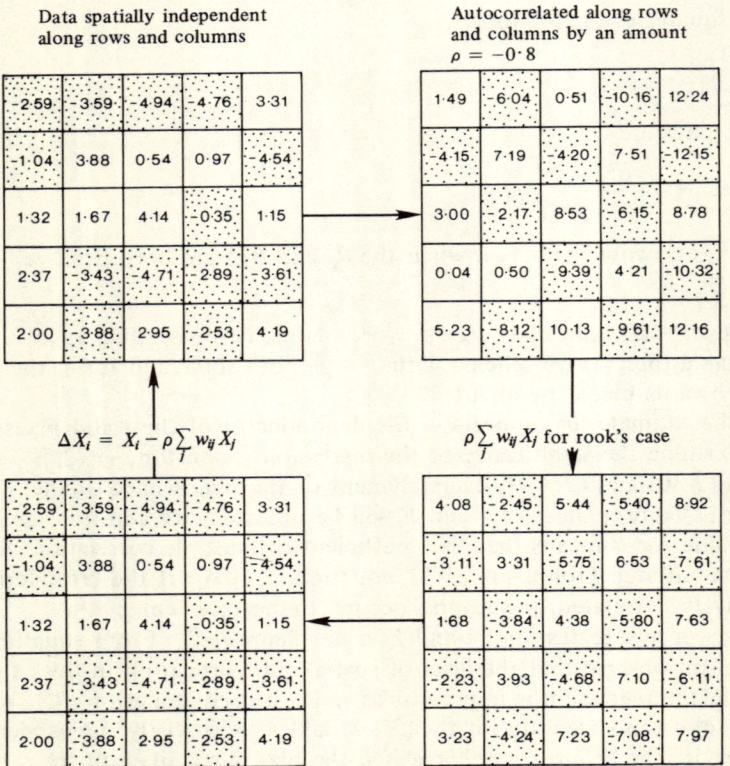

Figure 7.3. Use of the difference operator given in equation (7.15) to eliminate first-order spatial autocorrelation among contiguous quadrats on a lattice.

If variate differencing is used, instead of modified models, as a means of satisfying the independence assumption in analyses where hypothesis testing is the aim, the limited results available to date suggest that the following guidelines may be employed:

(a) as noted in sections 7.2 and 7.3, if $\rho \leq 0$, use of the original observations will result in tests of hypotheses which are conservative, in the sense that the true risk of a type I error will be less than the nominal (tabulated) value;

(b) if ρ is slightly greater than zero, use of the original observations will result in a slightly liberal test, whereas variate differences may produce a very conservative test;

(c) as ρ approaches one, the use of data which have been differenced may result in a slightly conservative test, but the conclusions will be much more reliable than those based upon the original observations. Tests based upon the original observations will carry a risk of a type I error greatly in excess of the nominal level.

These guidelines imply that in hypothesis testing, a strategy which will always tend to err on the safe side, in terms of the risk of a type I error, is to use the original observations when $\rho \leq 0$, and to use variate differences when $\rho > 0$.

It should be noted that spatial differencing cannot be used for a test of the mean because, when $\sum_j w_{ij} = 1$, the mean of the ΔX_i is *zero*, no matter what the common mean of the X_i may be.

The only paper in which spatial variate differencing is used, and in which substantial empirical results are given, is that of Martin (1974). Martin carried out a Monte Carlo study of regression estimates when the errors were autocorrelated, as described in section 7.3. He then reestimated the parameters after applying the difference filter given in equation (7.15). His results show that the difference filter renders significant improvement whenever the autocorrelations among the x and among the ϵ each exceed 0.4 or thereabouts.

In the spirit of this section, a possible approach to the hypothesis testing problem of section 7.2 is to use quasi-differences (cf Nerlove, 1964) such as

$$\Delta X_i = X_i - \rho_0 \sum w_{ij} X_j , \qquad (7.16)$$

where ρ_0 is specified *a priori* and could differ between the two samples. The means for the resulting ΔX_i are $\mu(1 - \rho_0)$ when $\sum_j w_{ij} = 1$ for all j, so that tests may be performed for any $\rho_0 \neq 1$. Despite the arbitrary element in the choice of ρ_0, when spatial autocorrelation is high and positive virtually any value of ρ_0 in the interval (0, 1) will be better than the implicit standard value $\rho_0 = 0$!

Lebart (1969) used the difference filter (7.15) prior to carrying out a factor analysis on spatial data. However, as Cliff and Ord (1975b, page 302) have noted, it is not clear that using the same filter for all variables in the study necessarily improves the results. Clearly, this is an area where much more work is required.

7.5 Tests using join counts
Earlier tests using join counts have concentrated upon a single statistic such as that for *BW* joins (section 1.3.1). Dacey (1965) has suggested an alternative approach of considerable interest which we shall now describe. Consider a two-colour choropleth map. We could count the number of counties of a given colour, *B*, say, in the map which are joined to *d* other counties ($d = 0, 1, 2, ...$) either of the same colour (when we obtain a frequency distribution of the number of counties in the map with $d = 0, 1, 2, ...$ *BB* joins), or of a different colour (yielding a frequency distribution of the number of counties in the map with $d = 0, 1, 2, ...$ *BW* joins). If we could derive the expected frequency distributions for *d* under the null hypothesis of no spatial autocorrelation, we could construct a $2 \times (R+1)$ contingency table of the sort given in table 7.2. Here *R* is the maximum value of *d* in the county system.

Dacey (1965) has shown that the expected frequency distributions in table 7.2 can be derived under H_0. He illustrated the approach for the distribution of *BW* joins; Cliff, Martin and Ord (1975) extended the results to cover *BB* joins and *k*-colour lattices. We now give the results for *BW* joins.

Table 7.2. Contingency table for *BB* or *BW* joins.

		Number of joins, *d*				
		0	1	2	3	... R
Number of counties with *d* joins	observed (O_d)					
	expected (E_d) under H_0					

7.5.1 Expectations under H_0
As with the *BW* statistic of section 1.3.1, the expectations may be computed under either of two assumptions, namely:
(a) *free sampling*, where each county is independently colour coded black with probability *p*, and white with probability $q = 1-p$; or
(b) *nonfree sampling*, where each county is coded by exhaustive random sampling without replacement from a population of n_1 black counties and n_2 ($= n - n_1$) white counties.

In an *n*-county map we use m_r to denote the number of counties which have *r* joins with other counties. Then the total number of joins, *A*, in the map is given by

$$A = \tfrac{1}{2} \sum_{r=1}^{R} rm_r , \qquad (7.17)$$

where $R = \max\{r | m_r > 0\}$. For a given county, *s*, say, with *r* joins in all, the probability of d joins coded BW is

$P(d|r) = P(s\text{th county is } B)P(d \text{ neighbours are } W|s \text{ is } B)$
$\qquad + P(s\text{th county is } W)P(d \text{ neighbours are } B|s \text{ is } W).$

The expected number of counties with d joins coded BW, n_d say, is the sum of $P(d|r)$ over all n counties,

$$n_d = \sum_{r=1}^{R} m_r P(d|r), \qquad d = 0, 1, ..., r. \qquad (7.18)$$

For free sampling, expression (7.18) becomes

$$n_d = \sum_{r=1}^{R} m_r \binom{r}{d} \{p^{r+1-d}q^d + p^d q^{r+1-d}\}. \qquad (7.19)$$

For nonfree sampling, the expected number of cells with d joins coded BW becomes

$$n_d = \sum_{r=1}^{R} m_r \binom{r}{d} \{n_1^{(d)} n_2^{(r+1-d)} + n_1^{(r+1-d)} n_2^{(d)}\} / n^{(r+1)}. \qquad (7.20)$$

Unfortunately, the observed and expected frequencies may not be compared by a standard X^2 test, because the observed counts are not independent; each cell appears as a neighbour for several other cells. Moreover, each cell appears both as a reference cell and as a neighbour, thus inducing an element of double counting. As a result both of theoretical and of Monte Carlo studies, Cliff, Martin, and Ord (1975) reached the conclusion that the familiar statistic,

$$X^2 = \sum \frac{(O_d - E_d)^2}{E_d} \qquad (7.21)$$

should be modified to $\tilde{X}^2 = \beta X^2$ where, for free sampling,

$$\beta = \frac{1}{2}\left\{1 - \frac{\frac{1}{2}D(1-4pq)}{A(1-2pq) + D(1-4pq)}\right\}, \qquad (7.22)$$

where

$$A = \tfrac{1}{2}\sum r m_r = \tfrac{1}{2} S_0, \qquad \text{and} \qquad D = \tfrac{1}{2} \sum_{i=1}^{n} w_{i.}^2 - A. \qquad (7.23)$$

The correction factor, β, is maximised if $p = \tfrac{1}{2}$, when $\beta = \tfrac{1}{2}$. This choice of p coincides with the best choice of p when the BW test is used as a test statistic (section 6.4.4). For nonfree sampling, or free sampling with p estimated by n_1/n, the simple choice $\beta = \tfrac{1}{2}$ appears to be adequate.

These authors showed that $\tilde{X}^2$ may be treated as having a chi-squared distribution under H_0 provided that n is not too small. The numerical studies indicate that the appropriate degrees of freedom are either
(a) R, when free sampling is used with p known, or
(b) between $R-1$ and R, when free sampling is used with p estimated, or nonfree sampling is appropriate.

In case (b), the suggested decision rule takes the form (for a test of size α)

if $\tilde{X}^2 < \chi^2_{(R-1)}(\alpha)$, do not reject H_0

if $\tilde{X}^2 > \chi^2_{(R)}(\alpha)$, reject H_0

if $\chi^2_{(R-1)}(\alpha) \leq \tilde{X}^2 \leq \chi^2_{(R)}(\alpha)$, defer judgement.

In practice, the area of indeterminacy is fairly small.

Finally, it is worth noting that an alternative approach to the problem would be to avoid double counting by selecting only a subset of the sites, such as those marked with a cross in figure 7.4. This represents the use of a method of coding (cf Besag, 1974), and the resulting X^2 statistics could be used without scaling. However, it is not known how much power might be lost by such a procedure.

```
X   •   X   •   X   •
•   X   •   X   •   X
X   •   X   •   X   •
```

Figure 7.4. Procedure for selecting sites to avoid double counting of joins by examining only sites marked with a cross.

7.6 Conclusions

In this chapter we have explored the consequences of applying statistical models, which assume independent observations, to variables which are spatially autocorrelated. The conclusion to be drawn is clear: serious inferential error is probable if observations are autocorrelated and modified models are not used. Ways of overcoming the difficulties, either by adapting the models to allow for spatial autocorrelation or by filtering the data to remove the spatial autocorrelation, have been considered. Clearly, the former approach should be the long term aim, but until the appropriate theory has been developed, the latter will ensure that major errors are unlikely, and it is certainly better than taking no action at all.

8
The analysis of regression residuals

8.1 Introduction
In the earlier chapters of this monograph, we have considered the case of spatial autocorrelation among sample data where the data have not been modified by any statistical procedure. On many occasions, however, we wish to carry out a regression analysis and look for autocorrelation, not in the original data, but among the residuals from the regression.

If autocorrelation, temporal or spatial, is detected among the regression residuals, then it can imply:
1. the presence of nonlinear relationships between the dependent and independent variables;
2. the omission of one or more regressor variables;
3. that the regression model should have an autoregressive structure.

The presence of autocorrelation among the population error terms leads to biased estimates of the residual variance and inefficient estimates of the regression coefficients when the method of ordinary least squares (OLS) is used. Therefore some check for autocorrelation should always be applied and remedial action taken when necessary.

When situation 1, above, is believed to be important, different models can be specified and interaction terms among the independent variables included (see section 8.6). If situation 2 is the main cause of autocorrelation, additional variables may be suggested by plotting the residuals on a map and searching for regular patterns in these residuals (see the discussion in sections 8.6 and 8.7). Finally, if situation 3 is thought to be the main cause, some kind of transformation must be carried out (see section 9.2).

In this chapter, the problems caused when population error terms are autocorrelated will be outlined in section 8.2; the I statistic forms the basis of an appropriate test of autocorrelation among regression residuals. In section 8.3, we evaluate its mean and variance, and then in section 8.4, we explore the alternative BLUS procedure. In section 8.5 some asymptotic distribution theory is developed.

In sections 8.6 to 8.8 some empirical studies are reported in which the I statistic has been used. Specifically in sections 8.6 and 8.7 we shall look at the work of O'Sullivan (1968) on the economic effects of road accessibility in Eire and the study by Taaffe et al (1963) on the degree of internal accessibility in underdeveloped countries. Finally, in section 8.8 we shall examine the residuals from a trend-surface analysis of agricultural land values in Iowa.

A test for autocorrelation will tell us whether a given model is adequate, or whether a different form is required. We have no desire to remove spatial autocorrelation as such, but simply to make allowances for it so that valid estimating procedures can be adopted.

Much of the material in sections 8.2–8.5 is based upon Cliff and Ord (1972), and this paper should be consulted for further details.

8.2 The linear model and alternative hypotheses

8.2.1 The classical regression model

Suppose that, for each of the n counties in a study area, the observation y_i has been recorded on the variate Y_i ($i = 1, ..., n$). In vector notation, let $y = (y_1, ..., y_n)^T$ be the ($n \times 1$) vector of observed values corresponding to the variate vector $Y = (Y_1, ..., Y_n)^T$. Y is assumed to have the mean vector $X\beta$ and covariance matrix $\sigma^2 V$; that is,

$$E(Y) = X\beta, \tag{8.1}$$

and

$$\text{var}(Y) = \sigma^2 V, \tag{8.2}$$

where X is an ($n \times k$) matrix of nonstochastic regressor variables,

$$X = \begin{bmatrix} 1 & X_{12} & ... & X_{1k} \\ 1 & X_{22} & ... & X_{2k} \\ \vdots & \vdots & \ddots & \vdots \\ 1 & X_{n2} & ... & X_{nk} \end{bmatrix}, \tag{8.3}$$

with the elements in the first column corresponding to the constant term. The vector β is of order ($k \times 1$), and contains the parameters β_i ($i = 1, ..., k$); that is $\beta = (\beta_1, ..., \beta_k)^T$, and V is an ($n \times n$) matrix with elements $\{v_{ij}\}$.

The linear regression model is defined as

$$Y = X\beta + \epsilon, \tag{8.4}$$

where ϵ is an ($n \times 1$) vector of random error terms. It follows from equation (8.1) and the assumption of nonstochastic X variables that

$$E(\epsilon) = 0. \tag{8.5}$$

Further, if

$$\text{var}(\epsilon) = \sigma^2 V \tag{8.6}$$

reduces to

$$\text{var}(\epsilon) = \sigma^2 I_n, \tag{8.7}$$

where I_n ($\equiv I$) is the unit matrix of order n, and X is of rank k (that is, full rank), then the OLS estimators b,

$$b = (X^T X)^{-1} X^T Y, \tag{8.8}$$

are the best linear unbiased estimators for β (cf Johnston, 1972, pages 123–127).

It follows from these assumptions that the unobservable error terms ϵ can be estimated by

$$e = Y - Xb ,\qquad(8.9)$$

and the best unbiased estimator for σ^2 is given by

$$\hat{\sigma}^2 = (n-k)^{-1}e^T e = (n-k)^{-1}\sum_{i=1}^{n} e_i^2 = (n-k)^{-1}(Y^T Y - b^T X^T Y) .\qquad(8.10)$$

Further properties of the linear model are given in Johnston (1972, chapter 5).

8.2.2 Consequences of autocorrelated errors

Equation (8.7) embodies the crucial assumption, H_0: that there is no autocorrelation among the error terms[10]. If, as often happens, this assumption is violated, the estimators (8.8) are inefficient, and the variance estimator (8.10) is downwards biased, thereby inflating the observed value of R^2.

To protect against such difficulties we may either
(a) fit a model which incorporates an autocorrelated structure for the error terms, or
(b) test the residuals from OLS analysis for spatial autocorrelation and fit the more complicated model only if this proves necessary.
Approach (b) is open to objection if formal significance tests are required at a later stage. Nevertheless this practice is widely followed and often saves a great deal of computational effort. Therefore, we can now proceed to develop a statistic for testing

$$H_0: \text{var}(\epsilon) = \sigma^2 I_n ,$$

against

$$H_1: \text{var}(\epsilon) = \sigma^2 V \neq \sigma^2 I_n ,$$

where $V = (I - \rho W)^{-1}$, or $V = [(I - \rho W)^T (I - \rho W)]^{-1}$, depending on whether we use the conditional scheme or the simultaneous scheme described in section 6.2. Clearly, we must take $\rho \neq 0$ under hypothesis H_1.

8.2.3 Choice of test statistic

As noted in section 8.2.1, the error terms ϵ_i ($i = 1, ..., n$) are unobservable and so any test for spatial autocorrelation must be based upon the calculated residuals e_i ($i = 1, ..., n$). It is evident that we could use the BB or BW join-count statistics by adopting a convention such as $x_i = 1$ if $e_i \geq 0$, and $x_i = 0$ if $e_i < 0$. However, such procedures will involve a considerable loss of efficiency, as was noted in section 6.4.

---

[10] Equation (8.7) also implies that the error terms are homoscedastic; that is they have equal variances. In the remainder of this chapter, we shall accept the assumption of homoscedasticity without further comment. The reader should consult Johnston (1972, pages 214–221) for details of this assumption.

For a finite number of regressor variables, the asymptotic power results for the different statistics given in section 6.4 will continue to hold. Therefore, the I statistic is chosen as the best simple statistic available, although Haining's (1978a) likelihood ratio statistic can be expected to have a slight edge for the joint scheme.

8.2.4 Correlation among the sample residuals under H_0

Under the usual assumptions of regression analysis, the least squares estimators e, of the error terms ϵ, have zero means and covariance matrix

$$E(ee^T) = \sigma^2[I - X(X^TX)^{-1}X^T]$$
$$= \sigma^2 M, \quad \text{say.} \tag{8.11}$$

That is to say, the sample residuals e are correlated under H_0, whether or not the population errors ϵ are autocorrelated. It follows that, for regression residuals, the expected values of the join-count statistics for free sampling under H_0 will differ from the values given in equations (1.25) and (1.27). The same argument holds for nonfree sampling. The correct moments under H_0 could be evaluated, but this would involve the evaluation of $\frac{1}{2}na$ bivariate integrals, where a is the average number of joins with nonzero weights per county.

For the I and c statistics, we could operate either under assumption N or under assumption R (see section 2.3). To make assumption R operational, we must consider the $n!$ random permutations of the n vectors $(y_i, X_{2i}, ..., X_{ki})$, which is equivalent to considering the randomisation of the $\{e_i\}$. This means that the moments given in section 2.3.2 apply (with e_i replacing z_i). Thus the test ignores autocorrelation among the regressor variables and it seems to the authors that these $n!$ random permutations are not the appropriate reference set for testing the residuals. However, Geary (1954) has evaluated the first two moments of c under the assumption that "the error terms are normally distributed and the $n!$ random permutations are considered".

We can conclude that the only operational, efficient, tests available are those based on the I and c statistics under assumption N. For the reasons given earlier, we chose to concentrate upon the I statistic. The moments for I in this case were first given by Cliff and Ord (1972), but the presentation here is somewhat different in that we make greater use of matrix algebra so as to obtain expressions that are more usable for computational purposes.

8.3 The moments of the I statistic

The methods used in this section are the same as those in section 2.3, although the moments are more involved because of the correlation of the variates under H_0. The linear regression model is assumed, as outlined in section 8.2.1, and it is also assumed that the random error terms are normally distributed.

The test statistic I, in terms of the residuals, is given by

$$I = \frac{n}{S_0} \frac{e^T W e}{e^T e} . \tag{8.12}$$

We can now proceed to evaluate the moments of I, using the Pitman-Koopmans theorem given in section 2.3.1. Direct application of this theorem implies that

$$E(I^p) = \left(\frac{n}{S_0}\right)^p \frac{E[(e^T W e)^p]}{E[(e^T e)^p]} , \qquad p = 1, 2, ..., \tag{8.13}$$

since I and $e^T e$ are independently distributed. Further, since $e^T e$ is distributed as chi-squared with $(n-k)$ degrees of freedom, we have that

$$E(e^T e) = n - k , \tag{8.14}$$

and

$$E[(e^T e)^2] = (n-k)(n-k+2) . \tag{8.15}$$

We shall now concentrate upon the quadratic form, $Q = e^T W e$, and in the course of our development, we shall make considerable use of traces; that is if $\mathbf{A}$, a square matrix, is given by $\mathbf{A} = \{a_{ij}\}$, then its trace, $\text{tr}(\mathbf{A})$, is $\sum_i a_{ii}$.

8.3.1 Mean of I

It follows from equations (8.8) and (8.9) that

$$e = [\mathbf{I} - \mathbf{X}(\mathbf{X}^T\mathbf{X})^{-1}\mathbf{X}^T] Y$$
$$= \mathbf{M} Y . \tag{8.16}$$

Hence, from equation (8.4), the sample error terms may be described by the random variables

$$\mathbf{M} Y = \mathbf{M} \mathbf{X} \beta + \mathbf{M} \epsilon = \mathbf{M} \epsilon . \tag{8.17}$$

Thus, from equation (8.17) we can write

$$Q = e^T W e = \epsilon^T \mathbf{M} \mathbf{W} \mathbf{M} \epsilon = \text{tr}(\epsilon^T \mathbf{M} \mathbf{W} \mathbf{M} \epsilon)$$
$$= \text{tr}(\mathbf{M} \mathbf{W} \mathbf{M} \epsilon \epsilon^T) ,$$

using the reordering properties of the trace operator that $\text{tr}(\mathbf{AB}) = \text{tr}(\mathbf{BA})$ for any conformable matrices $\mathbf{A}$ and $\mathbf{B}$. Since $E(\epsilon \epsilon^T) = \sigma^2 \mathbf{I}_n$ under H_0, it follows that

$$E(Q) = \sigma^2 \text{tr}(\mathbf{MWM}) = \sigma^2 \text{tr}(\mathbf{M}^2 \mathbf{W}) = \sigma^2 \text{tr}(\mathbf{MW}) , \tag{8.18}$$

since $\mathbf{M}^2 = \mathbf{M}$ ($\mathbf{M}$ is idempotent). Substituting back the form for $\mathbf{M}$ given in equation (8.11), we find that

$$E(Q) = \text{tr}(\mathbf{W}) - \text{tr}[\mathbf{X}(\mathbf{X}^T\mathbf{X})^{-1}\mathbf{X}^T\mathbf{W}]$$
$$= -\text{tr}[(\mathbf{X}^T\mathbf{X})^{-1}\mathbf{X}^T\mathbf{W}\mathbf{X}] , \tag{8.19}$$

since tr(**W**) = 0, and again making use of the reordering property. We shall set

$$\mathbf{A} = (\mathbf{X}^T\mathbf{X})^{-1}\mathbf{X}^T\mathbf{W}\mathbf{X} \ , \tag{8.20}$$

where **A** is a ($k \times k$) matrix, so that from equation (8.13),

$$E(I) = -\frac{n \operatorname{tr}(\mathbf{A})}{(n-k)S_0} \ . \tag{8.21}$$

When $k = 1$ (constant term only), expression (8.21) reduces to the usual value of $-(n-1)^{-1}$, as it should. For single variable regression, when $k = 2$, we find that

$$E(I) = -\frac{1+I_x}{n-2} \ , \tag{8.22}$$

where I_x is the spatial autocorrelation coefficient for the regressor variable x, evaluated in the usual way.

8.3.2 Variance of I

We need to evaluate [cf equation (2.4)]

$$E(Q^2) = \tfrac{1}{2}\sum_{(2)}(w_{ij}+w_{ji})^2 E(e_i^2 e_j^2) + \sum_{(3)}(w_{ij}+w_{ji})(w_{ik}+w_{ki})E(e_i^2 e_j e_k)$$
$$+ \sum_{(4)} w_{ij} w_{kl} E(e_i e_j e_k e_l) \ . \tag{8.23}$$

Since the errors are no longer identically distributed, we cannot use the reduced formulae given in section 2.1, and instead we must work from expression (8.23). Let the ith column vector of $\mathbf{M} = \{m_1, ..., m_n\}$ be denoted by m_i. Then, from equation (8.17),

$$E(e_i^2 e_j^2) = m_i^T E(\epsilon \epsilon^T m_j m_i^T \epsilon \epsilon^T) m_j \ . \tag{8.24}$$

Since the $\{\epsilon_i\}$ are independent and identically normally distributed it follows that

$$E(\epsilon_i^4) = 3\sigma^4 \ , \quad E(\epsilon_i^2 \epsilon_j^2) = \sigma^4 \ , \quad \text{and } E(\epsilon_i \epsilon_j \epsilon_k \epsilon_l) = 0 \text{ for all other cases.}$$

After some algebra and setting $\mathbf{M} = \{m_{ij}\}$, we find that

$$\left.\begin{array}{l} E(e_i^2 e_j^2) = \sigma^4(2m_{ij}^2 + m_{ii} m_{jj}) \ , \\ E(e_i^2 e_j e_k) = \sigma^4(2m_{ij} m_{ik} + m_{ii} m_{jk}) \ , \\ E(e_i e_j e_k e_l) = \sigma^4(m_{ij} m_{kl} + m_{ik} m_{jl} + m_{il} m_{jk}) \ . \end{array}\right\} \tag{8.25}$$

Substituting these results into equation (8.23), we obtain

$$E(Q^2) = [\operatorname{tr}(\mathbf{UM})]^2 + 2\operatorname{tr}[(\mathbf{UM})^2] \ , \tag{8.26}$$

where $\mathbf{U} = \tfrac{1}{2}(\mathbf{W}+\mathbf{W}^T)$. Clearly $\mathbf{U} = \mathbf{W}$ when the weights are symmetric. If we use **A**, as defined in equation (8.20), and set

$$\underset{(k \times k)}{\mathbf{B}} = 4(\mathbf{X}^T\mathbf{X})^{-1}\mathbf{X}^T\mathbf{U}^2\mathbf{X} \ , \tag{8.27}$$

then equation (8.26) reduces to
$$E(Q^2) = S_1 + [\operatorname{tr}(\mathbf{A})]^2 + 2\operatorname{tr}(\mathbf{A}^2) - \operatorname{tr}(\mathbf{B}) . \tag{8.28}$$
Finally, we arrive at
$$\operatorname{var}(I) = \frac{n^2}{S_0^2(n-k)(n-k+2)}\left\{S_1 + 2\operatorname{tr}(\mathbf{A}^2) - \operatorname{tr}(\mathbf{B}) - \frac{2[\operatorname{tr}(\mathbf{A})]^2}{n-k}\right\} . \tag{8.29}$$
When $k = 1$ (the constant term only), this reduces to equation (2.35) as we would expect.

8.3.3 Evaluation of the variance
We have expressed the variance in the form of equation (8.29) since this reduces the computational effort involved. The best computing procedure is as follows:
step 1 form $\mathbf{U} = \frac{1}{2}(\mathbf{W}+\mathbf{W}^\mathrm{T})$, if the matrix $\mathbf{W}$ is not symmetric already;
step 2 evaluate $\mathbf{Z} = \mathbf{U}\mathbf{X}$, an $(n \times k)$ matrix;
step 3 compute the $(k \times k)$ cross-product matrices
$$\mathbf{C}_1 = \mathbf{X}^\mathrm{T}\mathbf{Z}, \quad \text{and} \quad \mathbf{C}_2 = \mathbf{Z}^\mathrm{T}\mathbf{Z} ;$$
step 4 evaluate $\mathbf{G} = (\mathbf{X}^\mathrm{T}\mathbf{X})^{-1}$, and then form
$$\mathbf{A} = \mathbf{G}\mathbf{C}_1, \quad \mathbf{B} = \mathbf{G}\mathbf{C}_2, \quad \text{and} \quad \mathbf{A}^2 ;$$
step 5 evaluate the traces of the matrices formed in *step 4*.
The mean and variance then follow directly from equations (8.21) and (8.29).

8.4 The BLUS procedure
The reader will recall, from section 8.2.4, that the original test statistics could not be used because of the correlation among the sample residuals under null hypothesis H_0; in making allowances for this autocorrelation, the moments of the statistics became more complicated. To avoid these complications raised by the correlation among the sample residuals under H_0, we could construct different residuals which are uncorrelated (and hence independent if ϵ is taken to be normally distributed). Theil (1965) has suggested the use of BLUS estimators (Best Linear Unbiased estimators with a Scalar covariance matrix), and we shall next explore this approach.

8.4.1 Derivation of the estimates
Given that there are k regressor variables in equation (8.4), we nominate $(n-k)$ counties and estimate the $(n-k)$ vector of residuals z, where $z = (z_1, z_2, ..., z_{n-k})^\mathrm{T}$ say, such that
$$E(z) = 0, \quad \text{and} \quad \operatorname{var}(z) = \sigma^2 \mathbf{I}_{n-k} .$$
The counties may be renumbered so that the residuals are estimated for the first $(n-k)$ counties only. These estimators, the BLUS estimators, are

related to the OLS estimators by the relation,

$$z = Be, \qquad (8.30)$$

where B is an $(n-k) \times n$ matrix such that $BMB^T = I_{n-k}$, M being defined by equation (8.11). B^T is the matrix of eigenvectors corresponding to the $(n-k)$ eigenvalues equal to one in the idempotent matrix M. For details of the calculation of z, see Theil (1965).

Given the BLUS estimates, an I, c, or join-count test may be carried out in the usual way, working with the reduced county system of size $(n-k)$ and adjusted weights, as we shall now show. For example, consider the test statistic

$$I' = \frac{n-k}{S'_0} \frac{\sum'_{(2)} \omega_{ij} z_i z_j}{\sum' z_j^2}, \qquad (8.31)$$

where

$$\sum'_{(2)} \equiv \sum_{\substack{i=1 \\ i \neq j}}^{n-k} \sum_{j=1}^{n-k}, \qquad \sum' \equiv \sum_{j=1}^{n-k}, \qquad S'_0 = \sum'_{(2)} \omega_{ij},$$

and ω_{ij} represent the revised weights. These revised weights might be simply $\omega_{ij} = w_{ij}$, or $\omega_{ij} = w_{ij}/\sum' w_{ij}$.

8.4.2 Moments of I'

If we apply the methods of section 5.4, it follows that

$$E(I') = 0, \qquad (8.32)$$

and, putting $t_{ij} = \omega_{ij} + \omega_{ji}$, we also have

$$\text{var}(I') = \frac{(n-k)\sum'_{(2)} t_{ij}^2}{2(S'_0)^2 (n-k+2)}, \qquad (8.33)$$

with

$$\mu_3(I') = \frac{(n-k)^2 \sum'_{(3)} t_{ij} t_{jm} t_{mi}}{2(S'_0)^3 (n-k+2)(n-k+4)}, \qquad (8.34)$$

and

$$\mu_4(I') = \frac{(n-k)^3 \left[\frac{9}{2}\sum'_{(2)} t_{ij}^4 + 9\sum'_{(3)} t_{ij}^2 t_{im}^2 + \frac{3}{4}\left(\sum'_{(2)} t_{ij}^2\right)^2 + 3\sum'_{(4)} t_{ij} t_{jm} t_{ml} t_{li} \right]}{(S'_0)^4 (n-k+2)(n-k+4)(n-k+6)}, \qquad (8.35)$$

where $\sum'_{(3)} = \sum_{i \neq j \neq m}$, i, j, and m are defined over 1 to $(n-k)$, and $\sum'_{(4)}$ is defined similarly. From expressions (8.32-8.35) it is clear that, once the z_i have been computed, the analysis is simpler because of the zero correlations among these residuals.

The principal difficulty with the Theil procedure lies in the choice of the k counties to be dropped. In time-series studies, the first k time periods are often dropped, while for spatial data several rules come to mind, such as: choose the k counties

(1) on the borders of the study area;
(2) at random;
(3) in one particular subarea.

8.4.3 Relative efficiency of BLUS procedures

Abrahamse and Koerts (1969) reported extensive Monte Carlo comparisons of the Durbin-Watson statistic and the BLUS statistic for testing the presence of first-order autocorrelation in time series. Their results indicate that the exact Durbin-Watson test is more efficient, but this conclusion is qualified by the impracticality of producing tables for the exact test. Conversely, the BLUS procedure leads to the von Neumann statistic which is tabulated, and this benefit may outweigh any loss of efficiency.

To date, no comparable studies have been performed for the spatial case, but we believe that the results of any such studies would be rather similar. This belief is reinforced by our work on asymptotic relative efficiency (cf section 6.4) reported in Cliff and Ord (1975c). In all BLUS procedures the choice of which counties to omit raises some problems, but the ARE studies suggest, tentatively, that we should discard the least well connected counties; that is, those with the smallest values of $\sum_j w_{ji}$.

8.5 Distributions of the test statistics

Our treatment of the distributions of the test statistics considered in sections 8.3 and 8.4 falls into two parts. First, we demonstrate the asymptotic normality of the statistics under general conditions, and second, we outline a randomisation procedure for use with small samples or in situations where assumption N cannot be employed.

8.5.1 Asymptotic normality of the test statistics

The asymptotic normality of the I and c statistics was proved in section 2.4, subject to condition (2.51) on the eigenvalues of the matrix

$$\mathbf{T} = \mathbf{MWM}, \tag{8.36}$$

where

$$\mathbf{M} = \left(\mathbf{I}_n - \frac{1}{n}\mathbf{1}\mathbf{1}^T\right).$$

These results continue to hold with $\mathbf{M}$ replaced by the more general expression given in equation (8.11); that is

$$\mathbf{M} = \mathbf{I}_n - \mathbf{X}(\mathbf{X}^T\mathbf{X})^{-1}\mathbf{X}^T.$$

Since the number of regressor variables, k, is fixed and we allow n to go to infinity, the resulting conditions for asymptotic normality are precisely the same as those given in section 2.4.1. That is, if $\lambda_1, ..., \lambda_n$ are the eigenvalues of $\mathbf{W}$, then sufficient conditions are that either

$$\max_i \frac{\lambda_i^2}{S_1} \text{ be } O(n^{-1}),$$

or

$$\max_i w_{i.} \leq c < \infty, \quad \text{and} \quad 0 < \lim_{n \to \infty} \frac{S_1}{n} = \gamma^2 < \infty \,..$$

Exceptions to this result, like the star lattice, remain awkward.

The argument for the BLUS statistics is essentially the same. However, the simpler form of the higher order moments suggests that it may be possible to use better approximations, such as that based upon the beta distribution, in accord with the results for the d statistic (Durbin and Watson, 1971).

Given these asymptotic results, we shall proceed to use a normal approximation in the empirical work in sections 8.6 to 8.8. The validity of this approximation has not been checked empirically for small values of n.

8.5.2 A random permutations procedure
From equations (8.16) and (8.17) it is known that

$$e = M\epsilon \, . \tag{8.37}$$

Given M for a particular regression analysis, and that ϵ_i are independently distributed, the parameters of the distribution can be estimated from the observed residuals. A set of n drawings can then be generated from this distribution, and a 'dummy' vector e_g constructed from equation (8.37). If the ϵ_i are uncorrelated but not independent, a single drawing from a specified n variate distribution will have to be made; otherwise the procedure is unchanged. Usually, it will be assumed that ϵ_i is N(0, σ^2). No estimate for σ^2 is required, since the test statistics are scale free functions of ϵ. To achieve full comparability between the observed and generated e vectors, the generated values should be adjusted to have zero mean. Given e_g, the selected test statistic may be evaluated and the value recorded. By following the procedure of section 2.7, generate m such values. On the assumption that the upper tail is the critical region, the null hypothesis H_0, of no spatial autocorrelation, is rejected at the $100(j+1)/(m+1)\%$ level if the observed value of the statistic exceeds at least the $(m-j)$ smallest generated values. This Hope procedure is the same as that given in section 2.7, except that the generated variables must be transformed by equation (8.37) before the statistic is computed.

In accordance with the results given earlier, m set equal to 49 or 99 is suggested as a suitable number of generated values.

8.6 Eire: the economic effects of road accessibility
O'Sullivan determined the accessibility on the arterial road network of each of the twenty-six counties of Eire shown in figure 8.1. The accessibility of a vertex i, in a road network N, was defined by O'Sullivan as

$$A(i, N) = \sum_{j=1}^{n} d_{ij} \, , \tag{8.38}$$

where n is the number of vertices in the network, and d_{ij} is the distance in miles by road on the shortest path between the ith and jth vertices. The road network taken by O'Sullivan was all T (trunk) class roads in Eire, and the vertices were the thirty-one towns of over 5000 population in the 1961 census. Equation (8.38) was applied to each vertex. County accessibility values were approximated by taking the value of the largest town in the county if it was one of the thirty-one, or alternatively, by interpolating a value from an isopleth map of accessibility. With the accessibility on the arterial road network (ARA) defined in this way, it is evident that the ARA of eastern and southern counties will be 'better', and the ARA of western counties 'worse', than in reality because most of the thirty-one towns of over 5000 population are located in counties bordering the south and east coasts of Eire. There is no doubt that the eastern and southern counties of Eire do have considerably better transport services than the western counties (O'Sullivan, 1969) but, as constructed, the ARA index emphasises these differences.

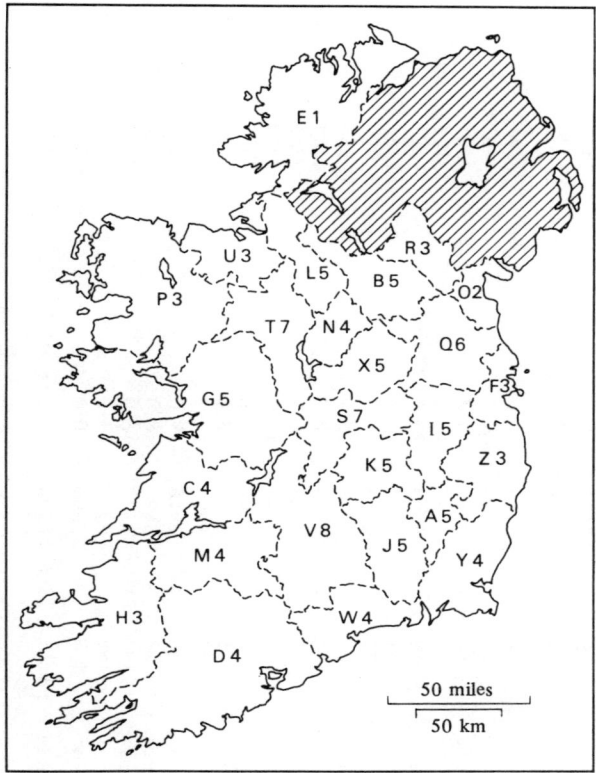

Figure 8.1. The counties of Eire. The serial letters correspond with table 8.1, and the numbers give the number of counties contiguous to each reference county.

Table 8.1. Eire regression data and regression residuals by county.

County serial letter and name	1961 population as % of 1926 population[a]	ARA	%, in value terms, of gross agricultural output of each county consumed by itself[b]	Value of retail sales[c] (£000)	Total personal income[d] (£000)	Regression residuals			
						1(a)	1(b)	2	3
A Carlow	97	3664	8.6	2962	7185	1.44	0.007	−1.93	1473
B Cavan	69	5000	15.0	4452	9459	−12.75	−0.057	−3.01	1733
C Clare	78	4321	19.0	3460	12435	−10.77	−0.044	4.59	−867
D Cork	90	4118	9.0	28402	65901	−0.87	−0.006	−4.34	−4823
E Donegal	75	7500	27.0	7478	17626	19.10	0.088	−4.26	345
F Dublin	142	3078	9.4	89424	164631	40.38	0.125	1.17	2838
G Galway	88	4537	21.9	8972	26950	1.46	0.022	6.44	−3200
H Kerry	78	5140	17.0	6341	20510	−2.30	0.003	−1.76	−2351
I Kildare	111	3200	9.0	4803	14703	10.64	−0.029	0.53	−750
J Kilkenny	87	3708	8.0	4321	13585	−8.11	−0.037	−3.17	−628
K Laoighis	87	3455	10.3	3128	9280	−10.72	−0.056	0.18	506
L Leitrim	60	5000	23.1	1885	5709	−21.75	−0.118	4.99	1193
M Limerick	95	4018	11.4	10786	27395	3.10	−0.023	−1.81	−1627
N Longford	77	4250	19.0	1960	5297	−12.50	−0.054	4.96	1491
O Louth	107	3948	10.1	7059	12156	14.37	0.069	−2.44	2883
P Mayo	71	6815	30.0	6758	19201	8.02	0.039	2.37	−1226
Q Meath	103	4008	8.7	3356	14512	10.99	0.057	−3.76	−2094
R Monaghan	72	4500	13.0	3960	8396	−15.92	−0.067	−2.36	1816
S Offaly	98	4108	14.3	3817	10320	7.03	0.042	1.71	633
T Roscommon	71	4500	23.0	2821	10223	−14.92	−0.073	7.64	−311
U Sligo	75	5997	22.0	3535	9461	3.56	−0.028	−1.30	815
V Tipperary	88	3926	9.0	9226	26424	−4.85	−0.017	−3.32	−2662
W Waterford	91	3691	8.0	7526	15696	−4.28	−0.019	−3.08	1436
X Westmeath	93	3872	16.0	3822	10842	−0.41	0.003	3.97	356
Y Wexford	87	3940	8.6	8231	15582	−5.71	−0.021	−3.40	2203
Z Wicklow	102	3600	10.2	4865	11921	5.78	0.024	−0.59	816

[a] Census of Population, 1966, Preliminary Report, table 8 (Central Statistics Office, 1967).
[b] Attwood and Geary (1963), table 3, $\left(\dfrac{\text{column } 12}{\text{column } 11}\right) \times 100$.
[c] Attwood and Geary (1963), table 6, column 2.
[d] Attwood and Geary (1963), table 2, column 6.

O'Sullivan then used these county accessibility values as the regressor variable in several regressions in which various measures of intensity of economic activity in the counties of Eire formed the dependent variable.

From O'Sullivan's work[11], the following least squares regressions were selected:

Regression (1)
  y = the 1961 population as a percentage of the 1926 population by county,
  x = the ARA,
  1(a) $y = 133 \cdot 45 - 0 \cdot 0103x$, $R^2 = 0 \cdot 40$,
        $(0 \cdot 0026)$
  1(b) $\log y = 4 \cdot 19 - 0 \cdot 6210 \log x$, $R^2 = 0 \cdot 52$.
        $(0 \cdot 1225)$

Note that here and elsewhere in this chapter, the number in brackets below each estimate is its estimated standard error, and the logarithms are to the base 10.

Regression (2)
  y = the percentage, in value terms, of the gross agricultural output of each county consumed by itself;
  x = the ARA,
  $y = -8 \cdot 49 + 0 \cdot 0053x$, $R^2 = 0 \cdot 70$.
       $(0 \cdot 0007)$

In addition, as an example where x was not the ARA, we took

Regression (3)
  y = the value of retail sales (£000) by county;
  x = the total personal income (£000) by county;
  $y = -2393 \cdot 8 + 0 \cdot 5405x$, $R^2 = 0 \cdot 987$.
       $(0 \cdot 0126)$

The county values for all variables, and the regression residuals, are given in table 8.1. The county serial letters refer to figure 8.1. The residuals were tested for spatial autocorrelation using I as defined in equation (8.12). In this equation we tried three forms for **W**:

1. $w_{ij} = 1$ if counties i and j had a length of county boundary in common, and $w_{ij} = 0$ otherwise (binary weights);
2. unstandardised boundary length/distance between county centres weights, defined as $w_{ij} = d_{ij}^{-1}\beta_{i(j)}$ (see section 1.3.4);
3. the standardised form of 2 given by $w_{ij}^* = w_{ij}/w_{i.}$, so that $\sum_{j} w_{ij}^* = 1$ for all i.

The moments of I were evaluated under assumption N [equations (8.21) and (8.29)], and under assumption R (but see section 8.2.4), when equations (1.37) and (1.39) were used. Positive spatial autocorrelation was suspected and so the upper tails were used as critical regions. The

---

[11] The authors wish to thank Professor O'Sullivan for making the full computer results of his study available to them.

Table 8.2. Results of tests for spatial autocorrelation in regression residuals from O'Sullivan (1969).

Value of I	Binary weights: regression				Unstandardised boundary length/county centre distance weights: regression				Standardised boundary length/county centre distance weights: regression			
	1(a)	1(b)	2	3	1(a)	1(b)	2	3	1(a)	1(b)	2	3
Assumption N												
I	0·1908	0·1301	0·3968	0·2297	0·2467	0·1890	0·4539	0·2281	0·1553	0·0803	0·4361	0·2928
$E(I)$	−0·0556	−0·0581	−0·0556	−0·0421	−0·0575	−0·0602	−0·0575	−0·0418	−0·0589	−0·0616	−0·0589	−0·0419
$\sigma(I)$	0·1137	0·1129	0·1137	0·1173	0·1438	0·1430	0·1438	0·1456	0·1391	0·1388	0·1391	0·1444
Standard deviate	2·17*	1·67*	3·98**	2·32*	2·12*	1·74*	3·56**	1·85*	1·54	1·02	3·56**	2·32*
Assumption R												
I	0·1908	0·1301	0·3968	0·2297	0·2467	0·1890	0·4539	0·2281	0·1553	0·0803	0·4361	0·2928
$E(I)$	−0·0400	−0·0400	−0·0400	−0·0400	−0·0400	−0·0400	−0·0400	−0·0400	−0·0400	−0·0400	−0·0400	−0·0400
$\sigma(I)$	0·1108	0·1157	0·1175	0·1158	0·1384	0·1448	0·1471	0·1450	0·1363	0·1427	0·1449	0·1428
Standard deviate	2·08*	1·47	3·72**	2·33*	2·07*	1·58	3·36**	1·85*	1·43	0·84	3·28**	2·33**

* Significant at $\alpha = 0·05$ level (one-tailed test)
** Significant at $\alpha = 0·01$ level (one-tailed test)

results of the analysis are given in table 8.2. The weighting matrices used are given in the appendix to this chapter.

From table 8.2 we note that in all cases the residuals are positively spatially autocorrelated, and that 10 of the 12 values of the standardised deviate are lower under assumption R than under N. This difference arises because under R we are assuming, falsely, that the sample residuals are uncorrelated under H_0, whereas by using assumption N we are allowing for the correlation among the sample residuals under H_0. The amount and direction of the bias introduced when we make assumption R depends critically on the correlation structure among the regressor variables. However, the general tendency is for the test under R to be more conservative than under N, if the residuals are positively spatially autocorrelated. Various empirical studies suggest that this conservatism increases considerably as the number of regressor variables increases. The *ad hoc* procedure under assumption R of taking $z = [I - E_N(I)]/\sigma_R(I)$ seems to work better, but it lacks any justification. It is, however, similar in spirit to Geary's (1954) suggestion for the use of the c statistic with regression residuals.

As regards the regression equations individually, several points emerge.
Regression (1) As shown in table 8.2, whatever the form of **W**, the positive spatial autocorrelation among the residuals from regression 1(a) is stronger than among the residuals from regression 1(b). One of the causes of autocorrelated disturbances mentioned in section 8.1 is that of making an incorrect specification of the form of the relation between the dependent and independent variables. For example, suppose we specify a linear relation between y and x when the true relation is, say, a quadratic. Even though the disturbance term in the true relation may not be autocorrelated, the residual in the linear relation will contain a term in x^2. If there is any autocorrelation in the x^2 values, then we shall have autocorrelation in the composite disturbance term. In the case of regression (1), the relation between y and x was originally specified by the simple linear equation 1(a). However, if we plot y against x (figure 8.2), it appears that the true relation is probably better described by a power curve of the form

$$y = \alpha x^\beta + \gamma, \tag{8.39}$$

with $\beta < 0$. By using regression equation 1(a), rather than equation (8.39), the disturbance term associated with it will contain a term in x^β. Also, x is highly spatially autocorrelated ($I = 0.335$, with the standard deviate equal to 3.42). This suggests that the cause of autocorrelated disturbances mentioned above may account in part for the result for regression 1(a) in table 8.2. Now if we assume $\gamma = 0$, a relation like that in equation (8.39) can be reduced to a linear form by taking logarithms of both sides, as done by O'Sullivan in regression 1(b). We are now specifying a linear relation between $\log y$ and $\log x$. As a result, there is a drop in the spatial autocorrelation in the residuals from this regression as compared with 1(a).

Although it is not attempted here, it appears that if equation (8.39) were fitted using nonlinear methods, better results than those obtained for regression 1(b) would be achieved.

Regression (2) This model is a measure of subsistence. The more remote and inaccessible Irish counties (essentially the western littoral and west-central counties) would be expected to consume more of their own agricultural produce than the less isolated eastern counties. Although a high value for R^2 is obtained, the striking feature is the degree of positive spatial autocorrelation in the residuals of all forms of **W** (table 8.2). The regression model overestimated the amount of their own agricultural produce consumed internally in (a) all counties on and south of a line from Limerick (M) to Wicklow (Z), (b) counties O, Q, and R, which form a block in the northeast part of the Irish Republic, and (c) U and E. Elsewhere, underestimation occurred. This spatial pattern of residuals suggests several alternative independent variables which might improve the model.

The areas of overestimation are counties which are served by the principal railways from Dublin, whereas the areas of underestimation are poorly linked by rail to Dublin, the latter being the chief market for agricultural produce in the country. We examined three independent variables which might measure the importance of rail transport to each county;

(1) rail accessibility measured by applying equation (8.38) to the largest town in each county—we defined d_{ij} as the distance in miles by rail on the shortest route between the *i*th and *j*th vertices;

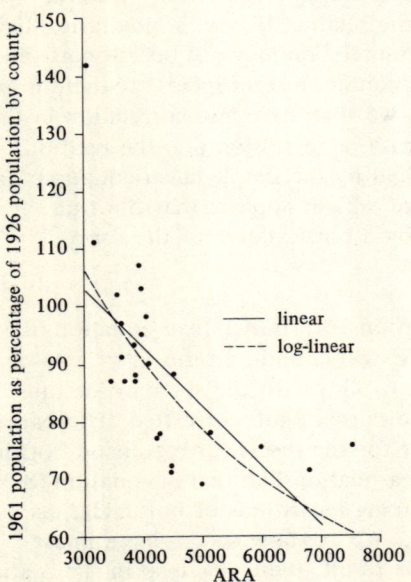

Figure 8.2. Graph of data and regression lines for O'Sullivan's regression 1(a) and 1(b).

(2) a binary variable, with a county being coded 1 if it was linked by a through freight train service to Dublin, and 0 otherwise;
(3) a measure of cost accessibility by rail of each county, which we defined for the ith county as

$$c_i = \sum_{j=1}^{n} c_{ij} , \qquad (8.40)$$

where c_{ij} is the cost of shipping one ton of goods by freight train between the ith and jth counties (data from Coras Iompair Eireann, 1966).

The areas of overestimation by the regression model are also those which are closest to the chief ports of Eire for the export of agricultural goods. From the Statistical Abstract of Ireland for 1961 (Central Statistics Office, 1962), the following ports each exported more than 180000 tons of cargo: Cork (including Cohb and Whitegate), Drogheda, Dublin (including Dun Laoghaire), Sligo, and Waterford. All others exported less than 26000 tons each. We therefore tried
(4) distance in miles between the largest town in each county and the nearest of these major ports.

O'Sullivan (1969, page 18) suggests that the areas where the amount of their own agricultural produce consumed internally is overestimated, are the chief commercial farming areas of Eire with large farm units, whereas the areas of underestimation coincide broadly with the subsistence farming areas with small farm units. To quantify this pattern, we looked at two variables:
(5) the number of agricultural holdings above 100 acres, June 1960, as a percentage of the total number of holdings by county [source: Statistical Abstract of Ireland 1961 (Central Statistics Office, 1962, table 80)];
(6) the number of males aged 18 years and over, who were members of the family engaged on their own family farm, as a percentage of the total male population, June 1960, by county [source: Statistical Abstract of Ireland 1961 (Central Statistics Office, 1962, tables 8 and 75)]. We hoped that variable (5) would pick out the areas of Eire in which the larger commercial farming units were located, whereas (6) would pick out the subsistence farming areas.

We tried a series of two-regressor variable multiple regressions, with y as the percentage, in value terms, of gross agricultural output of each county consumed by itself, x_1 as the ARA, and x_2 set equal, in turn, to each of the independent variables (1)–(6) described above. The values of R^2 obtained are shown in table 8.3.

The new independent variables postulated produced some increase in the value of R^2. However, even for the best, (6), the sign of the residual from the regression was changed in only four counties when compared with the signs for the original simple regression. All the sets displayed positive spatial autocorrelation when tested using I. Thus we were unable to break up the pattern of spatial autocorrelation in the residuals.

It therefore appeared that an autoregressive model might be worth considering, since such a model would allow for the persistence of regional variations such as those caused by inertia. We shall return to this approach in section 9.2.

Regression (3) Despite the very high R^2 value, the regression residuals display a high degree of spatial autocorrelation. First, the y variable (gross value of retail sales) is a commonly used surrogate variable for x (gross personal income), and this accounts in part for the inflated value of R^2. Second, the regression provides an example of the way in which the position of a regression line fitted by least squares can be dominated by outlier observations (in this case, the large values of counties Cork and Dublin; see figure 8.3). When this happens, a high R^2 value is obtained, although the regression equation may be a poor estimator of the relationship between nonextreme x and y values. The limited usefulness of the equation as a predictor of retail sales for nonextreme county values is revealed in the autocorrelated residuals. When this problem arises, the difficulty can be mitigated by using a criterion to fit the regression line which places less emphasis upon extreme observations. Minimum absolute deviations, where the line is fitted to minimise $\Sigma|e_i|$, rather than Σe_i^2, and linear programming techniques are possibilities (Wagner, 1959; 1962). Alternatively the regression may be fitted in the usual way, but omitting extreme data points (Tukey, 1962). In this particular example, the latter strategy was adopted. The extreme values for counties Cork and Dublin were dropped. The revised regression, based on twenty-four observations, was

$$y = 361 \cdot 82 + 0 \cdot 3489x, \quad R^2 = 0 \cdot 80.$$

If the residuals from this regression are tested for spatial autocorrelation, we obtain, with binary weights, $I = 0 \cdot 043$, $E(I) = -0 \cdot 059$, $\sigma(I) = 0 \cdot 117$, and the nonsignificant standard deviate of $0 \cdot 87$. According to Geary's (1954) arguments, the revised regression would be judged, because the residuals are uncorrelated, to have accounted for all the systematic spatial variation in y.

Table 8.3. Values of simple correlation coefficients and R^2 for new independent variables tried in O'Sullivan's *regression (2)*.

New independent variable added as x_2	r_{y, x_1}	r_{y, x_2}	r_{x_1, x_2}	R^2
(1)	0·835	0·292	0·591	0·759*
(2)	0·835	−0·599	−0·566	0·720*
(3)	0·835	0·312	0·579	0·741*
(4)	0·835	0·289	0·033	0·765*
(5)	0·835	−0·725	−0·574	0·787*
(6)	0·835	0·757	0·606	0·796*

* Increase in R^2 judged significant at 0·05 level using the F test.

The revised regression might therefore be regarded as more successful from a geographical point of view, despite the lower R^2. Because of the problems noted, the original regression model consistently overestimates retail sales in the poorer counties west of a line from Cork (D), Tipperary (V), and Mayo (P).

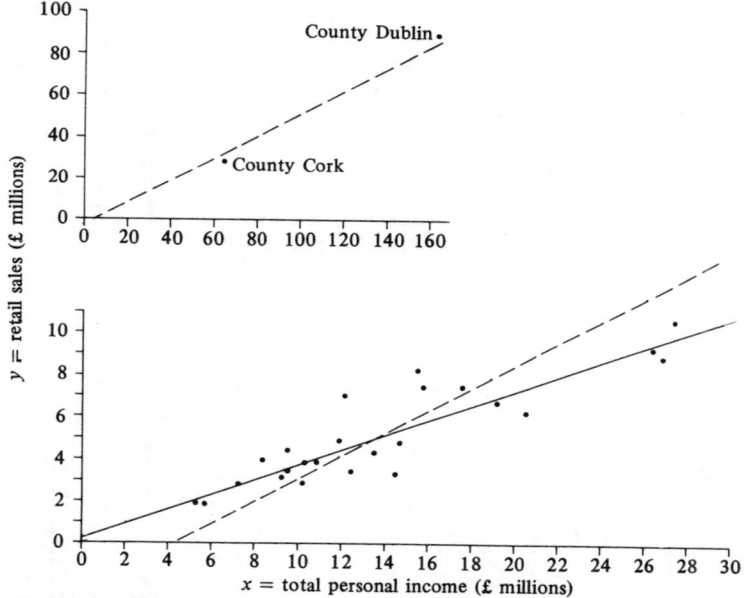

Figure 8.3. O'Sullivan's *regression (3)*. The OLS line based on all twenty-six observations is pecked. The solid line is the regression omitting counties Cork and Dublin, and the lower graph shows the position of these twenty-four counties at a larger scale.

8.7 Road accessibility in developing countries

In a well-known paper, Taaffe et al (1963) examined the degree of internal accessibility of underdeveloped countries as measured by the extent of their road networks. They postulated that this accessibility would be a function of such factors as population level, physical environment, rail competition, and degree of commercialisation. Regression analysis was used to quantify the relationship between the size of the road network and these factors. Ghana and Nigeria were used as examples[12]. The data are reproduced here in tables 8.4 and 8.5. The locations of the recording units are shown in figures 8.4 and 8.5, and the identity numbers of the recording units correspond with those given in tables 8.4 and 8.5, respectively. The basic regression equations obtained by Taaffe et al were

[12] The authors wish to thank Professor Taaffe and his colleagues for supplying us with the data used in their work.

Table 8.4. Basic regression data for Ghana.

Recording unit		y [a]	x_1 [b]	x_2	Recording unit		y [a]	x_1 [b]	x_2
Identity number	name	highway mileage	population (in thousands)	area (in tens of square miles)	Identity number	name	highway mileage	population (in thousands)	area (in tens of square miles)
1	Accra	284	225	92	21	Kumasi 1–2	259	157	39
2	Axim	168	74	151	22	Kumasi 3	205	70	81
3	Sekondi	140	106	37	23	Kumasi 4	287	84	395
4	Akwapim [c]	115	114	40	24	Kumasi 5	139	65	76
5	Kibi	286	219	225	25	Mampong E.	83	13	460
6	Mpraeso N.	16	6	200	26	Mampong W.	262	90	236
7	Mpraeso S.	129	73	82	27	Wenchi	359	73	681
8	Oda	162	72	94	28	Sunyani	238	109	248
9	Cape Coast	435	269	208	29	Dagomba E.	424	121	672
10	Dunkwa	103	40	93	30	Dagomba W.	308	104	289
11	Winneba	333	187	92	31	Gonja E.	226	52	578
12	Ho	432	173	246	32	Gonja W.	307	33	869
13	Keta	205	190	116	33	Krachi	194	32	338
14	Ada	90	114	80	34	Fra Fra	132	164	79
15	Sefwi	146	65	270	35	Gambaga	239	78	279
16	Volta River	201	165	146	36	Kusasi	266	148	124
17	Enchi	52	10	125	37	Navrongo	178	142	155
18	Tarkwa	256	121	346	38	Lawra	169	89	110
19	Bekwai	247	91	128	39	Tumu	214	30	273
20	Obuasi	169	67	94	40	Wa	333	85	340

[a] Source: *Road Map of the Gold Coast*, 1:500000, Department of Surveys, Accra, 1950. Only first and second class roads were used and no weighting system was applied.
[b] Source: *The Gold Coast, Census of Population, 1948, Report and Tables*, The Government Printing Department, Accra, Gold Coast (Ghana), 1950.
[c] Includes New Juaben.

Table 8.5. Basic regression data for Nigeria.

Recording unit		y^a highway mileage	x_1^b population (in thousands)	x_2 area (in tens of square miles)	Recording unit		y^a highway mileage	x_1^b population (in thousands)	x_2 area (in tens of square miles)
Identity number	name				Identity number	name			
1	Adamawa	545	799	1856	26	Kontagora	534	251	1322
2	Muri	444	260	1101	27	Minna	289	142	576
3	Numan	131	121	221	28	Emaa[d]	390	417	396
4	Bauchi	680	512	1452	29	Lowland	129	194	480
5	Gombe	402	477	648	30	Pankshin	192	279	380
6	Katagum	385	434	512	31	Argungu	213	171	336
7	Idoma	204	319	375	32	Gwandu	375	489	751
8	Lafia	144	132	395	33	Sokoto	1059	2021	2561
9	Nasarawa	251	162	556	34	Zaria	780	806	1649
10	Tiv	491	719	986	35	Abeokuta	448	630	427
11	Wukari	107	137	622	36	Benin	772	901	846
12	Bedde–Potiskum	111	160	367	37	Colony	101	510	135
13	Bornu	1339	1006	3299	38	Delta	315	590	644
14	Dikwa	330	265	515	39	Ibadan	617	1661	452
15	Biu	210	164	392	40	Ijebu	211	348	247
16	Borgu	312	100	1091	41	Ondo	695	946	816
17	Ilorin	339	460	265	42	Oyo	644	783	970
18	Lafiaga	170	70	416	43	Bamenda	252	429	693
19	Igala	353	361	498	44	Calabar	730	1841	625
20	Kabba[c]	310	303	497	45	Kumba–Victoria	211	224	532
21	Katsina	823	1483	947	46	Mamfe	136	100	432
22	Kano	1135	2933	1293	47	Ogoja	571	1082	749
23	Northern	302	424	370	48	Onitsha	770	1768	489
24	Abuja	189	101	395	49	Owerri	860	2080	387
25	Bida	299	221	574	50	Rivers	207	747	701

[a] Source: *Mobil Road Map of Nigeria, 1957*, from the Federal Survey Department, Lagos, at the scale of 1:750000. Maps of the Survey Department at a scale of 1:500000 were used to enumerate additional local roads in Onitsha and Owerri.
[b] Source: *Population Census of Nigeria, 1952–1953*, Government Statistician, Lagos, 1954.
[c] Includes the Kabba, Igbirra, Koto–Kaaifi districts of Kabba Province.   [d] Includes the Emaa, Jos, and southern districts of Plateau Province.

as follows:

Ghana
(a) $y = 130.636 + 0.8654 x_1$ $R^2 = 0.284$
          (0.2230)
(b) $\log y = 1.2774 + 0.5263 \log x_1$ $R^2 = 0.491$
                (0.0869)
(c) $y = 166.993 + 0.2298 x_2$ $R^2 = 0.195$
          (0.0757)
(d) $\log y = 1.7326 + 0.2461 \log x_2$ $R^2 = 0.100$
                (0.1199)
(e) $\log y = 0.1625 + 0.6293 \log x_1 + 0.4118 \log x_2$ $R^2 = 0.756$
                (0.0630) (0.0647)

Nigeria
(a) $y = 194.178 + 0.3742 x_1$ $R^2 = 0.657$
          (0.0390)
(b) $\log y = 1.0213 + 0.5805 \log x_1$ $R^2 = 0.650$
                (0.0615)

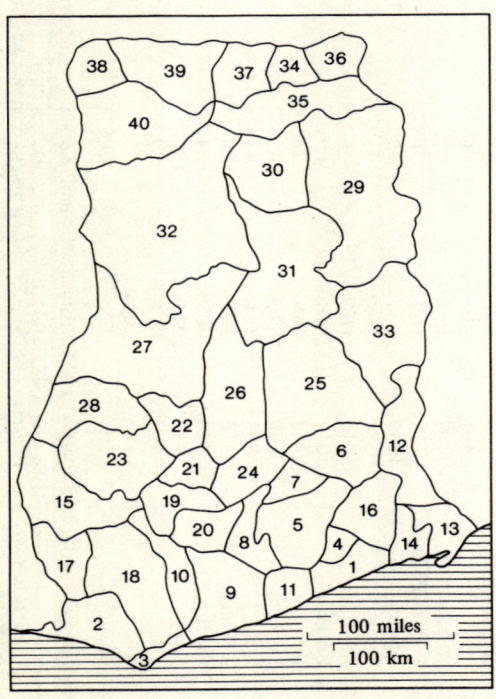

Figure 8.4. Location and identity numbers used in table 8.4 for recording units in Ghana.

(c) $y = 159.941 + 0.3628 x_2$ $\qquad R^2 = 0.541$
    $\qquad\quad (0.0482)$
(d) $\log y = 0.4272 + 0.7596 \log x_2$ $\qquad R^2 = 0.487$
    $\qquad\qquad\quad (0.1124)$
(e) $y = 60.733 + 0.2892 x_1 + 0.2510 x_2$ $\qquad R^2 = 0.883$
    $\qquad\quad (0.0247) \quad\; (0.0264)$
(f) $\log y = 0.0104 + 0.4512 \log x_1 + 0.4848 \log x_2$ $\quad R^2 = 0.817$
    $\qquad\qquad\quad (0.0491) \qquad\quad (0.0743)$

Regressions (b) and (e) appear to be the best for Ghana, whereas (e) and (f) are best for Nigeria. The residuals from them are shown in table 8.6.

We tested the residuals from all the regressions for spatial autocorrelation using the following statistics:
(1) I as defined in equation (8.12) under assumption N [using moments (8.21) and (8.29)] and under assumption R [using moments (1.37) and (1.39)].
(2) BB and BW as defined in equations (1.12) and (1.13). Here we coded a county B if it had a positive residual and W if it had a negative residual. The moments of BB and BW were evaluated both under free sampling and under nonfree sampling. Throughout the analysis we put $w_{ij} = 1$ if the

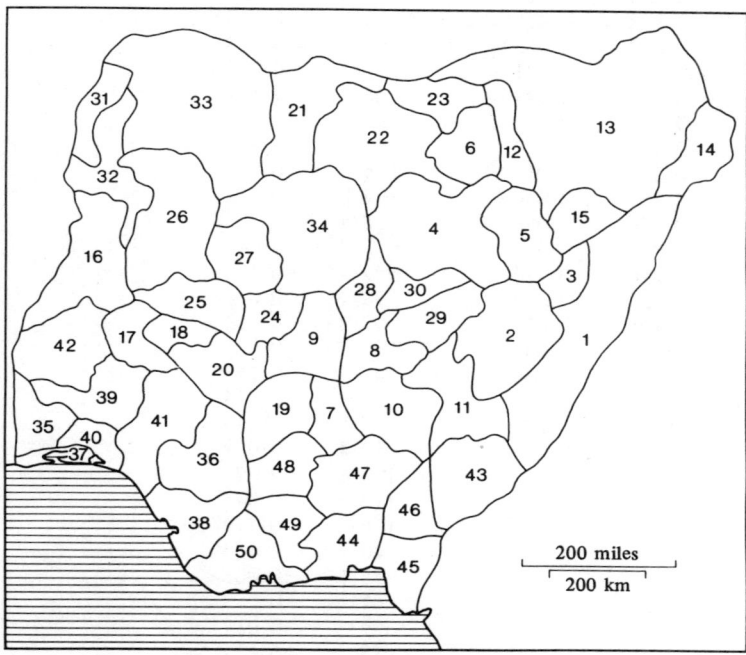

Figure 8.5. Location and identity numbers used in table 8.5 for recording units in Nigeria.

Table 8.6. Residuals from regressions for Ghana and Nigeria.

Recording unit	Regression for Ghana		Regression for Nigeria	
	(b)	(e)	(e)	(f)
1	−0·062	0·004	−212·9	−0·169
2	−0·036	−0·011	31·6	0·072
3	−0·197	0·063	−20·2	0·030
4	−0·299	−0·055	106·6	0·066
5	0·053	−0·148	40·6	0·022
6	−0·483	−0·396	70·2	0·071
7	−0·148	−0·013	−43·2	−0·079
8	−0·045	0·065	−54·1	−0·068
9	0·082	−0·008	3·8	0·061
10	−0·108	0·032	−25·3	−0·060
11	0·049	0·122	−149·6	−0·300
12	0·180	0·079	−88·2	−0·203
13	−0·165	−0·135	159·0	0·056
14	−0·406	−0·286	63·3	0·100
15	−0·067	−0·146	3·4	0·055
16	−0·141	−0·146	−51·6	0·108
17	−0·088	0·060	78·7	0·143
18	0·035	−0·110	−15·4	0·118
19	0·084	0·135	62·8	0·076
20	−0·011	0·103	36·8	0·054
21	−0·020	0·214	95·6	0·031
22	0·063	0·202	−98·6	−0·029
23	0·168	0·015	25·7	0·039
24	−0·089	0·065	−0·1	0·103
25	0·055	−0·041	30·2	0·070
26	0·112	0·049	68·7	0·121
27	0·297	0·032	42·6	0·141
28	0·027	−0·054	109·2	0·139
29	0·254	−0·011	−108·4	−0·232
30	0·150	0·046	−44·8	−0·082
31	0·174	−0·026	18·4	0·085
32	0·411	0·016	−15·7	−0·044
33	0·218	0·136	−229·3	−0·130
34	−0·323	−0·217	72·1	0·010
35	0·105	0·018	97·9	0·102
36	0·005	0·034	238·7	0·125
37	−0·160	−0·168	−141·1	0·261
38	−0·076	−0·002	−78·1	−0·124
39	0·276	0·237	−37·6	0·039
40	0·230	0·103	−12·4	0·007
41			155·8	0·077
42			113·3	0·045
43			−106·8	−0·174
44			−20·1	0·024
45			−48·1	−0·068
46			−62·1	−0·057
47			9·3	−0·017
48			75·2	0·107
49			100·6	0·172
50			−245·6	−0·371

ith and jth recording units shared a length of common boundary, and $w_{ij} = 0$ otherwise. The results are shown in table 8.7.

Given the assumption of independent, identically normally distributed residuals, the test based upon the coefficient I evaluated under assumption N is the efficient test, and we may judge the performance of the other coefficients using these results as our benchmark. As we wished to detect positive spatial autocorrelation, one-tailed critical regions were used. We note that:

(a) I evaluated under assumption R is on all occasions more conservative than I evaluated under assumption N in detecting positive spatial autocorrelation, and more liberal in detecting negative spatial autocorrelation. However, at the nominal $\alpha = 0.05$ level of significance, we would reach the same decision whether or not to accept the null hypothesis, H_0, of no spatial autocorrelation among the regression residuals.

(b) For the join counts, neither BB nor BW provides a strictly valid test of spatial autocorrelation among regression residuals for the reasons discussed in sections 6.4 and 8.2.3. However, despite this limitation, we found that for BW both under free and under nonfree sampling we would reach the same decision as for I under N as to whether or not to accept H_0 at the $\alpha = 0.05$ level for ten of the eleven regressions. There is also little to choose between the results for BW under free and nonfree sampling.

Table 8.7. Results of tests for spatial autocorrelation in the Taaffe et al (1963) regression residuals.

Standard deviate	Regression residuals					
	(a)	(b)	(c)	(d)	(e)	(f)
	Ghana					
I_N	2·59**	3·39**	−0·35	0·67	0·81	
I_R	2·48**	3·36**	−0·47	0·62	0·62	
Free sampling						
BB	0·60	0·36	−0·36	−1·33	0·84	
BW	−2·04*	−2·45**	0·20	−0·20	0·00	
Nonfree sampling						
BB	0·94	1·11	−0·73	−0·62	0·75	
BW	−2·39**	−2·85**	−0·50	−0·06	−0·16	
	Nigeria					
I_N	1·14	2·37**	2·37**	3·89**	0·90	−0·24
I_R	1·06	2·24**	2·34**	3·85**	0·72	−0·34
Free sampling						
BB	0·28	−0·23	2·08*	1·38	0·88	2·98**
BW	0·44	−2·55**	−2·91**	−2·91**	−0·26	−3·26**
Nonfree sampling						
BB	−1·14	0·88	0·92	1·47	1·66*	2·22*
BW	−0·53	−2·85**	−2·65**	−3·13**	−0·50	−2·80**

* Significant at $\alpha = 0.05$ level (one-tailed test).
** Significant at $\alpha = 0.01$ level (one-tailed test).

Conversely, for *BB* we would reach the same inferential decision as for *I* under N on only 6/11 occasions for free sampling, and on 4/11 occasions for nonfree sampling. There are also quite dramatic differences between the results under free and nonfree sampling, and with no consistent pattern to these differences.

However, the overall conclusion must be that the join-count statistics are essentially unreliable indicators of spatial autocorrelation among regression residuals.

(c) As regards the analysis carried out by Taaffe et al we note that, for Nigeria, the multiple regression produced a significant drop in the degree of spatial autocorrelation among the regression residuals compared with the results using either x_1 or x_2 alone. For Ghana, the multiple regression did not reduce autocorrelation significantly compared with x_2 alone. For both Nigeria and Ghana, the high values of R^2 for the multiple regressions, and the very small degree of spatial autocorrelation among the residuals, would suggest that these models are acceptable descriptions of the data. Taaffe et al found that little was gained by adding to the regressions the further independent variables postulated in their paper.

8.8 Trend-surface analysis: agricultural land values in Iowa, 1977-8

In trend-surface analysis, we assume that the value of some variate, Y, has been measured at several spatial locations, and that the x_1 (east-west) and x_2 (north-south) cartesian coordinates of the locations are known. The locations of the points are taken to provide information about the spatial pattern of variation in Y, leading to the polynomial regression,

$$Y = \sum_{i=0}^{p} \sum_{j=0}^{q} \beta_{ij} x_1^i x_2^j + \epsilon \ . \qquad (8.41)$$

Here, as before, the error terms, ϵ, are assumed to be independently normally distributed with zero means and variances, σ^2. The model is estimated by ordinary least squares, and is treated like any conventional multiple regression; all the assumptions about the errors discussed in sections 8.2.1 and 8.2.2 therefore apply. The model has been extensively used in geography, geology, and archaeology, and reviews are provided by Haggett et al (1977), Whitten (1974), and Hodder and Orton (1976). The values selected for p and q fix the order of the surface fitted. If $p = q = 1$, a linear (shed roof) surface is defined; $p = q = 2$ enters quadratic terms; and $p = q = 3$ defines a cubic surface. This is commonly the highest order of surface fitted.

To illustrate the use of the model, we have analysed data on the average $ value of land per acre in each of the 99 counties of Iowa for 1977, 1978, and the change from 1977-8. Here, 'change' is defined as

$$Y = 100 \times \left(\frac{1978 \text{ value}}{1977 \text{ value}} - 1 \right) \ .$$

The actual (observed) patterns of land values are shown in figures 8.6(a) and 8.7(a) (see over). Land values generally fall from north to south across the state. Highest values are found in the north-central region and in the east-central part of the state (Scott County). In the latter area, values are affected by development in the Quad Cities area of Davenport, Battendorf, Moline, and Rock Island. Areas exhibiting the biggest increase in values show a different pattern [figure 8.7(a)]. Greatest gains were made in counties in the north-east and south-west parts of the state.

Linear, quadratic, and cubic trend surfaces were fitted to the data. The centroids of each county were used to define x_1 and x_2. The surfaces are illustrated in figures 8.6(a) and 8.7(a), and reflect the features of the data described above. The values of the coefficients $\{\hat{\beta}_{ij}\}$ in equation (8.41) obtained for the best fit surfaces are recorded in table 8.8, along with the associated values of t and R^2. The analysis of variance is given in table 8.9. For the 1977 and 1978 land-values map, steady increases in R^2 are produced as quadratic and cubic terms are added. In the case of the change data, the quadratic surface is a dramatic improvement over the linear, whereas the cubic terms produce only a small increase in R^2. The relative unimportance of east–west variation alone, as opposed to north–south change, is confirmed by the nonsignificant t values for x_1^2 and x_1^3 in table 8.8, while the large values of the coefficients for $x_1 x_2$, $x_1^2 x_2$, and $x_1 x_2^2$ reflect the higher gains in the north east and south west.

In checking the overall goodness-of-fit of a trend surface, the residuals should, as we have noted, be tested for spatial autocorrelation and, when it is detected at a significant level, appropriate action taken. Indeed, recalling section 8.6, Geary (1954) has suggested that assessment of

Table 8.8. Trend-surface analysis of Iowa agricultural land-value data, 1977–8.

Terms in regression	1977		1978		% change, 1977–8	
	coefficient	t value	coefficient	t value	coefficient	t value
Constant	1955·63		2292·43		24·758	
x_1	−282·015	−3·75*	−315·427	−3·74*	−1·769	−13·45*
x_2	−367·935	−3·91*	−437·888	−4·15*	−1·471	8·47*
x_1^2	12·144	1·84	12·029	1·62	0·059	10·42*
$x_1 x_2$	80·373	9·42*	91·175	9·53*	0·104	13·38*
x_2^2	52·332	4·02*	61·015	4·18*	0·044	3·88*
x_1^3	0·139	0·70	0·223	1·00		
$x_1^2 x_2$	−3·138	−10·37*	−3·459	−10·19*		
$x_1 x_2^2$	−2·611	−6·52*	−3·015	−6·71*		
x_2^3	−2·361	−3·88*	−2·710	−3·97*		
Degrees of freedom	89		89		93	
R^2	0·85		0·86		0·72	

* Significant at $\alpha = 0·05$ level (two-tailed test).

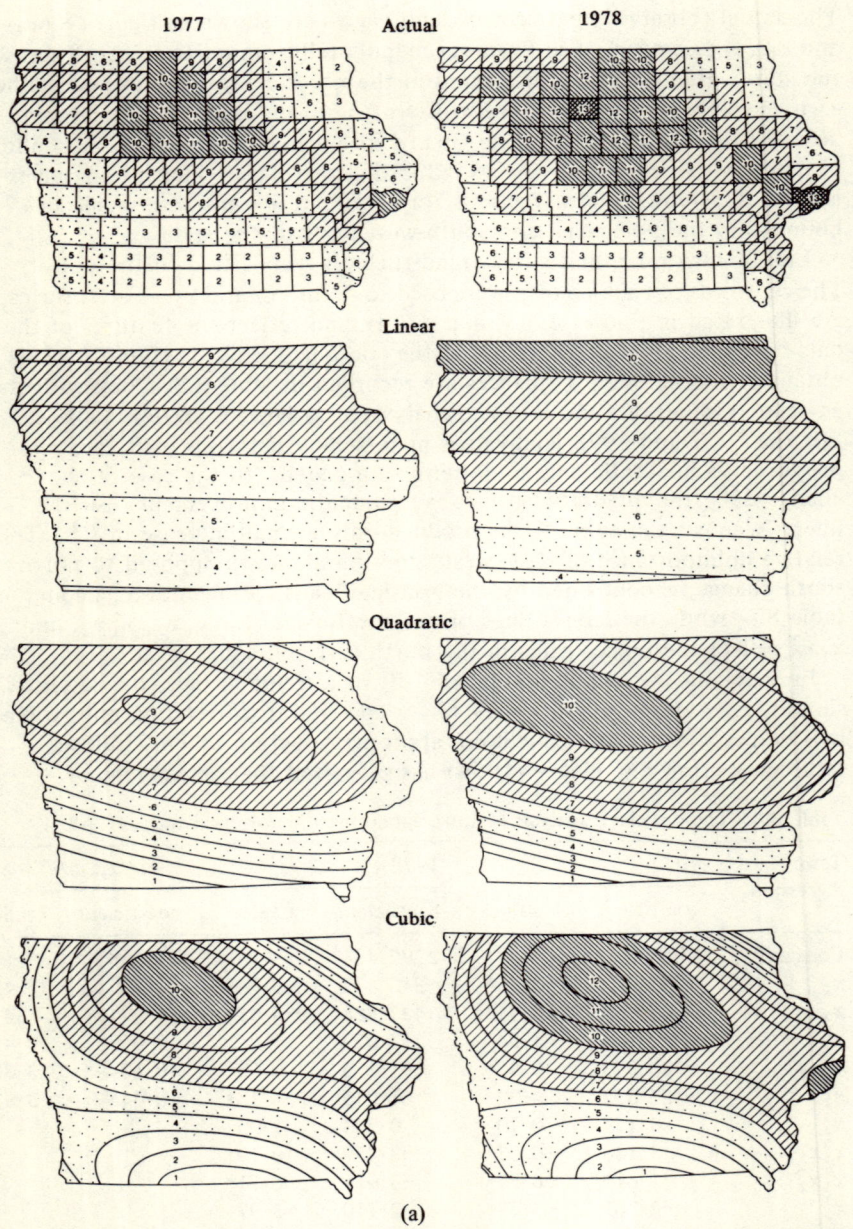

(a)

Figure 8.6. Average agricultural land values ($) per acre by county, Iowa, 1977 and 1978. (a) Actual land values and linear, quadratic, and cubic trend surfaces. (b) Residuals from trend surfaces and standardised deviate for I evaluated under assumption N. (Data source: *Des Moines Register*, December 15, 1978).

The analysis of regression residuals

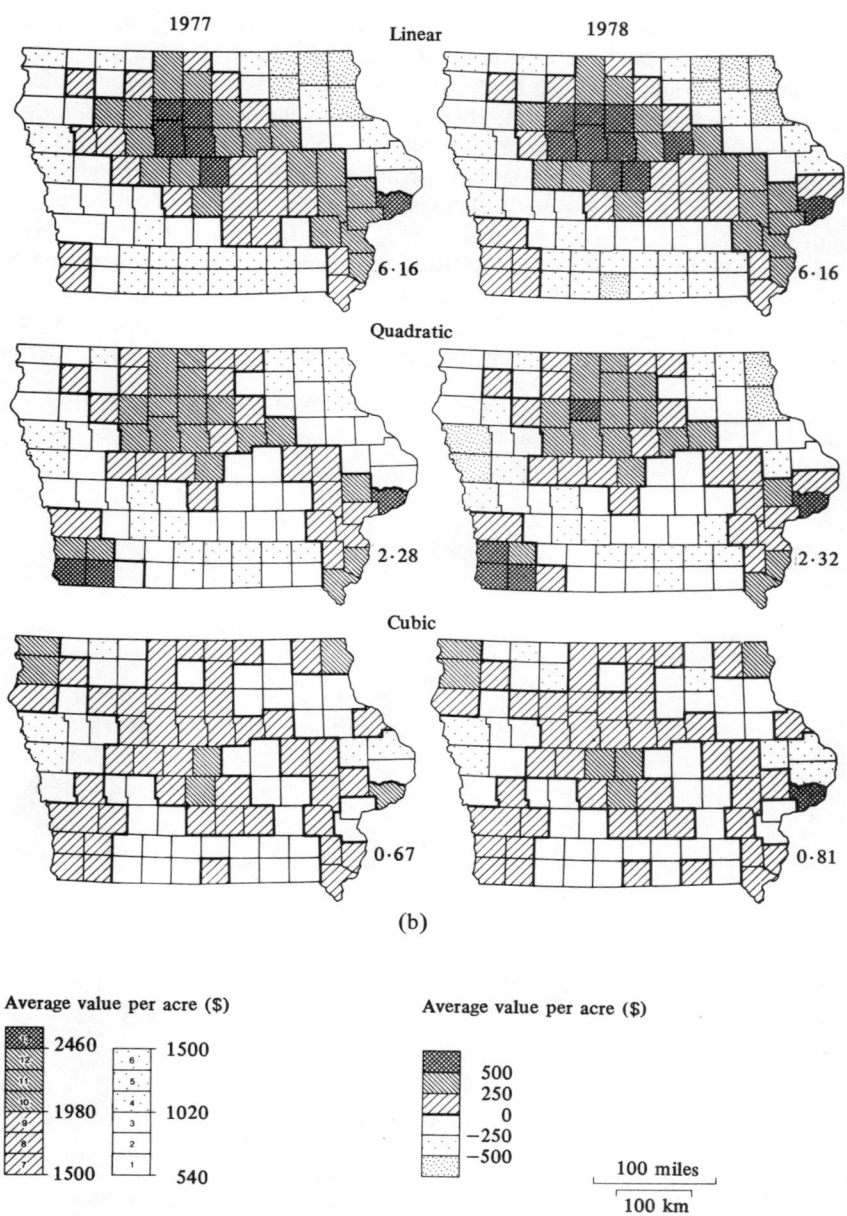

Figure 8.6 (continued)

goodness-of-fit should not be based on quantities like R^2 alone. He has argued that the purpose of a regression is to account for the systematic variation in some variable, Y, by a series of regressor variables. If the residuals lack autocorrelation, then all the systematic variation in Y has been accounted for, whatever the value in R^2, and there is no point in attempting to extend the analysis further by including more regressors.

We examined the residuals from all the trend surfaces for spatial autocorrelation using the coefficient, I, given in equation (8.12). We have tested I for significance by using the moments under assumption N [equations (8.21) and (8.29)] and by ignoring distributional issues [assumption R, equations (1.37) and (1.39)]. In these equations, we took $w_{ij} = 1$ if counties i and j were physically adjacent, and $w_{ij} = 0$ otherwise.

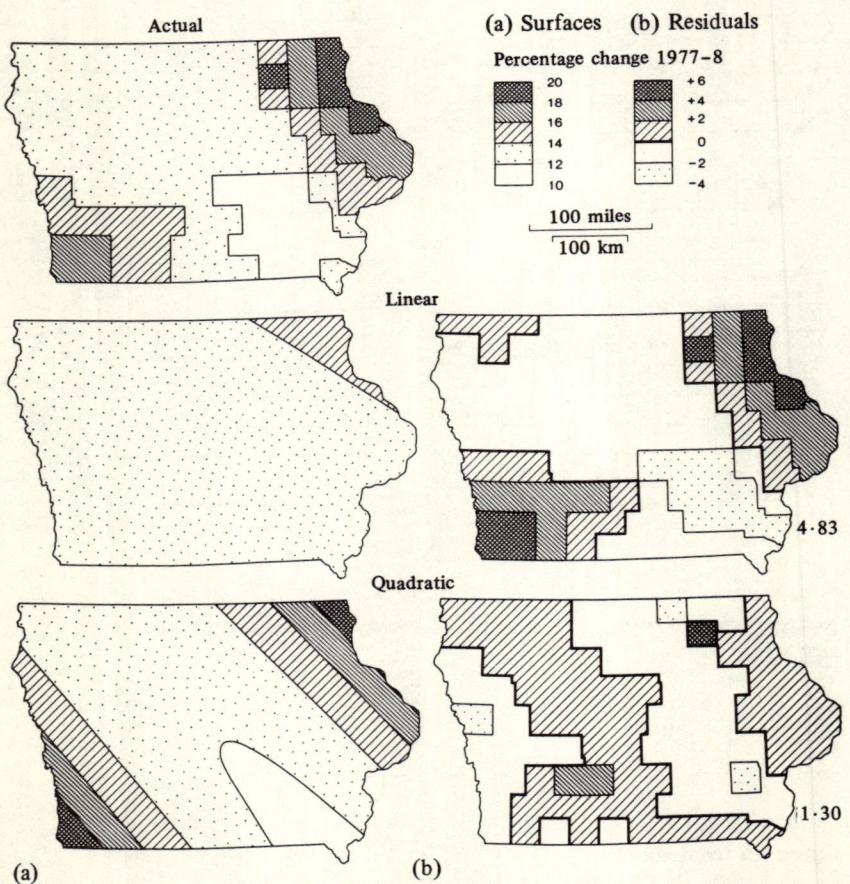

Figure 8.7. Percentage change by county in agricultural land values ($) per acre, Iowa, 1977-8. (a) Observed percentage change, linear, and quadratic trend surfaces. (b) Residuals from trend surfaces and standardised deviate for I evaluated under assumption N.

Table 8.9. Analysis of variance of trend-surface results for Iowa agricultural land-value data, 1977-8.

Year [a]	Source	Degrees of freedom	Sums of squares [a]	Observed F values	Increase in R^2
1977	Linear	2	6·11	$F_{2, 96} = 25·36*$	0·35
	AQ	3	4·54	$F_{3, 93} = 20·01*$	0·26
	AC	4	4·44	$F_{4, 89} = 38·09*$	0·24
	Residual	89	2·59		
1978	Linear	2	7·90	$F_{2, 96} = 26·93*$	0·36
	AQ	3	5·18	$F_{3, 93} = 18·03*$	0·24
	AC	4	5·64	$F_{4, 89} = 38·50*$	0·26
	Residual	89	3·26		
% change 1977-8	Linear	2	4·86	$F_{2, 96} = 0·56$	0·01
	AQ	3	300·74	$F_{3, 93} = 79·37*$	0·71
	AC	4	41·04	$F_{4, 89} = 11·95*$	0·10
	Residual	89	76·41		

* Significant at $\alpha = 0·05$ level.
[a] For 1977 and 1978, the sums of squares have been multiplied by 10^{-6}.
AQ ≡ addition of quadratic. AC ≡ addition of cubic.

Table 8.10. Results of tests for spatial autocorrelation in residuals from trend surfaces of Iowa agricultural land-value data.

Statistic	1977 data, trend surface			1978 data, trend surface		
	linear	quadratic	cubic	linear	quadratic	cubic
I	0·308	0·091	−0·004	0·308	0·093	0·004
Using assumption N						
$E(I)$	−0·022	−0·032	−0·041	−0·022	−0·032	−0·041
$\sigma(I)$	0·054	0·054	0·056	0·054	0·054	0·056
SD	6·16*	2·28*	0·67	6·16*	2·32*	0·81
Using assumption R						
$E(I)$	−0·010	−0·010	−0·010	−0·010	−0·010	−0·010
$\sigma(I)$	0·056	0·056	0·056	0·056	0·056	0·056
SD	5·71*	1·81	1·10	5·70*	1·85	0·25

	% change, 1977-8, trend surface		
	linear	quadratic	cubic
I	0·237	0·038	0·021
Using assumption N			
$E(I)$	−0·022	−0·032	−0·041
$\sigma(I)$	0·054	0·054	0·056
SD	4·83*	1·30	1·12
Using assumption R.			
$E(I)$	−0·010	−0·010	−0·010
$\sigma(I)$	0·056	0·056	0·053
SD	4·43*	0·87	0·59

* Significant at $\alpha = 0·05$ level (two-tailed test).   SD ≡ standard deviate.

The results are given in table 8.10. The spatial patterns of residuals and the standardised scores for statistic I under assumption N are shown in figures 8.6(b) and 8.7(b). The scores for the 1977 and 1978 data reflect the value of fitting third-order surfaces to those data, and confirm that the quadratic is adequate for the change maps. The cubic terms permit modelling of the high values both in the north-central and in the southeast parts of the 1977 and 1978 maps; a quadratic surface does not contain enough inflexions to allow this to be done.

From table 8.10 it is apparent that I evaluated under assumption R is again generally more conservative in assessing levels of positive autocorrelation than I evaluated under assumption N.

8.9 Conclusions
In this chapter, the theory of tests for spatial autocorrelation in residuals from regression analysis has been developed, and important differences in the distribution theory compared with tests using raw data have been indicated. The methods described have been illustrated by application to regression data on the economy of Eire, the structure of the transport networks of Ghana and Nigeria, and agricultural land values in Iowa. We stress that it is necessary to examine regression residuals for autocorrelation, and that in so doing, the formally correct tests are the only ones that should be used.

APPENDIX

Tables of weights for the counties of Eire, determined as functions of the distance between centres and the length of common boundary.

Table A8.1. Unstandardised weighting matrix for the twenty-six counties of Eire.

County	Contiguous counties and weights
A	I, 0·0039 J, 0·0142 K, 0·0031 Y, 0·0112 Z, 0·0119
B	L, 0·0099 N, 0·0061 Q, 0·0089 R, 0·0113 X, 0·0006
C	G, 0·0134 H, 0·0018 M, 0·0103 V, 0·0028
D	H, 0·0125 M, 0·0084 V, 0·0008 W, 0·0030
E	L, 0·0186
F	I, 0·0101 Q, 0·0246 Z, 0·0105
G	C, 0·0087 P, 0·0068 S, 0·0010 T, 0·0079 V, 0·0013
H	C, 0·0021 D, 0·0175 M, 0·0046
I	A, 0·0028 F, 0·0038 K, 0·0062 Q, 0·0059 S, 0·0059 Z, 0·0090
J	A, 0·0111 K, 0·0088 V, 0·0086 W, 0·0051 Y, 0·0044
K	A, 0·0025 I, 0·0074 J, 0·0090 S, 0·0185 V, 0·0033
L	B, 0·0106 E, 0·0008 N, 0·0055 T, 0·0079 U, 0·0098
M	C, 0·0080 D, 0·0077 H, 0·0030 V, 0·0095
N	B, 0·0079 L, 0·0067 T, 0·0105 X, 0·0202
O	Q, 0·0324 R, 0·0134
P	G, 0·0094 T, 0·0049 U, 0·0100
Q	B, 0·0066 F, 0·0051 I, 0·0046 O, 0·0071 R, 0·0005 S, 0·0005 X, 0·0071
R	B, 0·0319 O, 0·0109 Q, 0·0023
S	G, 0·0012 I, 0·0044 K, 0·0134 Q, 0·0006 R, 0·0010 V, 0·0062 X, 0·0088
T	G, 0·0088 L, 0·0049 N, 0·0054 P, 0·0039 S, 0·0009 U, 0·0054 X, 0·0015
U	L, 0·0100 P, 0·0133 T, 0·0089
V	C, 0·0016 D, 0·0006 G, 0·0012 J, 0·0047 K, 0·0018 M, 0·0071 S, 0·0046 W, 0·0055
W	D, 0·0051 J, 0·0071 V, 0·0140 Y, 0·0023
X	B, 0·0006 N, 0·0179 Q, 0·0092 S, 0·0120 T, 0·0022
Y	A, 0·0165 J, 0·0083 W, 0·0032 Z, 0·0108
Z	A, 0·0116 F, 0·0076 I, 0·0119 Y, 0·0072

In some examples we excluded county F (Dublin) from the analysis. If this is done the weights of counties contiguous to Dublin are changed, and the revised weights for these counties are as follows.

I	A, 0·0034 K, 0·0081 Q, 0·0063 S, 0·0068 Z, 0·0110
Q	B, 0·0083 I, 0·0052 O, 0·0087 R, 0·0008 S, 0·0008 X, 0·0086
Z	A, 0·0139 I, 0·0146 Y, 0·0086

Table A8.2. Standardised weighting matrix for the twenty-six counties of Eire.

County	Contiguous counties and weights						
A	I, 0·0874	J, 0·3207	K, 0·0699	Y, 0·2540	Z, 0·2680		
B	L, 0·2690	N, 0·1658	Q, 0·2426	R, 0·3073	X, 0·0153		
C	G, 0·4808	H, 0·0617	M, 0·3590	V, 0·0985			
D	H, 0·5056	M, 0·3411	V, 0·0327	W, 0·1206			
E	L, 1·0000						
F	I, 0·2226	Q, 0·5442	Z, 0·2332				
G	C, 0·3392	P, 0·2639	S, 0·0394	T, 0·3076	V, 0·0499		
H	C, 0·0866	D, 0·7218	M, 0·1916				
I	A, 0·0820	F, 0·1123	K, 0·1839	Q, 0·1760	S, 0·1761	Z, 0·2697	
J	A, 0·2919	K, 0·2312	V, 0·2259	W, 0·1345	Y, 0·1166		
K	A, 0·0610	I, 0·1808	J, 0·2217	S, 0·4558	V, 0·0807		
L	B, 0·3057	E, 0·0229	N, 0·1604	T, 0·2277	U, 0·2833		
M	C, 0·2827	D, 0·2732	H, 0·1075	V, 0·3365			
N	B, 0·1628	L, 0·1387	T, 0·2169	X, 0·4816			
O	Q, 0·7080	R, 0·2920					
P	G, 0·3881	T, 0·2000	U, 0·4119				
Q	B, 0·2109	F, 0·1604	I, 0·1465	O, 0·2239	R, 0·0167	S, 0·0161	X, 0·2254
R	B, 0·7075	O, 0·2416	Q, 0·0508				
S	G, 0·0348	I, 0·1245	K, 0·3742	Q, 0·0177	T, 0·0289	V, 0·1748	X, 0·2451
T	G, 0·2871	L, 0·1586	N, 0·1748	P, 0·1269	S, 0·0306	U, 0·1749	X, 0·0471
U	L, 0·3114	P, 0·4126	T, 0·2760				
V	C, 0·0602	D, 0·0203	G, 0·0432	J, 0·1743	K, 0·0649	M, 0·2611	S, 0·1713
	W, 0·2045						
W	D, 0·1797	J, 0·2488	V, 0·4901	Y, 0·0814			
X	B, 0·0134	N, 0·4288	Q, 0·2196	S, 0·2861	T, 0·0521		
Y	A, 0·4254	J, 0·2146	W, 0·0810	Z, 0·2791			
Z	A, 0·3036	F, 0·1985	I, 0·3099	Y, 0·1880			

In some examples we excluded county F (Dublin) from the analysis. If this is done the weights of counties contiguous to Dublin are changed, and the revised weights for these counties are as follows.

I	A, 0·0959	K, 0·2274	Q, 0·1766	S, 0·1907	Z, 0·3095	
Q	B, 0·2560	I, 0·1624	O, 0·2685	R, 0·0233	S, 0·0249	X, 0·2649
Z	A, 0·3740	I, 0·3935	Y, 0·2325			

9

Models containing components both regressive and autoregressive

9.1 Introduction

In chapter 6 we developed autoregressive models to describe purely spatial processes. Although these schemes have the advantage that they take account of the covariance structure of the process, they require that the mean be constant. The natural way to handle a location dependent mean is to use regression analysis. However, as we have seen in the last chapter, the assumption of independence among the residuals may not be satisfied, so that ordinary least squares procedures become inefficient. Thus, we need to develop models which incorporate both a regression component for the mean and an autoregressive component to reflect the spatial structure. We describe the theory and estimation procedures for such models in section 2, and then go on in sections 3 and 4 to illustrate the application of these schemes.

9.2 Regression models with autoregressive components

There are two ways of defining a 'mixed' scheme; either we can specify a regression which has autoregressive terms, or we can formulate a regression model with spatially autocorrelated residuals. These alternatives represent different assumptions concerning the nature of the process and are not competitors; we shall develop them in turn. In each case, we assume that there are n sites with dependent variable Y_i and regressor variables $x_{i1}, x_{i2}, ..., x_{ik}$; where $i = 1, ..., n$, and $x_{i1} \equiv 1$. Error terms are denoted by ϵ_i and u_i. In matrix notation, these variables are written as Y, ϵ, u [all $(n \times 1)$ vectors] and X [an $(n \times k)$ matrix].

9.2.1 A regression model with autoregressive terms

If we use the simultaneous approach of section 6.2.3, we may write

$$Y = \rho WY + X\beta + \epsilon , \tag{9.1}$$

where W is the weighting matrix, and ρ and β ($k \times 1$ vector) are parameters. When $\rho = 0$, expression (9.1) reduces to a standard regression equation, while for $\beta = 0$, we obtain the purely spatial scheme of equation (6.14). To complete the specification of the model, we assume that the errors are normally distributed with zero means and equal variances, σ^2; that is

$$\epsilon \sim N(0, \sigma^2 I) .$$

In accordance with section 6.3.2, the log-likelihood may be written as

$$\mathcal{L} = \text{const} - \tfrac{1}{2} n \ln \omega + \ln|I - \rho W| - \frac{1}{2\omega}(Y - \rho WY - X\beta)^T(Y - \rho WY - X\beta) , \tag{9.2}$$

where $\omega = \sigma^2$.

If we let $Z = Y - \rho WY$, then the maximum likelihood estimators are given by (cf Ord, 1975; Bodson and Peeters, 1975)

$$\left.\begin{aligned}\hat{\beta} &= (X^TX)^{-1}X^TZ, \\ n\hat{\omega} &= Z^TZ - Z^TX\hat{\beta},\end{aligned}\right\} \quad (9.3)$$

and $\hat{\rho}$ is that value of ρ which maximises

$$-n\ln\hat{\omega} + 2\ln|I - \rho W|. \quad (9.4)$$

The value $\hat{\rho}$ must then be substituted back into equations (9.3). The objective function, (9.4), is of the same form as that for the purely autoregressive scheme given in equation (6.40). To simplify the numerical details, we may write $n\hat{\sigma}^2 = n\hat{\omega}$, as

$$n\hat{\omega} = S(y, y) - 2\rho S(y, y^*) + \rho^2 S(y^*, y^*),$$

where $y^* = Wy$, and

$$S(y, y^*) = \sum_{i=1}^{n} (y_i - \bar{y})(y_i^* - \bar{y}^*),$$

and so on. The large sample variance-covariance matrix is derived in the appendix to this chapter. In Cliff and Ord (1973), we suggested that the estimators could be calculated by an iterative procedure, by following the procedure of Cochrane and Orcutt (1949). However, the further experience of ourselves and others (Doreian, 1981) suggests that the best method is to evaluate $\hat{\rho}$ by a search on function (9.4), and then to solve equations (9.2) and (9.3) directly.

From the discussion in section 6.2.4, we might propose a conditional model as

$$\left.\begin{aligned}E(Y_i|y_i^*) &= \rho w_i^T y_i^* + x_i^T \beta, \\ \text{var}(Y_i|y_i^*) &= \sigma^2,\end{aligned}\right\} \quad (9.5)$$

where w_i^T and x_i^T are the ith rows of W and X respectively, and y_i^* denotes y after deletion of y_i. When the Y are taken to be normal, this specification yields the joint distribution,

$$Y \sim N(A^{-1}X\beta, \sigma^2 A^{-1}),$$

in which $A = I - \rho W$. Maximum likelihood estimation for this model is awkward, because the iterative procedure requires the inversion of A at each stage. However, the form of equations (9.5) is such that ordinary least squares provides consistent estimators for ρ and β.

The advantage of schemes (9.1) and (9.5) is that they offer a ready extension of the simultaneous and conditional models developed earlier. The drawback, as with the original simultaneous scheme, is that they do not necessarily correspond to any kind of spatiotemporal process. We shall now investigate this question further.

9.2.2 Spatiotemporal processes

The standard regression model involving first-order lags both in space and in time is

$$Y_t = HY_{t-1} + X_t\beta + \epsilon_t , \qquad t = 1, ..., T ; \tag{9.6}$$

where H is an $(n \times n)$ matrix dependent on one or more parameters. A moment's reflection (or a lot of algebra!) is sufficient to demonstrate that we cannot hope to encapsulate this process by a purely spatial model, since the information carried by X_t varies at each time point. However, if the model is reformulated in terms of spatially autocorrelated errors as

$$\left. \begin{array}{l} Y_t = X_t\beta + \epsilon_t \\ \epsilon_t = H\epsilon_{t-1} + u_t \end{array} \right\} \tag{9.7}$$

or

$$Z_t = HZ_{t-1} + u_t \tag{9.8}$$

where

$$Z_t = Y_t - X_t\beta ,$$

then progress is possible. In particular, if we now assume that the errors u_{ti} are independent and identically normally distributed,

$$u_t \sim N(0, \sigma_u^2 I_n) ,$$

then it follows that

$$Z_t \sim N[0, \sigma_u^2(I - H^2)^{-1}] ,$$

provided that all the eigenvalues of H are less than one in modulus. Thus

$$Y_t \sim N[X_t\beta, \sigma_u^2(I - H^2)^{-1}] ; \tag{9.9}$$

representation (9.9) is possible because we are imposing the spatial structure upon the errors, rather than as a direct autoregressive component. Note that expression (9.9) gives a conditional form for the covariance matrix; see section 6.2.4. Although there is no simple correspondence between H^2 and the weighting matrices we might employ for purely spatial schemes, result (9.9) does suggest that the conditional version should be used rather than the simultaneous scheme. An important feature of (9.9) is that the vector β carries over from the spatiotemporal process to the spatial model, so that these parameters may be estimated from the purely spatial data.

When data are available for several time periods, either of models (9.6) or (9.7) may be fitted by standard regression methods (cf Bennett, 1979, chapter 9).

9.2.3 Regression with autocorrelated errors

Following on from the discussion of the last section, we now propose the spatial model

$$Y \sim N[X\beta, \sigma^2(I-G)^{-1}], \qquad (9.10)$$

where the time subscripts have been dropped for convenience and we have inserted a standard form of spatial covariance for the conditional scheme. From section 6.2.4, this may be written in conditional expectation form as

$$E(Y_i|) \equiv E(Y_i|y_j, j \neq i) = x_i^T\beta + \sum_{j \neq i} g_{ij}(y_j - x_j^T\beta), \qquad (9.11)$$

and

$$\text{var}(Y_i|Y_j = y_j, j \neq i) = \sigma^2,$$

where x_i^T denotes the ith row of X, $i = 1, ..., n$. Equation (9.11) could be used to develop least squares estimators, as shown by the following example.

Example 9.1
Suppose that

$$Y_i = \alpha + \beta x_i + \epsilon_i, \qquad (9.12)$$

and

$$\epsilon \sim N[0, \sigma^2(I - \rho W)^{-1}],$$

so that expression (9.11) becomes

$$E(Y_i|) = \alpha + \beta x_i + \rho \sum_{j \neq i} w_{ij} y_j - \rho \alpha w_{i.} - \rho \beta \sum w_{ij} x_j, \qquad (9.13)$$

where $w_{i.} = \sum_{j \neq i} w_{ij}$.

If we treat γ ($= \rho\beta$) as a fourth independent parameter, ordinary least squares can be used, although there will be no constant unless $w_{i.} = w_{k.}$ for all $k \neq i$. A more efficient approach would be to use constrained least squares, fitting the equation subject to $\gamma = \rho\beta$. Both these alternatives are likely to be more efficient than calibrating equation (9.12) by least squares, and ignoring the structure of the errors. Model (9.5) is easier to fit by least squares, although the likelihood procedure is simpler for model (9.10). Both schemes are conditional models, although the spatio-temporal pedigree of model (9.10) appears to make it more attractive.

Returning to model (9.10), when $G = \rho W$, it is possible to develop the maximum likelihood estimators from the log-likelihood ($\omega = \sigma^2$),

$$\mathcal{L} = \text{const} - \tfrac{1}{2}n \ln \omega + \tfrac{1}{2}\ln|I - \rho W| - \frac{1}{2\omega}(Y - X\beta)^T A(Y - X\beta), \qquad (9.14)$$

in which $A = I - G$.

The usual line of argument yields

$$\hat{\beta} = (X^T A X)^{-1} X^T A X, \qquad (9.15)$$

$$n\hat{\omega} = (Y - X\hat{\beta})^T A(Y - X\hat{\beta}), \qquad (9.16)$$

where $\hat{\rho}\mathbf{W}$ should be substituted for $\mathbf{G}$, and $\hat{\rho}$ is that value of ρ which maximises

$$-n \ln \hat{\omega} + \ln|\mathbf{I} - \rho\mathbf{W}| . \qquad (9.17)$$

Comparing functions (9.4) and (9.17) we see that the only difference is the factor of 2; this reflects the use of the conditional model in place of the simultaneous scheme. If the simultaneous scheme is used the only changes in the estimators are that

$\mathbf{A}$ becomes $(\mathbf{I} - \rho\mathbf{W})(\mathbf{I} - \rho\mathbf{W}^T)$,

and the second term in function (9.17) has coefficient 2. The large sample variances for both schemes are given in the appendix to the chapter. The best computational procedure is to evaluate $\hat{\rho}$ by a direct search on function (9.17) and then to evaluate equations (9.15) and (9.16). The steps of the procedure are as follows:

(1) Calculate

scalars $a_0 = y^T y$, $a_1 = y^T \mathbf{W} y$,

vectors $v_0^T = y^T \mathbf{X}$, $v_1^T = y^T \mathbf{W} \mathbf{X}$,

and

matrices $\mathbf{M}_0 = \mathbf{X}^T \mathbf{X}$, $\mathbf{M}_1 = \mathbf{X}^T \mathbf{W} \mathbf{X}$.

(2) Given a trial value of ρ, evaluate

$$n\hat{\omega} = a_0 - \rho a_1 - (v_0 - \rho v_1)^T (\mathbf{M}_0 - \rho \mathbf{M}_1)^{-1} (v_0 - \rho v_1) .$$

(3) Function (9.17) may now be written as

$$-n \ln \hat{\omega} + \sum \ln(1 - \rho \lambda_i) , \qquad (9.18)$$

where λ_i denotes the ith eigenvalue of $\mathbf{W}$.

(4) Determine $\hat{\rho}$ by a direct search procedure on function (9.18).

(5) Hence evaluate $\hat{\omega}$, and

$$\hat{\beta} = (\mathbf{M}_0 - \hat{\rho}\mathbf{M}_1)^{-1}(v_0 - \hat{\rho}v_1) .$$

For the joint scheme, we define

$a_2 = y^T \mathbf{W}^T \mathbf{W} y$, $v_2 = y^T \mathbf{W}^T \mathbf{W} \mathbf{X}$, and $\mathbf{M}_2 = \mathbf{X}^T \mathbf{W}^T \mathbf{W} \mathbf{X}$,

whence

$$n\hat{\omega} = a_0 - 2\rho a_1 + \rho^2 a_2$$
$$- (v_0 - 2\rho v_1 + \rho^2 v_2)(\mathbf{M}_0 - 2\rho\mathbf{M}_1 + \rho^2\mathbf{M}_2)^{-1}(v_0 - 2\rho v_1 + \rho^2 v_2) ;$$

also, we replace function (9.18) by (9.4).

For a discussion on the use of such models in the analysis of field experiments, see Bartlett (1978).

9.3 The Huk rebellion in the Philippines

Mitchell (1969) studied the pattern of insurgent control during the Huk rebellion in the Philippines, linking control to a variety of cultural and economic factors; the data refer to 57 municipalities in four provinces. Doreian and Hummon (1976) pointed out that control of any given area either by the government or by the insurgents has immediate relevance for the control in adjacent areas—we would expect the insurgency to spread, or to contract, through contiguous areas. Thus Doreian and Hummon proposed a regression model with a spatial component of the form set out in scheme (9.1). The regressor variables were:

P   the proportion of the population speaking the Pampangan dialect,
FMP   farmers as a percentage of the population,
OWN   owners as a percentage of farmers,
SGR   the percentage of cultivated land given over to sugar cane,
MNT   the presence of mountainous terrain (dummy),
SWP   the presence of swamps (dummy).

For further details on these variables, and on the rationale for their inclusion, see Mitchell (1969) or Doreian and Hummon (1976). In the regression equation, P is used multiplicatively with the other variables, so that the exogenous variables are P*FMP, P*OWN, P*SGR, P*MNT, and P*SWP.

Table 9.1 gives three sets of estimates for a (linear) model linking insurgent (Huk) control to the cultural, demographic, economic, and physical exogenous variables for each municipality. The nonspatial model is the standard linear model, fitted by ordinary least squares without any spatial autoregressive component. In the second part of the table, two spatial models are estimated. In the first of these, the simultaneous scheme (9.1) has been fitted by maximum likelihood, whereas in the second, ordinary least squares has been used to estimate the parameters of conditional model (9.5). The OLS results for the spatial model have not been identified previously in terms of a conditional scheme. The most noteworthy features of the results are the inflated parameter estimates and the greatly increased standard errors of these estimates induced by the nonspatial scheme. In addition, for the spatial schemes the proportion of variance explained increases from 73% to 80%. Both the spatial schemes indicate a strong element of geographical interaction and, despite their differences in formulation, they arrive at closely similar estimates for ρ. Thus the presence of spatial interaction is well established.

The estimated standard errors are somewhat higher for the conditional scheme than for the simultaneous version. How far this is due to the inefficient procedure used for the conditional model and how far it reflects a difference between the two formulations is not known. The only striking difference between the standard errors is for the spatial interaction coefficient itself. Here the use of the asymptotic expression (given in the appendix) may account for part of the difference. Further work is required to establish the adequacy of the asymptotic results for these models.

If we now turn to the regression element of the spatial model, it can be seen that only the variables P*FMP, P*OWN, and P*MNT are significant at the 10% level, whereas all variables appear significant in the nonspatial version. However, the sugar cane variable (P*SGR) alone has a coefficient less than its standard error, and it would appear that this term is acting as a surrogate for spatial interaction in the nonspatial model.

Table 9.1. Alternative estimations for the (multiplicative) model of Huk insurgent control. (Source: Doreian, 1981.)

Nonspatial model
OLS

$$Y = 1 \cdot 147 + 3 \cdot 794 \text{P*FMP} - 1 \cdot 912 \text{P*OWN} + 0 \cdot 461 \text{P*SGR} + 38 \cdot 38 \text{P*MNT}$$
$$(2 \cdot 94)^a \quad (0 \cdot 939) \quad\quad (0 \cdot 438) \quad\quad (0 \cdot 161) \quad\quad (7 \cdot 02)$$
$$+ 17 \cdot 17 \text{P*SWP}$$
$$(7 \cdot 94)$$

$$R^2 = 0 \cdot 73$$

Spatial model
Simultaneous: MLE [b]

$$Y = -1 \cdot 316 + 0 \cdot 571 \text{WY} + 1 \cdot 942 \text{P*FMP} - 0 \cdot 889 \text{P*OWN} + 0 \cdot 118 \text{P*SGR}$$
$$(2 \cdot 39) \quad (0 \cdot 008) \quad\quad (0 \cdot 762) \quad\quad (0 \cdot 355) \quad\quad (0 \cdot 132)$$
$$+ 28 \cdot 75 \text{P*MNT} + 11 \cdot 41 \text{P*SWP}$$
$$(5 \cdot 69) \quad\quad\quad (6 \cdot 44)$$

$$R^2 = 0 \cdot 80$$

Conditional: OLS

$$Y = -1 \cdot 382 + 0 \cdot 586 \text{WY} + 1 \cdot 892 \text{P*FMP} - 0 \cdot 862 \text{P*OWN} + 0 \cdot 108 \text{P*SGR}$$
$$(2 \cdot 62) \quad (0 \cdot 138) \quad\quad (0 \cdot 928) \quad\quad (0 \cdot 453) \quad\quad (0 \cdot 164)$$
$$+ 28 \cdot 49 \text{P*MNT} + 11 \cdot 26 \text{P*SWP}$$
$$(6 \cdot 51) \quad\quad\quad (7 \cdot 01)$$

$$R^2 = 0 \cdot 80$$

[a] The estimated standard errors for each coefficient are given in brackets. The asymptotic formulae (see the appendix) were used for the simultaneous scheme.
[b] WY is used to denote the regressor $w_i^T y_i^*$ for Y_i.

9.4 Consumption of agricultural output in Eire

In section 8.6, we mentioned the persistence of a pattern of positive spatial autocorrelation among the residuals from O'Sullivan's regression, which suggested the use of an autoregressive model. The reader will recall that O'Sullivan regressed the gross agricultural output of each county consumed by itself upon the index of arterial road accessibility (ARA).

Making use of the procedure outlined in section 9.2.3, we can specify the equations

$$Y_i = \alpha + \beta x_i + \epsilon_i , \qquad i = 1, ..., n , \tag{9.19}$$

and

$$\epsilon \sim N(0, \sigma^2 A^{-1}) ,$$

where **A** is defined either according to the simultaneous or to the conditional scheme. In the conditional case these equations yield the transformed relationship shown in (9.13). This should, however, be estimated subject to the constraint that $\gamma = \rho\beta$.

We have considered several different models and estimation procedures as follows.

A nonspatial model The original regression was fitted without any allowance for spatial autocorrelation among the residuals.

A spatial model, simultaneous version For this type of model we use standardised weights **W** such that $\sum_j w_{ij} = 1$ for all i. These weights are given in the appendix to chapter 8.

The forms tried were:

First differences Using first differences, the new variables

$$y_1 = (I - W)y \qquad \text{and} \qquad x_1 = (I - W)x ,$$

were computed and the relationship $Y_{1i} = \delta_1 + \delta_2 x_{1i} + \epsilon_i$ was fitted by OLS. This assumes *a priori* that $\rho = 1$, so that the spatial process is considered nonstationary.

Maximum likelihood In this case $\tilde{y} = (I - \hat{\rho}W)y$, and $\tilde{x}$ is similarly defined.

A spatial model, conditional version Using unstandardised weights, we tried:

Full regression The 'full' regression was fitted by least squares, ignoring the constraint $\gamma = \rho\beta$. The variable name WY is used to denote the vector **W**y.

Constrained regression The regression model was fitted subject to the constraint $\gamma = \rho\beta$.

The results of this work are presented in table 9.2, and the different coefficients are summarised in table 9.3. As for the estimates recorded in table 9.1, we can see that the nonspatial least squares scheme appears to overestimate the x coefficient substantially. The lack of constraints on the other least squares procedures produces estimates for the WY coefficient outside the feasible region. This problem also occurred in Haining's (1978b) simulation study of least squares estimators. All in all, it would seem that there is no real alternative to maximum likelihood estimation unless the sample size is large enough to permit a high degree of accuracy in the least squares estimators.

Returning to the original problem studied by O'Sullivan (1969), the high level of spatial autocorrelation suggests a strong structural component in the rural economy, so that the extent of 'home' consumption of agricultural output is determined by historical and other factors as well as by road accessibility. Nevertheless, it is evident from table 9.3 that the ARA remains an important explanatory variable.

Another application of this model is the interesting study of the Belgian labour market by Bodson and Peeters (1975). Their results also show a reduction in absolute value of most of the straight regression coefficients for the maximum likelihood (ML) scheme relative to the nonspatial least squares version. Again, the estimated standard errors are lower for the ML scheme.

Table 9.2. Results of autoregressive analyses of O'Sullivan's regression (2) data.

Regression	Equation	R^2	Explained variance [a]
Original variables	$y = -8.49 + 0.00527x$ (0.00070)	0.700	0.700
First differences (simultaneous)	$y = -0.038 + 0.00245x$ (0.00067)	0.356	0.880
Maximum likelihood (simultaneous)	$\tilde{y} = -1.32 + 0.00387\tilde{x}$ (0.00067)	—	—
Full regression (conditional)	$y = 0.625 + 0.00245x$ (0.00068) $-0.00265Wx + 1.018Wy$ $(0.00213)\quad(0.2325)$	0.882	0.882
Constrained regression (conditional)	$y = -0.036 + 0.00260x$ (0.00067) $-0.00265Wx - 1.018Wy$ $(0.00106)\quad(0.1143)$	0.881	0.881

Note: for the constrained regression, the standard errors of the coefficients of x and Wy were obtained by the asymptotic maximum likelihood formula. For the coefficient of Wx, the standard error was calculated using the asymptotic formula for the variance of a product. For the iterative method, the asymptotic maximum likelihood formula for the standard error was used.

[a] Explained variance is given by $\dfrac{\text{error variance}}{\text{variance}(y)}$.

Table 9.3. Coefficients implied by the autoregressive equations summarised in table 9.2.

Regression	x [a]	Wx [a]	Wy
Original variables	0.527	0	0
First differences (simultaneous)	0.245	−0.245	1
Maximum likelihood (simultaneous)	0.387	−0.195	0.505
Full regression (conditional)	0.245	−0.265	1.018
Constrained regression (conditional)	0.260	−0.265	1.018

[a] Coefficient multiplied by 100.

9.5 Residual autocorrelation in autoregressive schemes

Even after the scope of the spatial model has been extended to cover both regressive and autoregressive elements, there is still the possibility of an incomplete specification for the model, which may be reflected in a different pattern of autocorrelation among the residuals. Unfortunately, as noted by several analysts in the time-series case, the standard tests for autocorrelation cease to be valid when the model contains an autoregressive component. Durbin (1970) suggested a statistic for time series which enables the researcher to test for residual autocorrelation in the model, and his work has been extended by Wickens (1972) who used an instrumental variable technique. The greater complexity of the estimation procedures for spatial models is such that no satisfactory technique has been developed to handle the problem to date. This is one of many areas where further work is required and where greater practical experience (of the shortcomings) of existing methods would be useful.

9.6 Conclusions

In this chapter, the models described earlier in the book have been combined to produce a spatial autoregressive-regression scheme. Estimation procedures have been developed using ordinary least squares where appropriate, and also maximum likelihood. The models have been used to analyse data on the Huk rebellion in the Philippines and on the consumption of agricultural produce in Eire. In both cases, although it is noted that cultural and economic factors form an essential part of the underlying processes involved, it is also apparent that the inclusion of spatial components adds materially to an understanding of the problems analysed. More generally, we believe that a spatial dimension is likely to be important in a wide range of investigations, and it is our hope that this book has gone some way towards indicating how the presence of such spatial factors may be detected, suitably modelled, and estimated.

APPENDIX

Large sample variances for the regression-autoregressive models
A1. The regression model with autoregressive terms

The log-likelihood is given by equation (9.2). In accordance with Doreian (1981), we may evaluate the second derivatives of $\mathcal{L}$, setting $\omega = \sigma^2$ for convenience, as

$$\frac{\partial^2 \mathcal{L}}{\partial \omega} = \frac{n}{2\omega^2} - \frac{n\hat{\omega}}{\omega^3},$$

where $\hat{\omega} = \hat{\sigma}^2$ in function (9.4);

$$\frac{\partial^2 \mathcal{L}}{\partial \beta^2} = -\frac{\mathbf{X}^T \mathbf{X}}{\omega};$$

$$\frac{\partial^2 \mathcal{L}}{\partial \beta \partial \omega} = \frac{-\mathbf{X}^T(\mathbf{Z} - \mathbf{X}\beta)}{\omega^2};$$

$$\frac{\partial^2 \mathcal{L}}{\partial \omega \partial \rho} = \frac{-\mathbf{Y}^T \mathbf{W}(\mathbf{Z} - \mathbf{X}\beta)}{\omega^2};$$

$$\frac{\partial^2 \mathcal{L}}{\partial \beta \partial \rho} = \frac{-\mathbf{X}^T \mathbf{W} \mathbf{Y}}{\omega};$$

and

$$\frac{\partial^2 \mathcal{L}}{\partial \rho^2} = -\alpha - \frac{\mathbf{Y}^T \mathbf{W}^T \mathbf{W} \mathbf{Y}}{\omega},$$

where $\lambda_1 \geq \ldots \geq \lambda_n$ are the eigenvalues of $\mathbf{W}$, and

$$\alpha = \sum_{i=1}^{n} \frac{\lambda_i^2}{(1 - \rho \lambda_i)^2}.$$

The large sample covariance matrix is then given by

$$\mathbf{V} = \left[-\mathrm{E} \left(\frac{\partial^2 \mathcal{L}}{\partial \theta_r \partial \theta_s} \right) \right]^{-1},$$

where θ_r and θ_s are typical parameters; see Kendall and Stuart (1979, pages 59-60).

For our model,

$$\mathbf{V}(\omega, \beta, \rho) = \omega \begin{bmatrix} n/2\omega & \mathbf{0}^T & \mathrm{tr}(\mathbf{B}) \\ & \mathbf{X}^T \mathbf{X} & \mathbf{X}^T \mathbf{B} \mathbf{X} \beta \\ & & g \end{bmatrix}^{-1},$$

where the lower half is filled by symmetry;

$$\mathbf{B} = \mathbf{W}(\mathbf{I} - \rho \mathbf{W})^{-1},$$

$$g = \omega \, \mathrm{tr}(\mathbf{B}^T \mathbf{B}) + \beta^T \mathbf{X}^T \mathbf{B}^T \mathbf{B} \mathbf{X} \beta + \alpha \omega,$$

and
$$\mathrm{tr}(\mathbf{B}) = \sum \frac{\lambda_i}{1-\rho\lambda_i}.$$

When $\mathbf{W}$ is symmetric, $\mathrm{tr}(\mathbf{B}^T\mathbf{B}) = \alpha$.

For the spatial scheme without a regression component, the variance–covariance matrix reduces to

$$\mathbf{V} = \omega \begin{bmatrix} n/2\omega & \mathrm{tr}(\mathbf{B}) \\ & g \end{bmatrix}^{-1},$$

where $g = \mathrm{tr}(\mathbf{B}^T\mathbf{B}) + \alpha\omega$.

A2. Regression with autocorrelated residuals

By the same approach as that outlined above, we arrive at
the conditional scheme

$$\mathbf{V}(\omega, \beta, \rho) = \omega \begin{bmatrix} n/2\omega & \mathbf{0}^T & \tfrac{1}{2}\mathrm{tr}(\mathbf{B}) \\ & (\mathbf{X}^T\mathbf{A}\mathbf{X}) & \mathbf{0} \\ & & \tfrac{1}{2}\alpha\omega \end{bmatrix}^{-1},$$

where $\mathbf{A} = \mathbf{I} - \rho\mathbf{W}$, and $\mathbf{B} = \mathbf{W}(\mathbf{I} - \rho\mathbf{W})^{-1}$;
the simultaneous scheme

$$\mathbf{V}(\omega, \beta, \rho) = \omega \begin{bmatrix} n/2\omega & \mathbf{0}^T & \mathrm{tr}(\mathbf{B}) \\ & (\mathbf{X}^T\mathbf{A}\mathbf{X}) & \mathbf{0} \\ & & \alpha + \mathrm{tr}(\mathbf{B}^T\mathbf{B}) \end{bmatrix}^{-1},$$

where $\mathbf{A} = (\mathbf{I} - \rho\mathbf{W}^T)(\mathbf{I} - \rho\mathbf{W})$.

Glossary of notation

The glossary is divided into two parts: (a) mathematical notation and (b) terms used. In list (b) we have followed the convention that if a symbol is used *only* within the (sub)section where it is introduced and never again, then it does not appear in the glossary; the list is organised alphabetically A–Z, α–ω.

(a) Mathematical

$\hat{}$	the circumflex is used to denote the maximum likelihood estimator
$j \in J_i$	county j is a member of the set, J_i, of counties contiguous to county i
$n^{(j)}$	$n(n-1) \ldots (n-j+1)$
$\Gamma(x)$	gamma function, $\Gamma(x+1) = x\Gamma(x)$
$\sum_{(2)}, \sum_{(3)}, \sum_{(4)}$	$\sum_{\substack{i=1 \\ i \neq j}}^{n} \sum_{j=1}^{n}$, $\sum_{\substack{i=1 \\ i \neq j \neq k}}^{n} \sum_{j=1}^{n} \sum_{k=1}^{n}$, $\sum_{\substack{i=1 \\ i \neq j \neq k \neq l}}^{n} \sum_{j=1}^{n} \sum_{k=1}^{n} \sum_{l=1}^{n}$
5(1)10	the set of numbers 5,6,7,8,9,10
5(2)9	the set of numbers 5,7,9
→	approaches, goes to
$O(n^{-1})$	terms of order n^{-1}
$o(n^{-1})$	terms of smaller order than n^{-1}
∧, ∨	representing compounding and generalising of distributions

(b) Terms used

A	the total number of joins in the county system
$A(i, N)$	the accessibility of the ith vertex in the Nth road network; see section 8.6
ARA	arterial road accessibility; see section 8.6
ARE	asymptotic relative efficiency; see section 6.4
b_1	the sample coefficient of skewness, m_3^2/m_2^3; m_3 and m_2 are defined below
b_2	the sample coefficient of kurtosis, m_4/m_2^2; m_4 and m_2 are defined below
BB	the number of black–black joins in a county system
bishop's case	in a regular lattice with binary weights, $\delta_{ij} = 1$ if the ith and jth cells have a common vertex, and $\delta_{ij} = 0$ otherwise

BLUS estimators	the best linear unbiased estimators of the $(n-k)$ regression residuals which have a scalar covariance matrix
BW	the number of black–white joins in a county system
c	the spatial autocorrelation test statistic defined in equation (1.10)
$C(g)$	set of pairs of counties g steps apart
column-only case	in a regular lattice with binary weights, $\delta_{ij} = 1$ if the ith and jth cells have an edge in common on a column of the lattice, and $\delta_{ij} = 0$ otherwise
corr(x_1, x_2)	the sample correlation between the observations x_{i1}, x_{i2} ($i = 1, ..., n$) on the random variables X_1 and X_2
D	diameter of a graph
df	degrees of freedom
density function	for continuous variates, the probability in an elemental range dx, usually denoted by $f(x)dx$. Also, $f(x)$ is the derivative of the distribution function $F(x) = \text{prob}(X \leq x)$
d_{ij}	the 'distance' between the ith and jth counties
eigenvalue	for any matrix $\mathbf{A}$ the eigenvalues are the solution to the determinantal equation $\|\mathbf{A} - \lambda\mathbf{I}\| = 0$. Thus if $\mathbf{A} = \begin{bmatrix} 2 & 1 \\ 1 & 2 \end{bmatrix}$, $$\|\mathbf{A} - \lambda\mathbf{I}\| = \begin{vmatrix} 2-\lambda & 1 \\ 1 & 2-\lambda \end{vmatrix} = (2-\lambda)^2 - 1,$$ and $(2-\lambda)^2 - 1 = 0$ has roots $\lambda = 1$, and $\lambda = 3$. The corresponding row vectors u or column vectors v, for which $u\mathbf{A} = \lambda u$, or $\mathbf{A}v = \lambda v$, are called the eigenvectors of $\mathbf{A}$. The eigenvectors are determined up to an arbitrary scaling constant. Thus for the example, when $\lambda = 3$, $v = c_1 \begin{pmatrix} 1 \\ 1 \end{pmatrix}$, and when $\lambda = 1$, $v = c_2 \begin{pmatrix} 1 \\ -1 \end{pmatrix}$, c_1 and c_2 any constants. When $\mathbf{A}$ is symmetric, $u = v^T$ for each value of λ

Glossary of notation

eigenvector	see eigenvalue
$E(X)$	the expected, average or mean value of X, the first moment of X
F	Snedecor's F distribution
F as a subscript	refers to the free sampling assumption defined on page 12
$F, \hat{F}$	distribution function (DF) and empirical DF
$F(h)$	the asymptotic efficacy of a test based on h; see section 6.4.1
H_0	null hypothesis
H_1	alternative hypothesis
$\mathbf{I}$ or $\mathbf{I}_n$	the unit matrix (of order n)
I	the spatial autocorrelation test statistic defined in equation (1.9)
I'	a spatial autocorrelation test statistic; two uses (1) weighted form, equation (1.11), (2) BLUS regression residuals form, equation (8.31)
$I(g)$	the gth-order spatial autocorrelation using the test statistic I; see section 5.2
idempotent	a matrix $\mathbf{A}$ is said to be idempotent if $\mathbf{A}^2 = \mathbf{A}$
$I_{\text{s-t}}$	the space–time index (interaction coefficient) defined by equation (1.44)
J_{tot}	the total number of joins between counties of different colours
$K(r)$	distance-based statistic; see section 4.4.3
L	likelihood function
$\mathcal{L}$	log-likelihood function
likelihood ratio test	a test of a hypothesis H_0 against an alternative H_1 based on the ratio of two likelihood functions, one derived from each of H_0 and H_1
m_j, m_j'	sample moments corresponding to μ_j and μ_j', respectively, which are defined below
MAD	minimum absolute deviation
Markovian scheme	a stochastic process such that the conditional probability distribution for the state at any future instant, given the present state, is unaffected by any additional knowledge of the past history of the system

MCP	mean cross product
moment generating function	defined as $E[\exp(tX)]$ or $\sum_{j=0}^{\infty} t^j \mu'_j/j!$; exists only if all moments exist
MS	mean square
N	the normal distribution
N as a subscript	refers to assumption N defined on page 14
n	the total number of counties in the study area
n_1	the number of black cells in a two-colour lattice
n_2	the number of white cells in a two-colour lattice
n_r	the number of cells of colour r in a k-colour lattice
$N(\mu, \sigma^2)$	normal distribution with mean μ and variance σ^2
NF as a subscript	refers to the nonfree sampling assumption defined on page 12
NN distance	nearest neighbour (distance)
OLS	ordinary least squares
p	probability that a cell is coloured black in a two-colour lattice
$\hat{p}$	an estimate of p from the data, for example by n_1/n
p_r	probability that a cell is of colour r in a k-colour lattice
$P(h, \psi)$	power of a test based on the test statistic h when the parameter has the value ψ under H_1
PI distance	distance from a randomly selected point to nearest individual
positive definite	if the matrix $\mathbf{A}$ is a positive definite, all the eigenvalues of $\mathbf{A}$ will be positive
power of a test	$1 - \text{prob}(\text{type II error})$; the probability of rejecting H_0 when H_1 is true (cf size of a test)
$\text{prob}(X_i = 1)$ or $p(X_i = 1)$	probability that $X_i = 1$
$q = 1 - p$	probability that a cell is coloured white in a two-colour lattice
queen's case	in a regular lattice with binary weights, $\delta_{ij} = 1$ if the ith and jth cells have a common edge or vertex, and $\delta_{ij} = 0$ otherwise
R as a subscript	refers to assumption R defined on page 14

Glossary of notation

R^2	the coefficient of multiple correlation
R or R_i	random variable describing distance to an individual
r or r_{12}	the sample correlation between the observations x_{i1}, x_{i2} ($i = 1, ..., n$) on the random variables X_1 and X_2
$r_{12.3}$	the sample partial correlation between the observations x_{i1}, x_{i2} ($i = 1, ..., n$) on the random variables X_1 and X_2, after allowing for the effect of X_3
$r(k)$	the kth-order correlation
rook's case	in a regular lattice with binary weights, $\delta_{ij} = 1$ if the ith and jth cells have a common edge, and $\delta_{ij} = 0$ otherwise
row-only case	in a regular lattice with binary weights, $\delta_{ij} = 1$ if the ith and jth cells have a common edge on a row of the lattice, and $\delta_{ij} = 0$ otherwise
S	test using spacings between individuals in one dimension
S_0	$S_0 = \sum_{(2)} w_{ij}$, the sum of the weights ($= 2A$ when $w_{ij} = \delta_{ij}$); w_{ij} and δ_{ij} are defined below
S_1	$S_1 = \frac{1}{2} \sum_{(2)} (w_{ij} + w_{ji})^2 = 4A$ when $w_{ij} = \delta_{ij}$
S_2	$S_2 = \sum_{i=1}^{n} (w_{i.} + w_{.i})^2$; $w_{i.}$ and $w_{.i}$ are defined below
SCP	sum of cross products
SS	sum of squares
size of a test	prob(type I error), the probability of rejecting H_0 when it is true (cf power of a test)
T_0	$T_0 = \sum_{(2)} y_{ij}$; y_{ij} is defined below
T_1	$T_1 = \frac{1}{2} \sum_{(2)} (y_{ij} + y_{ji})^2$
T_2	$\sum_{i=1}^{n} (y_{i.} + y_{.i})^2$; $y_{i.}$ and $y_{.i}$ are defined below
TSS	total sum of squares
torus	the three-dimensional shape which results when the edges of a plane are joined together so that all end points (boundaries) are eliminated and the surface is continuous. The resulting object resembles a doughnut

tr(**T**)	the trace operator which sums elements on the leading diagonal of the matrix **T**
type I error	the error committed if, as the result of a statistical test, H_0 is rejected when it is true
type II error	the error committed if, as the result of a statistical test, H_0 is not rejected when it is false
$u = (u_1, u_2)$	point coordinates
V	covariance matrix with elements $\{v_{ij}\}$
v_{ij}	covariance between X_i and X_j
var(X)	the variance of X, the second moment of X
vâr	estimate of the variance of a statistic
w_{ij}	the weight assigned to the link between counties i and j (assumed to be nonnegative)
w'_{ij}	$w'_{ij} = w_{ij} + w_{ji}$
$w_{i.}, w_{.i}$	$w_{i.} = \sum_{j=1}^{n} w_{ij}, \quad w_{.i} = \sum_{j=1}^{n} w_{ji}$
W	the weighting matrix with elements w_{ij}
W(g)	weighting matrix for counties g steps apart
WW	the number of white–white joins in a county system
X or X_i	random variable
x or x_i	particular value taken on by X or X_i
$\bar{x}$	the average or mean value of the set of values
X	matrix of regressor variables
X^2	the test statistic, $X^2 = \sum_{i=1}^{n} [(O_i - E_i)^2 / E_i]$, which is approximately distributed as χ^2. Here O_i is the observed frequency in the ith cell and E_i is the expected frequency in that cell under H_0
y_{ij}	value assigned to the pair of counties i and j
$y_{i.}, y_{.i}$	$y_{i.} = \sum_{j=1}^{n} y_{ij}, \quad y_{.i} = \sum_{j=1}^{n} y_{ji}$
Y_i or $Y(x)$	random variable relating to location i or x
Y_{ij}	random variable relating to locations i and j

Glossary of notation

z_i, z_i^*	$z_i = x_i - \bar{x}$, $\quad z_i^* = \sum_{j=1}^{n} w_{ij} z_j$
Z_i	random variable equal to $X_i - \bar{X}$
α	the size of a statistical test (probability of type I error)
β	the beta distribution
β	vector of parameters in regression model
$\beta_{i(j)}$	the proportion of the perimeter of county i in common with county j
β_1	skewness coefficient, μ_3^2/μ_2^3; μ_2 and μ_3 are defined below
$\gamma(h)$	variogram; see section 5.3.3
Δ	minimum distance between two individuals; difference operator in section 7.4
δ, δ_i	random error
δ_{ij}	the special form of binary $\{w_{ij}\}$; that is $w_{ij} = 1$, or $w_{ij} = 0$
ϵ	vector of random errors
ϵ or ϵ_i	random error
θ	parameter for spatial moving-average process
κ_j	the jth cumulant. Cumulants are constants of a frequency distribution defined in terms of the moments by the identity in t, $$\exp\left(\sum_{r=1}^{\infty} \frac{\kappa_r t^r}{r!}\right) = \sum_{r=0}^{\infty} \frac{\mu_r' t^r}{r!}.$$ They are thus given by the coefficients in the expansion of a power series formed from the logarithm of the characteristic function of a variable, if such an expansion exists
λ	likelihood ratio statistic (sections 6.4–6.6 only)
λ or λ_i	eigenvalue or parameter of the Poisson distribution
$\lambda(u)$	parameter of the Poisson process at location u
$\mu(X)$ or μ	the expected, average or mean value of X, the first moment of X; $\mu = E(X)$
$\mu_2(X)$ or μ_2	the variance of X, the second moment of X
$\mu_j(X)$ or μ_j	the jth moment of X about the mean

$\mu'_j(X)$ or μ'_j	the jth moment of X about the origin
μ	vector of means
ρ	generally, autocorrelation parameter; also used in chapter 4 for intensity of point process
ρ_{12}	the population analogue of r_{12}
σ^2	the variance of X, the second moment of X
σ_{ij}	covariance between X_i and X_j
Σ	covariance matrix with elements σ_{ij}
φ	the index of dispersion given by m_2/m_1; m_2 and m_1 are defined above
χ^2	the chi-squared distribution
ω	variance ($= \sigma^2$)

References

The numbers in square brackets after each reference indicate the sections of the text in which the work is cited.

Abrahamse A P J, Koerts J, 1969 "A comparison between the power of the Durbin-Watson test and the power of the BLUS test" *Journal of the American Statistical Association* **64** 938-948 [8.4.3]

Abramowitz M, Stegun I A, 1965 *Handbook of Mathematical Functions* (Dover, New York) [6.3.1]

Alonso W, 1964 *Location and Land Use* (Harvard University Press, Cambridge, Mass) [5.4.4]

Anderson T W, 1948 "On the theory of testing serial correlation" *Skandanavisk Aktuarietidskrift* **31** 88-116 [6.4.2]

Arora S S, Brown M, 1977 "Alternative approaches to spatial autocorrelation: an improvement over current practice" *International Regional Science Review* **2** 67-78 [5.6]

Attwood E A, Geary R C, 1963 *Irish County Incomes in 1960* Paper Number 16, The Economic Research Institute, Dublin [8.6]

Bartko J J, Greenhouse S W, Patlak C S, 1968 "On expectations of some functions of Poisson variates" *Biometrics* **24** 97-102 [4.2.3]

Bartlett M S, 1963 "The spectral analysis of point processes" *Journal of the Royal Statistical Society, series B* **25** 264-296 [4.4.4]

Bartlett M S, 1964 "The spectral analysis of two-dimensional point processes" *Biometrika* **51** 299-311 [4.4.4]

Bartlett M S, 1971 "Physical nearest-neighbour models and non-linear time series" *Journal of Applied Probability* **8** 222-232 [5.3.1, 6.2.4]

Bartlett M S, 1975 *The Statistical Analysis of Spatial Pattern* (Chapman and Hall, Andover, Hants) [4.1, 4.3.1]

Bartlett M S, 1978 "Nearest neighbour models in the analysis of field experiments" (with discussion) *Journal of the Royal Statistical Society, series B* **40** 147-174 [9.2.3]

Barton D E, David F N, 1966 "The random intersection of two graphs" in *Research Papers in Statistics: Festschrift for J. Neyman* editor F N David (John Wiley, New York) pp 455-459 [2.4.4]

Bennett R J, 1974 "The representation and identification of spatio-temporal systems: an example of population diffusion in north-west England" *Transactions and Papers, Institute of British Geographers* **66** 73-94 [1.2.2]

Bennett R J, 1975a, 1975b, 1975c, 1975d "Dynamic systems modelling of the North-west region:
1. Spatio-temporal representation and identification
2. Estimation of the spatio-temporal policy model
3. Adaptive-parameter policy model
4. Adaptive spatio-temporal forecasts"
Environment and Planning A **7** 525-538, 539-566, 617-636, 887-898 [5.5, 5.5.3]

Bennett R J, 1979 *Spatial Time Series* (Pion, London) [9.2.2]

Besag J E, 1974 "Spatial interaction and the statistical analysis of lattice systems" (with discussion) *Journal of the Royal Statistical Society, series B* **36** 192-236 [6.2.4, 6.3.1, 6A, 7.5]

Besag J E, 1975 "Statistical analysis of non-lattice data" *The Statistician* **24** 179-196 [6.3.1]

Besag J E, 1977a "Discussion on Ripley (1977)" *Journal of the Royal Statistical Society, series B* **39** 193-195 [4.4.3]

Besag J E, 1977b "Errors in variables estimation for Gaussian lattice schemes" *Journal of the Royal Statistical Society, series B* **39** 73-78 [6.3]

Besag J E, Diggle P J, 1977 "Simple Monte Carlo tests for spatial pattern" *Applied Statistics* **26** 327-333 [2.6.3]

Besag J E, Gleaves J T, 1973 "On the detection of spatial pattern in plant communities" *Bulletin of the International Statistical Institute* **45**(1) 153-158 [4.3.1, 4.3.2]

Besag J E, Moran P A P, 1975 "On the estimation and testing of spatial interaction in Gaussian lattice processes" *Biometrika* **62** 555-562 [6.3.1]

Birch B P, 1967 "The measurement of dispersed patterns of settlement" *Tijdschrift voor Economische en Sociale Geografie* **58** 68-75 [4.2]

Bissell A F, 1972a "A negative binomial model with varying element sizes" *Biometrika* **59** 435-441 [4.2.2]

Bissell A F, 1972b "Another negative binomial model with varying element sizes" *Biometrika* **59** 691-693 [4.2.2]

Bissell A F, 1973 "Monitoring event rates using varying sample element sizes" *The Statistician* **22** 43-58 [4.2.2]

Bivand R, 1980 "A Monte Carlo study of correlation coefficient estimation with spatially autocorrelated observations" *Quaestiones Geographical* (forthcoming) [7.3.1]

Bodson P, Peeters D, 1975 "Estimation of the coefficients of a linear regression in the presence of spatial autocorrelation. An application to a Belgian labour-demand function" *Environment and Planning A* **7** 455-472 [9.2.1, 9.4]

Box G E P, Jenkins G M, 1970 *Time Series Analysis, Forecasting and Control* (Holden-Day, San Francisco) [5.5, 6.2.5]

Brandsma A S, Ketellapper R H, 1979 "A biparametric approach to spatial autocorrelation" *Environment and Planning A* **11** 51-58 [6.3.3]

Brown L A, Moore E G, 1969 "Diffusion research in geography: a perspective" *Progress in Geography* **1** editors C Board, R J Chorley, P Haggett, D R Stoddart (Edward Arnold, London) pp 119-157 [3.2.2]

Burghardt A F, 1959 "The location of river towns in the central lowland of the United States" *Annals of the Association of American Geographers* **49** 305-323 [4.1]

Burridge P, 1980 "On the Cliff-Ord test for spatial correlation" *Journal of the Royal Statistical Society, series B* **42** 107-108 [6.4.3]

Casetti E, Semple R K, 1969 "Concerning the testing of spatial diffusion hypotheses" *Geographical Analysis* **1** 154-159 [3.3.3]

Central Statistics Office, 1962 *Statistical Abstract of Ireland, 1961* (The Stationery Office, Dublin) [8.6]

Central Statistics Office, 1967 *Census of Population, 1966, Preliminary Report* (The Stationery Office, Dublin) [8.6]

Chan H, Hayya J C, Ord J K, 1977 "A note on trend removal methods: the case of polynomial regression versus variate differencing" *Econometrica* **45** 737-744 [5.5.4]

Clark P J, 1956 "Grouping in spatial distributions" *Science* **123** 373-374 [4.3.3]

Clark P J, Evans F C, 1954 "Distance to nearest neighbour as a measure of spatial relationships in populations" *Ecology* **35** 445-453 [4.3.2, 4.4.2]

Cliff A D, 1969 *Some Measures of Spatial Association in Areal Data* unpublished PhD thesis, University of Bristol, Bristol, Glos [2.2.2]

Cliff A D, Haggett P, 1980 "Mapping respiratory diseases" in *Scientific Foundations of Respiratory Medicine* editors J G Scadding, G Cumming (Heinemann, London) chapter 6 [1.1]

Cliff A D, Haggett P, Ord J K, Bassett K, Davies R B, 1975 *Elements of Spatial Structure: A Quantitative Approach* (Cambridge University Press, Cambridge) [4.3.2, 4.3.4, 5.2.2]

Cliff A D, Haggett P, Ord J K, Versey G R, 1981 *Spatial Diffusion: An Icelandic Example* (Cambridge University Press, Cambridge) forthcoming [6.2]

References

Cliff A D, Martin R L, Ord J K, 1975 "A test for spatial autocorrelation in choropleth maps based upon a modified X^2 statistic" *Transactions and Papers, Institute of British Geographers* **65** 109-129 [7.5]

Cliff A D, Ord J K, 1969 "The problem of spatial autocorrelation" in *London Papers in Regional Science 1. Studies in Regional Science* editor A J Scott (Pion, London) pp 25-55 [1.3.3, 2.3.2]

Cliff A D, Ord J K, 1971 "Evaluating the percentage points of a spatial autocorrelation coefficient" *Geographical Analysis* **3** 51-62 [2.5, 2.5.1, 2.5.2, 2.7]

Cliff A D, Ord J K, 1972 "Testing for spatial autocorrelation among regression residuals" *Geographical Analysis* **4** 267-284 [2.4.1, 8.1, 8.2.4]

Cliff A D, Ord J K, 1973 *Spatial Autocorrelation* (Pion, London) [2.4.1, 9.2.1]

Cliff A D, Ord J K, 1975a "The comparison of means when samples consist of spatially autocorrelated observations" *Environment and Planning A* **7** 725-734 [7.2.1, 7.2.2]

Cliff A D, Ord J K, 1975b "Model building and the analysis of spatial pattern in human geography" (with discussion) *Journal of the Royal Statistical Society, series B* **37** 297-348 [6.3.4, 7.4]

Cliff A D, Ord J K, 1975c "The choice of a test for spatial autocorrelation" in *Display and Analysis of Spatial Data* editors J C Davis, M J McCullagh (John Wiley, Chichester) pp 54-77 [6.4, 6.4.3, 6.4.4, 6.5.2, 8.4.3]

Cliff A D, Ord J K, 1980 "On statistical models for spatial diffusion processes" *Geographical Analysis* (forthcoming) [3.4.4]

Cliff A D, Robson B T, 1978 "Changes in the size distribution of settlements in England and Wales, 1801-1968" *Environment and Planning A* **10** 163-171 [4.3.2]

Cochrane D, Orcutt G H, 1949 "Applications of least squares regressions to relationships containing autocorrelated error terms" *Journal of the American Statistical Association* **44** 32-61 [9.2.1]

Coras Iompair Eireann, 1966 *Scale of Charges by Merchandise Trains* (The Stationery Office, Dublin) [8.6]

Cormack R M, 1979 "Spatial aspects of competition between individuals" in *Spatial and Temporal Analysis in Ecology* editors R M Cormack, J K Ord (op cit) pp 151- pp 151-212 [4.3.1, 4.3.2, 5.3.2]

Cormack R M, Ord J K (editors), 1979 *Spatial and Temporal Analysis in Ecology* International Statistical Ecology Program (International Co-operative Publishing House, Fairland, Md) [reference cited under individual contributors]

Cox D R, Lewis P A W, 1966 *The Statistical Analysis of Series of Events* (Methuen, London) [4.3.1]

Cox T F, Lewis T, 1976 "A conditional distance ratio method for analyzing spatial patterns" *Biometrika* **63** 483-489 [4.3.2]

Cresswell W L, Froggart P, 1963 *The Causation of Bus Driver Accidents—An Epidemiological Study* (Oxford University Press, London) [4.2.1, 6.6]

Cummings L P, Manly B J, Weinand H C, 1973 "Measuring association in link-node problems" *Geoforum* **13** 43-51 [3.2, 3.4]

Dacey M F, 1960 "The spacing of river towns" *Annals of the Association of American Geographers* **50** 59-61 [4.3.4]

Dacey M F, 1964 *Two Dimensional Random Point Patterns—A Review and Interpretation* Department of Geography, Northwestern University, Evanston (mimeo) [4.2]

Dacey M F, 1965 "A review of measures of contiguity for two and k-color maps" in *Spatial Analyses: A Reader in Statistical Geography* editors B J L Berry, D F Marble (Prentice-Hall, Englewood Cliffs) pp 479-495 [1.3.4, 2.2.2, 7.5]

Dacey M F, 1966a "A county seat model for the areal pattern of an urban system" *Geographical Review* **56** 527-542 [4.2]

Dacey M F, 1966b "A compound probability law for a pattern more dispersed than random with areal inhomogeneity" *Economic Geography* **42** 172-179 [4.2]

Dacey M F, 1968 "An empirical study of the areal distribution of houses in Puerto Rico" *Transactions and Papers, Institute of British Geographers* **45** 51-69 [4.2, 4.2.3]

Dacey M F, 1969a "Similarities in the areal distributions of houses in Japan and Puerto Rico" *Area* **3** 35-37 [4.2, 4.2.3]

Dacey M F, 1969b "Proportion of reflexive nth order neighbours in spatial distribution" *Geographical Analysis* **1** 385-388 [4.3.3]

Dacey M F, Tung T, 1962 "The identification of randomness in point patterns" *Journal of Regional Science* **4** 83-96 [4.3.2]

David F N, 1971a "Measurement of diversity" *Proceedings of the Sixth Berkeley Symposium on Mathematical Statistics and Probability* volume 1 (University of California Press: Berkeley) pp 631-648 [1.3.1]

David F N, 1971b "Measurement of diversity: multiple cell contents" *Proceedings of the Sixth Berkeley Symposium on Mathematical Statistics and Probability* volume 4 (University of California Press, Berkeley) pp 109-136 [1.3.1]

Delfiner P, Delhomme J P, 1975 "Optimum interpolation by Kriging" in *Display and Analysis of Spatial Data* editors J C Davis, M J McCullagh (John Wiley, Chichester) pp 96-114 [6.2.6]

Diggle P J, 1975 "Robust density estimation using distance methods" *Biometrika* **62** 39-48 [4.3.1, 4.3.5, 4.4.2]

Diggle P J, 1977a "A note on robust density estimation for spatial point patterns" *Biometrika* **64** 91-95 [4.3.5]

Diggle P J, 1977b "Detection of random heterogeneity in plant populations" *Biometrics* **33** 390-394 [4.3.2]

Diggle P J, 1978 "On parameter estimation for spatial point processes" *Journal of the Royal Statistical Society, series B* **40** 178-181 [4.4.3]

Diggle P J, 1979a "Statistical methods for spatial point patterns in ecology" in *Spatial and Temporal Analysis in Ecology* editors R M Cormack, J K Ord (op cit) pp 95-150 [4.3, 4.3.1, 4.4.2, 4.4.3]

Diggle P J, 1979b "On parameter estimation and goodness-of-fit testing for spatial point patterns" *Biometrics* **35** 87-101 [4.4.3]

Diggle P J, Besag J E, Gleaves J T, 1976 "Statistical analysis of spatial point patterns by means of distance methods" *Biometrics* **32** 659-667 [4.3.1]

Doreian P, 1981 "On the estimation of linear models with spatially distributed data" chapter 11 in *Sociological Methodology* editor Samuel S Leinhardt, The American Sociological Association, School of Urban and Public Affairs, Carnegie-Mellon University, Pittsburgh [9.2.1, 9.3, 9A]

Doreian P, Hummon N P, 1976 *Modelling Social Processes* (Elsevier, New York) [9.3]

Draper N R, Smith H, 1966 *Applied Regression Analysis* (John Wiley, New York) [6.2.1]

Durbin J, 1960 "Estimation of parameters in time series regression models" *Journal of the Royal Statistical Society, series B* **22** 139-153 [5.5.1, 6A]

Durbin J, 1965 "Discussion on Pyke (1965)" *Journal of the Royal Statistical Society, series B* **27** 437-438 [4.3.2]

Durbin J, 1970 "Testing for serial correlation in least squares regression when some of the regressors are lagged dependent variables" *Econometrica* **38** 410-421 [9.5]

Durbin J, Watson G S, 1950 "Testing for serial correlation in least squares regression I" *Biometrika* **37** 409-428 [1.3.2]

Durbin J, Watson G S, 1951 "Testing for serial correlation in least squares regression II" *Biometrika* **38** 159-178 [1.3.2, 2.4.1]

Durbin J, Watson G S, 1971 "Testing for serial correlation in least squares regression, III" *Biometrika* **58** 1-19 [1.3.2, 8.5.1]

References

Eagleson P S, 1967 "Optimum density of rainfall networks" *Water Resources Research* **3** 1021-1033 [6.2.6, 6.3.5]

Feller W, 1943 "On a general class of contagious distributions" *Annals of Mathematical Statistics* **14** 389-400 [4.2.1]

Fisher R A, 1941 "The negative binomial distribution" *Annals of Eugenics* **11** 182-187 [4.2.1]

Ford E D, 1976 "The canopy of a Scots pine forest: description of a surface of complex roughness" *Journal of Animal Ecology* **11** 215-244 [4.4.4]

Freeman G H, 1953 "Spread of diseases in a rectangular plantation with vacancies" *Biometrika* **40** 287-296 [1.3.1]

Gale S, 1971 "A simple device for the recognition of geographic patterns" *Geographical Analysis* **3** 187-194 [3.2]

Galton F, 1889 "Comments on Tylor, E.B. 'On a method of investigating the development of institutions, applied to laws of marriage and descent'" *Journal of the Royal Anthropological Institute of Great Britain and Ireland* **18** 245-272 [1.2.1]

Gantmacher F R, 1959 *Applications of the Theory of Matrices* (John Wiley, New York) [2.4.1]

Gatrell A C, 1979 "Autocorrelation in spaces" *Environment and Planning A* **11** 507-516 [5.6]

Geary R C, 1954 "The contiguity ratio and statistical mapping" *The Incorporated Statistician* **5** 115-145 [1.3.2, 2.3.1, 6.4.4, 8.2.4, 8.6]

General Register Office, 1961 *England and Wales: Preliminary Census Report, 1961* (HMSO, London) [2.7]

Getis A, 1964 "Temporal analysis of land use patterns with the use of nearest neighbor and quadrat methods" *Annals of the Association of American Geographers* **54** 391-399 [4.2]

Getis A, Boots B N, 1978 *Models of Spatial Processes* (Cambridge University Press, London) [4.1, 4.3.5]

Ginsberg N S, 1958 *The Pattern of Asia* (Prentice-Hall, Englewood Cliffs, NJ) [4.2.3, 4.2.4]

Gould P R, 1970 "Is *Statistix Inferens* the geographical name for a wild goose?" *Economic Geography* **46** 439-448 [1.2.1]

Green R F, 1979 "A graphical test to detect interference in selecting nest sites" in *Contemporary Quantitative Ecology and Related Ecometrics* editors G P Patil, M L Rosenzweig (International Co-operative Publishing House, Fairland, Md) pp 439-452 [4.4.5]

Greig-Smith P, 1964 *Quantitative Plant Ecology* (Butterworth, London) [4.2, 5.3.1]

Greig-Smith P, 1971 "Analysis of vegetation data: the user viewpoint" in *Statistical Ecology, volume 3* editors G P Patil, E C Pielou, W E Waters (Pennsylvania State University Press, University Park, Pa) pp 147-166 [5.3.1]

Griffith D A, 1978 "A spatially adjusted ANOVA model" *Geographical Analysis* **10** 296-301 [7.2.3]

Hägerstrand T, 1953 "On Monte Carlo simulation of diffusion" reprinted in *Quantitative Geography, Part I: Economic and Cultural Topics* editors W L Garrison, D F Marble, 1967 *Studies in Geography, 13* (Northwestern University, Evanston) pp 1-32 [3.1, 3.2.2, 3.3.1, 3.3.2, 3.5]

Hägerstrand T, 1967 *Innovation Diffusion as a Spatial Process* Postscript and translation by Allan Pred (University of Chicago Press, Chicago); translated from *Innovationsförloppet ur korologisk synpunkt* (Gleerup, Lund) 1953 [3.3.1]

Haggett P, 1972 "Contagious processes in a planar graph: an epidemiological application" in *Medical Geography* editor N D McGlashan (Methuen, London) pp 307-324 [1.1, 5.2.2]

Haggett P, 1976 "Hybridizing alternative models of an epidemic diffusion process" *Economic Geography* **52** 136-146 [1.7.2]

Haggett P, Cliff A D, Frey A E, 1977 *Locational Analysis in Human Geography* (Edward Arnold, London) [1.1, 1.3.4, 4.2.3, 4.3.1, 4.3.3, 4.3.5, 4.4.4, 5.2.1, 6.2, 8.8]

Haining R P, 1977 "Model specification in stationary random fields" *Geographical Analysis* **9** 107-129 [6.3.2, 6.4.2, 6.5.2]

Haining R P, 1978a "The moving average model for spatial interaction" *Transactions and Papers, Institute of British Geographers, New Series* **3** 202-225 [6.2.5, 6.3, 6.3.1, 6.3.4, 6.5.2, 8.2.3]

Haining R P, 1978b "Estimating spatial interaction models" *Environment and Planning A* **10** 305-320 [6.3.2, 9.4]

Haining R P, 1979 "Statistical tests and process generators for random field models" *Geographical Analysis* **11** 45-64 [6.2.4]

Haining R P, 1981a "Spatial interdependencies in population distributions—a study in univariate map analysis. 1 Rural population densities"; 1981b "2 Urban population distributions" *Environment and Planning A* **13** forthcoming) [6.2]

Harvey D W, 1966 "Geographical processes and the analysis of point patterns: testing models of diffusion by quadrat sampling" *Transactions and Papers, Institute of British Geographers* **40** 81-95 [4.2]

Harvey D W, 1968 "Some methodological problems in the use of the Neyman type A and negative binomial probability distributions for the analysis of spatial point patterns" *Transactions and Papers, Institute of British Geographers* **44** 85-95 [4.2.4]

Hepple L W, 1976 "A maximum likelihood model for econometric estimation with spatial data" in *London Papers in Regional Science 6. Theory and Practice in Regional Science* editor I Masser (Pion, London) pp 90-104 [6.3.3]

Hepple L W, 1978 "The econometric specification and estimation of spatio-temporal models" in *Timing Space and Spacing Time, Volume 3* editors T Carlstein, D Parkes, N Thrift (Edward Arnold, London) pp 66-80 [1.2.1]

Hill M O, 1973 "The intensity of spatial pattern in plant communities" *Journal of Ecology* **61** 225-235 [5.3.2]

Hodder I, Orton C, 1976 *Spatial Analysis in Archaeology* (Cambridge University Press, London) [8.8]

Hoeffding W, 1952 "The large-sample power of tests based on permutations of observations" *Annals of Mathematical Statistics* **23** 169-192 [2.4.1, 2.4.3]

Holgate P, 1965 "Tests of randomness based on distance methods" *Biometrika* **52** 345-353 [4.3.2]

Hope A C A, 1968 "A simplified Monte Carlo significance test procedure" *Journal of the Royal Statistical Society, series B* **30** 582-598 [2.7]

Hooper P M, Hewings G J D, 1980 "Some properties of space-time process" paper presented at the Annual Meeting, British Section of the Regional Science Association, University College London, September 3-5 [5.5.3]

Hopkins B, 1954 "A new method for determining the type of distribution of plant individuals" *Annals of Botany* **18** 213-226 [4.3.2]

Hubert L J, 1978 "Nonparametric tests for patterns in geographic variation: possible generalizations" *Geographical Analysis* **10** 86-88 [1.6.1]

Hubert L J, Golledge R G, Costanzo C M, 1981 "Generalized procedures for evaluating spatial autocorrelation" *Geographical Analysis* (forthcoming) [1.6.1]

Hudson J C, 1967 *Theoretical Settlement Geography* unpublished PhD thesis, University of Iowa, Iowa [4.2]

Huijbrechts C, 1975 "Regionalised variables and quantitative analysis of spatial data" in *Display and Analysis of Spatial Data* editors J C Davis, M J McCullagh (John Wiley, Chichester) pp 38-53 [6.2.6]

References

Jenkins G M, Watts D G, 1968 *Spectral Analysis and its Applications* (Holden-Day, San Francisco) [4.4.4]

Johnston J, 1972 *Econometric Methods* second edition (McGraw-Hill, London) [6.2.1, 7.3.2, 8.2.1, 8.2.2]

Jones E, Sinclair D J (Eds), 1968 *Atlas of London and the London Region* (Pergamon Press, Oxford) sheets 43-45 [5.1, 5.4.4]

Jumars P A, Thistle D, Jones M L, 1977 "Detecting two dimensional spatial structure in biological data" *Oecologia* 28 109-123 [1.1]

Jusatz H, 1977 "Cholera" in *A World Geography of Human Diseases* editor G M Howe (Academic Press, London) pp 131-143 [1.1]

Katti S K, Gurland J, 1962 "Some methods of estimation for the Poisson-binomial distribution" *Biometrics* 18 42-51 [4.2.1]

Kemp C D, 1967 "On a contagious distribution suggested for accident data" *Biometrics* 23 241-255 [4.2.1]

Kendall M G, 1976 *Time-Series* (Griffin, London) [1.3.1, 5.2.1]

Kendall M G, Stuart A, 1975 *The Advanced Theory of Statistics* Volume 3, third edition (Griffin, London) [5.3.1]

Kendall M G, Stuart A, 1977 *The Advanced Theory of Statistics* Volume 1, fourth edition (Griffin, London) [4.2.1]

Kendall M G, Stuart A, 1979 *The Advanced Theory of Statistics* Volume 2, fourth edition (Griffin, London) [5.4.1, 5.5.3, 6.2.1, 6.2.2, 6.3.1, 6.4, 6.4.1, 6.4.3, 9A]

Kershaw K A, 1957 "The use of cover and frequency in the detection of patterns in plant communities" *Ecology* 38 291-299 [5.3.1]

Kershaw K A, 1964 *Quantitative and Dynamic Ecology* (Edward Arnold, London) [4.2]

Knox E G, 1964 "The detection of space-time interactions" *Applied Statistics* 13 25-29 [1.6.1, 2.4.4]

Kooijman S A L M, 1976 "Some remarks on the statistical analysis of grids especially with respect to ecology" *Annals of Systems Research* 5 113-132 [1.4.2, 5.2.1, 6.2.6, 6.3.5]

Kooijman S A L M, 1977 *Inference about Dispersal Patterns* Unpublished PhD thesis, University of Leiden, Holland [4.4.1, 4.4.3]

Kooijman S A L M, 1979 "The description of point patterns" in *Spatial and Temporal Analysis in Ecology* editors R M Cormack, J K Ord (op cit) pp 305-332 [4.4.1, 4.4.3, 6.2.1, 6.2.2]

Koopmans T C, 1942 "Serial correlation and quadratic forms in normal variables" *Annals of Mathematical Statistics* 13 14-33 [2.3.1]

Krishna Iyer P V A, 1949 "The first and second moments of some probability distributions arising from points on a lattice, and their applications" *Biometrika* 36 135-141 [1.3.1, 2.2.2]

Lankford P M, 1974 "Testing simulation models" *Geographical Analysis* 6 295-302 [3.2, 3.4]

Lebart L, 1969 "Analyse statistique de la contiguïté" *Publications de l'Université de Paris* 18 81-112 [7.4]

Ludwig J A, 1979 "A test of different quadrat variance methods for the analysis of spatial pattern" in *Spatial and Temporal Analysis in Ecology* editors R M Cormack, J K Ord (op cit) pp 289-304 [5.3.2]

McConnell M, 1966 *Quadrat Methods in Map Analysis* Department of Geography, University of Iowa (mimeo) [4.2]

McGuire J U, Brindley T A, Bancroft T A, 1957 "The distribution of European cornborer larvae, *Pyrusta nubilalis* (H.B.N.) in field corn" *Biometrics* 13 65-78 [4.2.1]

Malm R, Olsson G, Wärneryd O, 1966 "Approaches to simulations of urban growth" *Geografiska Annaler, B* 48 9-22 [4.2]

Mantel N, 1967 "The detection of disease clustering and a generalized regression approach" *Cancer Research* **27** 209-220 [1.6.1, 4.4.5]
Marriott F H C, 1979 "Monte Carlo tests: how many simulations?" *Applied Statistics* **28** 75-77 [2.7, 3.2.2]
Martin R L, 1974 "On autocorrelation, bias and the use of first spatial differences in regression analysis" *Area* **6** 185-194 [7.4]
Martin R L, Oeppen J E, 1975 "The identification of regional forecasting models using space-time correlation functions" *Transactions and Papers, Institute of British Geographers* **66** 95-118 [5.5, 5.5.3]
Matérn B, 1960 "Spatial variation" *Meddlanden Frán Statens Skogsforskningsinstitut* **49** 1-144 [4.1, 4.3.1, 4.4.3]
Matérn B, 1971 "Doubly stochastic Poisson processes in the plane" in *Statistical Ecology* Volume 1, editors G P Patil, E C Pielou, W E Waters (Pennsylvania State University Press, University Park, Pa) pp 195-213 [4.3.1]
Matérn B, 1979 "The analysis of ecological maps as mosaics" in *Spatial and Temporal Analysis in Ecology* editors R M Cormack, J K Ord (op cit) pp 271-288 [4.1, 4.3.5]
Matheron G, 1971 *The Theory of Regionalised Variables* (Centre de Morphologie Mathématique, Fontainebleau) [5.3.3, 6.2.6, 6.3.5]
Matui I, 1932 "Statistical study of the distribution of scattered villages in two regions of the Tonami Plain, Toyama Prefecture" *Japanese Journal of Geology and Geography* **9** 251-266 [4.2.3, 4.2.4, 5.3.1]
Mead R, 1967 "A mathematical model for the estimation of interplant competition" *Biometrics* **23** 189-205 [6.3.1]
Mead R, 1974 "A test for spatial pattern at several scales using data from a grid of contiguous quadrats" *Biometrics* **30** 295-307 [5.3.1]
Mitchell E J, 1969 "Some econometrics of the Huk rebellion" *American Political Science Review* **63** 1159-1171 [9.3]
Moellering H, Tobler W R, 1972 "Geographical variances" *Geographical Analysis* **4** 34-50 [4.2.3, 5.3.1]
Mollison D, 1975 "Comments on Cliff, A.D. and Ord, J.K. (1975)" *Journal of the Royal Statistical Society, series B* **37** 334-335 [3.3.3]
Mollison D, 1977 "Spatial contact models for ecological and epidemic spread" (with discussion) *Journal of the Royal Statistical Society, series B* **39** 283-326 [3.3.3]
Mollison D, 1978 "Structural choices for spatial models" (abstract) *Advances in Applied Probability* **10** 491-492 [4.4.5]
Moore P G, 1954 "Spacing in plant populations" *Ecology* **35** 222-227 [4.3.2]
Moran P A P, 1948 "The interpretation of statistical maps" *Journal of the Royal Statistical Society, series B* **10** 243-251 [1.3.1, 2.2.2, 2.4.2]
Moran P A P, 1950 "Notes on continuous stochastic phenomena" *Biometrika* **37** 17-23 [1.3.2]
Moran P A P, 1973 "A Gaussian Markovian process on a square lattice" *Journal of Applied Probability* **10** 54-62 [6.3.1]
Mountford M D, 1961 "On E.C. Pielou's index of non-randomness" *Journal of Ecology* **49** 271-276 [4.3.2]
Nerlove M, 1964 "Spectral analysis of seasonal adjustment procedures" *Econometrica* **32** 241-286 [7.4]
Neyman J, 1939 "On a new class of contagious distributions applicable in entomology and bacteriology" *Annals of Mathematical Statistics* **10** 35-57 [4.2.1]
Neyman J, Scott E L, 1958 "A statistical approach to problems of cosmology" *Journal of the Royal Statistical Society, series B* **20** 1-29 [4.3.1]
Noether G E, 1970 "A central limit theorem with non-parametric applications" *Annals of Mathematical Statistics* **41** 1753-1755 [2.4.2]

Olsson G, 1966 "Central place systems, spatial interaction, and stochastic processes" *Papers, Regional Science Association* **18** 13-45 [4.2]

Ord J K, 1972 *Families of Frequency Distributions* (Griffin, London) [2.6.4, 4.2.1, 4.2.2]

Ord J K, 1975 "Estimation methods for models of spatial interaction" *Journal of the American Statistical Association* **70** 120-126 [6.3, 6.3.1, 6.3.3, 9.2.1]

Ord J K, 1977 "How many trees in a forest?" *Mathematical Scientist* **3** 23-33 [4.3.5]

Ord J K, 1981 "Tests of significance using non-normal data" *Geographical Analysis* **13** (forthcoming) [2.3]

O'Sullivan P M, 1968 "Accessibility and the spatial structure of the Irish economy" *Regional Studies* **2** 195-206 [8.1]

O'Sullivan P M, 1969 *Transport Networks and the Irish Economy* London School of Economics and Political Science Geographical Papers, Number 4 (Weidenfeld and Nicolson, London) [8.6, 9.4]

Persson O, 1971 "The robustness of estimating density by distance measurements" in *Statistical Ecology* Volume 2, editors G P Patil, E C Pielou, W E Waters (State University Press, University Park, Pa) pp 175-190 [4.3.5]

Pfeiffer P E, Deutsch S J, 1980 "A three-stage iterative procedure for space-time modeling" *Technometrics* **22** (forthcoming) [5.5.3]

Pielou E C, 1959 "The use of point-to-plant distances in the study of the pattern of plant populations" *Journal of Ecology* **47** 607-613 [4.3.2]

Pielou E C, 1975 *Ecological Diversity* (John Wiley, New York) [1.3.1]

Pielou E C, 1977 *Mathematical Ecology* second edition (John Wiley, New York) [4.1, 4.3.2, 4.3.3, 4.3.5, 5.3.2]

Pitman E J G, 1937 "The 'closest estimates' of statistical parameters" *Proceedings of the Cambridge Philosophical Society* **33** 212-222 [2.3.1]

Pitman E J G, 1948 *Lecture Notes on Non-parametric Inference* unpublished notes used by the author in his lecture course at Stanford University, Calif. [6.4]

Polya G, 1931 "Sur quelques points de la théorie des probabilités" *Annales de l'Institut de H. Poincaré* **1** 117-162 [4.2.1]

PP 1850, XXI: Report of the General Board of Health on the Epidemic Cholera of 1848 and 1849 *British Parliamentary Papers* [1.1, 1.7.3]

PP 1850, XXII: Report by the General Board of Health on the Supply of Water to the Metropolis *British Parliamentary Papers* [1.1]

PP 1854-5, XXI: Report and Appendix to Report of the Committee for Scientific Inquiries in Relation to the Cholera-Epidemic of 1854 *British Parliamentary Papers* [1.1]

Pyke R, 1965 "Spacings" (with discussion) *Journal of the Royal Statistical Society, series B* **27** 395-449 [4.3.2]

Quenouille M H, 1949 "A relationship between the logarithmic, Poisson and negative binomial series" *Biometrics* **5** 162-164 [4.2.1]

Rayner J N, Golledge R G, 1972 "Spectral analysis of settlement patterns in diverse physical and economic environments" *Environment and Planning* **4** 347-371 [4.4.4]

Rayner J N, Golledge R G, 1973 "The spectrum of U.S. Route 40 re-examined" *Geographical Analysis* **5** 338-350 [4.4.4]

Ripley B D, 1977 "Modelling spatial patterns" (with discussion) *Journal of the Royal Statistical Society, series B* **39** 172-212 [4.3.1, 4.4.2, 4.4.3]

Ripley B D, Kelly F P, 1977 "Markov point processes" *Journal of the London Mathematical Society* **15** 188-192 [4.3.1]

Roach S A, 1968 *The Theory of Random Clumping* (Methuen, London) [4.1]

Robinson A H, 1956 "The necessity of weighting values in correlation of areal data" *Annals of the Association of American Geographers* **46** 233-236 [5.4.1]

Robinson J E, 1975 "Frequency analysis, sampling and errors in spatial data" in *Display and Analysis of Spatial Data* editors J C Davis, M J McCullagh (John Wiley, Chichester) pp 78-95 [4.4.4]

Rogers A, 1965 "A stochastic analysis of the spatial clustering of retail establishments" *Journal of the American Statistical Association* **60** 1094-1102 [4.2]

Rogers A, 1974 *Statistical Analysis of Spatial Dispersion: The Quadrat Method* (Pion, London) [1.3.1, 4.2.1]

Ross-Parker M, 1975 "Inter-plant competition models and their sampling distributions" *Advances in Applied Probability* **7** 453-454 [6.3.2]

Sen A K, 1976 "Large sample-size distribution of statistics used in testing for spatial correlation" *Geographical Analysis* **8** 175-184 [2.4.1]

Shumway R, Gurland J, 1960 "Fitting the Poisson-Binomial distribution" *Biometrics* **16** 522-533 [4.2.1]

Siegel S, 1956 *Nonparametric Statistics for the Behavioral Sciences* (John Wiley, New York) [1.3.1]

Skellam J G, 1951 "Random dispersal in theoretical populations" *Biometrika* **38** 196-218 [4.3.2]

Sokal R R, 1979 "Ecological parameters inferred from spatial correlograms" in *Contemporary Quantitative Ecology and Related Ecometrics* editors G P Patil, M L Rosenzweig (International Co-operative Publishing House, Fairland, Md) pp 167-196 [1.5.2]

Sokal R R, Oden N L, 1978a "Spatial autocorrelation in biology, 1. Methodology" *Biological Journal of the Linnean Society* **10** 199-228 [1.1]

Sokal R R, Oden N L, 1978b "Spatial autocorrelation in biology, 2. Some biological applications of evolutionary and ecological interest" *Biological Journal of the Linnean Society* **10** 229-249 [1.1]

Sprott D A, 1958 "The method of maximum likelihood applied to the Poisson binomial distribution" *Biometrika* **14** 97-106 [4.2.1]

Stephan F F, 1934 "Sampling errors and the interpretation of social data ordered in time and space" *Journal of the American Statistical Association* **29** 165-166 [1.2.1]

Stetzer F C, 1977 *The Application of Nonstationary Spatial Models* Unpublished PhD thesis, Department of Geography, University of Iowa, Iowa [6.3]

Strauss D J, 1975 "A model for clustering" *Biometrika* **62** 467-475 [4.4.2]

Strauss D J, 1977 "Clustering on coloured lattices" *Journal of Applied Probability* **14** 135-143 [2.4.3]

Student, 1914 "The elimination of spurious correlation due to position in time or space" *Biometrika* **10** 179-180 [6.2.1]

Taaffe E J, Morrill R L, Gould P R, 1963 "Transport expansion in underdeveloped countries: A comparative analysis" *Geographical Review* **53** 503-509 [6.5.1, 8.1, 8.7]

Theil H, 1965 "The analysis of disturbances in regression residuals" *Journal of the American Statistical Association* **60** 1067-1079 [8.4, 8.4.1]

Thomas M, 1949 "A generalisation of Poisson's binomial limit for use in ecology" *Biometrika* **36** 18-25 [4.2.1]

Thomas E N, Anderson D W, 1965 "Additional comments on weighting values in correlation analysis of areal data" *Annals of the Association of American Geographers* **55** 492-505 [5.4.1]

Thompson H R, 1955 "Spatial point processes, with applications to ecology" *Biometrika* **42** 102-115 [4.3.5]

Tinline R R, 1971 "Linear operators in diffusion research" in *Regional Forecasting* Proceedings of 22nd Colston Symposium, editors M Chisholm, A Frey, P Haggett (Butterworth, London) pp 71-91 [3.3.3]

Tobler W R, 1965 "Computation of the correspondence of geographical patterns" *Papers, Regional Science Association* **15** 131-139 [3.2]
Tobler W R, 1969a "Geographical filters and their inverses" *Geographical Analysis* **1** 234-253 [3.3.3]
Tobler W R, 1969b "The spectrum of U.S. 40" *Papers, Regional Science Association* **23** 45-52 [4.4.4]
Tobler W R, 1970 "A computer movie simulating urban growth in the Detroit Region" *Economic Geography* Supplement **46** 234-240 [1.2.1]
Tukey J W, 1962 "The future of data analysis" *Annals of Mathematical Statistics* **33** 1-67 [8.6]
Usher M B, 1969 "The relation between mean square and block size in the analysis of similar patterns" *Journal of Ecology* **57** 505-514 [5.3.2]
Usher M B, 1975 "Analysis of pattern in real and artificial plant populations" *Journal of Ecology* **63** 569-586 [5.3.2]
Von Neumann J, 1941 "Distribution of the mean square successive difference to the variance" *Annals of Mathematical Statistics* **12** 367-395 [1.3.2]
Wagner H M, 1959 "Linear programming techniques for regression analysis" *Journal of the American Statistical Association* **54** 206-212 [8.6]
Wagner H M, 1962 "Non-linear regression with minimal assumptions" *Journal of the American Statistical Association* **57** 572-578 [8.6]
Wallis K F, 1972 "Testing for fourth order autocorrelation in quarterly regression equations" *Econometrica* **40** 617-636 [6.4]
Warren W G, 1962 *Contributions to the Study of Spatial Point Processes* University of North Carolina, Institute of Statistics Series number 337 (mimeo) [4.3.1]
Warren W G, 1971 "The centre-satellite concept as a basis for ecological sampling" in *Statistical Ecology* Volume 2, editors G P Patil, E C Pielou, W E Waters (Pennsylvania State University Press, University Park, Pa) pp 87-118 [4.3.1]
Warren W G, Batcheler C L, 1979 "The density of spatial patterns: robust estimation through distance methods" in *Spatial and Temporal Processes in Ecology* editors R M Cormack, J K Ord (op cit) pp 247-270 [4.3.1, 4.3.5]
Watson G S, 1971 "Trend surface analysis" *Mathematical Geology* **3** 215-226 [6.2.1]
Watson G S, 1972 "Trend surface analysis and spatial correlation" *Geological Society of America, Special Paper* **146** 39-46 [6.2.1]
Whitten E H T, 1974 "Scale and directional field and analytical data for spatial variability studies" *Mathematical Geology* **6** 183-198 [6.2.1, 8.8]
Whitten E H T, 1975 "The practical use of trend-surface analysis in the geological sciences" in *Display and Analysis of Spatial Data* editors J C Davis, M J McCullagh (John Wiley, Chichester) pp 282-297 [6.2.1, 8.8]
Whittle P, 1954 "On stationary processes in the plane" *Biometrika* **41** 434-449 [1.2.2, 6.2.3, 6.3, 6.3.1, 6.3.2]
Whittle P, 1963 "Stochastic processes in several dimensions" *Bulletin of the International Statistical Institute* **40** 974-994 [4.1]
Wickens M R, 1972 "Testing for serial correlation in equations with lagged dependent variables" Technical Report, Department of Economics, University of Bristol [9.5]
Yapa L, 1976 "On the statistical significance of the observed map in spatial diffusion" *Geographical Analysis* **8** 255-268 [3.2, 3.4]
Yule G U, Kendall M G, 1965 *An Introduction to the Theory of Statistics* fourteenth edition (Griffin, London) [5.4.1]
Zahl S, 1974 "Applications of the S-method to the analysis of spatial pattern" *Biometrics* **30** 513-524 [5.3.2]
Zellner A, 1962 "An efficient method of estimating seemingly unrelated regressions and tests for aggregation bias" *Journal of the American Statistical Association* **57** 348-368 [5.4.3]

Index

Aggregation, *see* spatial scale
Agricultural land values 197, 222
Agricultural subsidy 71
Analysis of variance 125, 223
   hierarchical 118, 123-134
Approximations to sampling distributions
   Geary statistic 56
   join-count statistics 52, 61-63
   Moran statistic 54
   Moran statistic for regression residuals 205
Asymptotic relative efficiency (ARE) 163-165, 168-172 passim
Asymptotic sampling theory 34, 39, 46, 51, 205
Autocorrelation
   causes and consequences of 184-197 passim
   interpretation 21
Autocorrelation measures
   for nominal data, *see* join counts
   for ordinal and interval data, *see* Geary statistic, Moran statistic
Autocorrelation, partial, *see* correlogram, partial
Autocorrelation, spatial, *see* spatial autocorrelation
   in time 9, 14
Autoregressive model
   in space 8, 134, 141, 145-149, 152-160, 165, 166, 167, 179-183, 185, 197
   in time 9, 145
   in time and space 10, 233
   see also regression

Beta distribution 50, 62
BLUS procedure 203
   relative efficiency 205
Bounds for I 21
Bronchitis 5, 6, 7, 24

c statistic, *see* Geary statistic
Centre-satellite process 102
Chi-squared approximation 57
Chi-squared test 195
Choice of test statistic 14
Cholera outbreaks 1, 7, 27
Colour lattices, *see* join counts

Combinations of tests 172
Conditional autoregressive model 134, 146, 148, 152, 153, 154-159, 165, 166, 167, 179-183, 232, 236-239 passim
Consumption of agricultural produce 209, 212, 237
Contagion, true or apparent 90
Contagious distributions, *see* generalised and compound distributions
Contingency tables 194
Correlation 118, 127, 131
   tests for 189
Correlogram 16, 18, 118, 120, 126, 134-139, 158, 161
   computation of 135
   partial 118, 134, 137
   spatiotemporal 136
Covariance function 141, 146, 150, 151, 153, 162

Dacey statistic 16, 43, 46
Data, mapped vs random 87
Data sets used:
   Asby, Sweden, acceptance of agricultural subsidy 71
   Cornwall, England, measles notifications 24, 118, 120
   Eire
      road accessibility 197, 206
      consumption of agricultural produce 237
   Ghana transport network 173, 215
   Iowa, USA, agricultural land values 222
   London, England
      cholera 1, 27
      bronchitis 5, 24
      land use 118, 131, 137
   Mississippi Valley, USA, location of towns 86, 107
   Nebraska and Kansas, USA
      population values 161
      wheat yields 158
   Nigeria, transport network 173, 215
   Phillipines, insurgent control 236
   Redwood seedlings, locations 113
   Southwest England, measles notifications 1, 24

Data set used (continued)
  Tonami Plain, Japan, house locations 92, 124
  Welsh counties
    bronchitis deaths 6, 24
    population change 63
    tuberculosis deaths 6, 24
Diffusion processes 1, 24, 31, 66, 68–85, 143
Directional bias 137
Dirichlet cells 110
Dispersion, index of 90, 97
Distributions, *see* sampling distributions, individual distributions by name
Diversity, *see* spatial diversity
Durbin's method for partial autocorrelations 135

Efficacy 164
Efficiency of tests 163
Epidemics 1–5, 24–33, 120
Errors
  type I 67, 70, 163, 177, 185, 193
  type II 67, 163, 177
Estimation for spatial models 153
Exponential distribution 101

Free sampling models 12, 19, 37–38, 51, 55, 84, 170, 173, 194, 200, 219
Functional relationship 128

Gamma distribution 91, 101
Geary statistic 13, 17, 21, 34, 42, 56, 167, 174
  for regression residuals 200
Generation of random variables 152
Graphs 24, 118
Greig-Smith's method for quadrats 123

Hägerstrand model 71
Hammersley-Clifford theorem 149, 181
Heterogeneity 89, 103, 106
Hoeffding's theorem 47
Hope procedure 63, 70, 80, 206
Huk rebellion 236

I statistic, *see* Moran statistic
Identification of model 134
Inhibition processes 103

Insurgent control 236
Intensity, estimation of 109, 111, 126

Join-count statistics 11, 15, 17, 19, 24, 34, 36–41, 51–52, 56–63, 66, 74 170, 173, 174, 194
  for regression residuals 199, 219

Lagrange multiplier test 170
Land use 131, 137, 222
Least squares, in AR models 153–160 passim, 179–183
Likelihood ratio test 166, 170, 174, 177
Limitations of measures 15
Limiting distributions 46
Limiting form of moments 39
Location
  of houses 92, 124
  of seedlings 113
  of towns 107
Logistic curve 32, 84
Log-series distribution 91, 102

Map comparisons 66
Markov point processes 99
Matérn process 103, 116
Maximum likelihood estimation 139, 153–161 passim
  autoregressive-regression model 231, 236, 238, 241
  regression with autocorrelated residuals 234, 242
Mean Information Field 71, 81–84
Measles 1, 7, 22, 24, 118, 120
Model identification 134
  for spatial processes 141–183
  see also autoregressive model, covariance function, moving-average model, regionalised variables
Moments
  Geary statistic 14, 21, 42–46
  join counts 12, 19, 36–41
  Moran statistic 14, 21, 42–46
    for observations about the origin 68
    for ranks 46
    for regression residuals 200, 203, 204
  space-time interaction coefficient 34

Monte Carlo studies 51, 54, 56, 63, 71,
    80, 126, 164, 174, 185, 189, 193,
    195
  sampling variation 65, 70
Moran statistic 13, 17, 21, 34, 42-51, 54,
    96, 118, 119, 167, 170-178 passim,
    187
  maximised 120, 139
  measuring observations about natural
      origin 67
  for regression residuals 200-206,
      209, 219, 221, 226
Moving-average model, in space 141, 149,
    152, 153, 160, 174, 177
Multidimensional scaling 140
Multivariate normal distribution 142, 146,
    179

Nearest neighbour distances 86,
    99-117 passim
Negative binomial distribution 88, 89,
    90-96
Neyman type A distribution 88, 89, 96
Noether's theorem 51
Nonfree sampling models 12, 20, 37, 39,
    52, 56-61, 74, 84, 171, 173, 194, 200,
    221
Normal approximations 47, 57
Normality assumption 14, 21, 42, 52,
    165, 200, 209, 221

Outliers, in regression 214

Partial correlogram, see correlogram
Permutations tests 63, 70, 80, 206
  comparison with approximations 63
Pitman-Koopmans theorem 43, 201
Point patterns, see spatial point patterns
Poisson approximation 24, 52, 61
Poisson distribution 88, 93, 100
  compound 89, 96
  generalised 88, 96
Poisson process 87, 100
  doubly stochastic 103
  trend-surface method for 145
Population changes 63
Population density 161, 162

Power
  of autocorrelation measures 164-172,
      174-178
  of a test 59, 163, 200
Pseudolikelihood estimator 153

Quadrat counts 86-99 passim, 109
Quadrats
  choice of size 86, 92
  contiguous 118, 123, 125
  Greig-Smith's method for 123
  other methods for 125

Randomisation assumption 14, 21, 23,
    42, 52, 200, 209, 221
Randomised model
  for Moran statistic with observations
      about natural origin 67-68
  for Moran and Geary statistics 21, 42
  for regression residuals 206
Ranks
  in Moran statistic 23, 46
  test, efficiency of 172
Reflexive pairs 106-109
Regionalised variables 151, 153, 162
Regression analysis 141, 145, 189, 198
  autocorrelation among residuals
      197-228, 231, 234, 240
  correlation among residuals under H_0
      200
  estimation of parameters in presence of
      autocorrelation 190, 197, 199
  models in space and time 233
    with autocorrelated errors 234
    with autoregressive terms 155, 214,
      213-242
  estimation of parameters under
      constraints 234, 237
  residuals 66, 173, 197-228
  seemingly unrelated equations 131
  weighted 129
Road accessibility 206, 215

Sampling distributions 34-65
  Geary statistic 42, 47, 56
  join-count statistics 36, 51, 52, 56, 61
  $I_{s \cdot t}$ 52
  Monte Carlo studies 53-65

Sampling distributions (continued)
　Moran statistic 42, 47, 54
　　among regression residuals 205
Scale components 127-134
Sen's conditions 49
Settlement patterns 92, 99, 107
Simultaneous autoregressive scheme 146, 152, 153, 159, 165, 185, 231, 236, 238, 242
Space-time index 7, 23, 31, 34, 37, 62
Space-time interactions 1, 9, 22, 27, 117
Spatial autocorrelation
　and correlation 189
　among counts 91, 92
　definition of 5
　effect on test statistics 184-196
　interpretation of 21
　and regression 190
　specification of alternative hypothesis 165
　tests for, *see* test of hypotheses
　and tests of means 184
Spatial diversity of species 13
Spatial models, *see* conditional auto-regressive models, moving-average models, simultaneous autoregressive models
Spatial point patterns 86
Spatial scale 85, 87, 118, 123, 127, 131-134 passim, 139
Spectral analysis 116
Stationarity
　in space 137, 142, 143, 147, 150, 151
　strict sense 142
　wide sense 142
Structural relationship 128

Tests of hypotheses 19
　asymptotic relative efficiency (ARE) of 164
　chi-squared 195
　choice of test statistic 163, 199
　combinations 172
　comparison of tests for spatial auto-correlation 163-178
　effects of spatial autocorrelation upon 184-196
　　for correlation 189
　　for two means 184
　for nominal data 11, 14, 36
　for ordinal and interval data 13, 42
　for regression residuals 199, 203
　for space-time interaction 24, 34
　using distances 104
　using random permutations 63
Topological invariance 15
Transport networks 206, 215
Trend removal 139
Trend-surface analysis 131, 145, 153, 222-228
Tuberculosis 6, 24

Variance components 124, 127, 129, 131
　negative estimates for 130
Variate differencing 192
Variogram 126, 152, 153, 162
von Neumann ratio 205

Weights
　Eire 229-230
　estimation of 118, 139
　standardisation of 18
　structure for 7, 17-19
Wheat yields 158
White noise process 152